T0254800

Praise for An Introduction to Stochastic Dynamics

"Jinqiao Duan's book introduces the reader to the actively developing theory of stochastic dynamics through well-chosen examples that provide an overview, useful insights, and intuitive understanding of an often technically complicated topic."

– P. E. Kloeden, Goethe University, Frankfurt am Main

"Randomness is an important component of modeling complex phenomena in biological, chemical, physical, and engineering systems. Based on many years teaching this material, Jinqiao Duan develops a modern approach to the fundamental theory and application of stochastic dynamical systems for applied mathematicians and quantitative engineers and scientists. The highlight is the staged development of invariant stochastic structures that underpin much of our understanding of nonlinear stochastic systems and associated properties such as escape times. The book ranges from classic Brownian motion to noise generated by α-stable Levy flights."

– A. J. Roberts, University of Adelaide

AN INTRODUCTION TO STOCHASTIC DYNAMICS

The mathematical theory of stochastic dynamics has become an important tool in the modeling of uncertainty in many complex biological, physical, and chemical systems and in engineering applications – for example, gene regulation systems, neuronal networks, geophysical flows, climate dynamics, chemical reaction systems, nanocomposites, and communication systems. It is now understood that these systems are often subject to random influences, which can significantly impact their evolution.

This book serves as a concise introductory text on stochastic dynamics for applied mathematicians and scientists. Starting from the knowledge base typical for beginning graduate students in applied mathematics, it introduces the basic tools from probability and analysis and then develops for stochastic systems the properties traditionally calculated for deterministic systems. The book's final chapter opens the door to modeling in non-Gaussian situations, typical of many real-world applications. Rich with examples, illustrations, and exercises with solutions, this book is also ideal for self-study.

JINQIAO DUAN is professor and director of the Laboratory for Stochastic Dynamics at Illinois Institute of Technology. During 2011–13, he also served as professor and associate director of the Institute for Pure and Applied Mathematics (IPAM) at UCLA. An expert in stochastic dynamics, stochastic partial differential equations, and their applications in engineering and science, he has been the managing editor for the journal *Stochastics and Dynamics* for more than a decade. He is also a coauthor of the research monograph *Effective Dynamics of Stochastic Partial Differential Equations* (2014).

Cambridge Texts in Applied Mathematics

All titles listed below can be obtained from good booksellers or from Cambridge University Press. For a complete series listing, visit www.cambridge.org/mathematics.

Flow, Deformation and Fracture
G. I. BARENBLATT

The Mathematics of Signal Processing
STEVEN B. DAMELIN AND WILLARD MILLER, JR.

Nonlinear Dispersive Waves
MARK J. ABLOWITZ

Complex Variables: Introduction and Applications (2nd Edition)
MARK J. ABLOWITZ AND ATHANASSIOS S. FOKAS

Scaling
G. I. R. BARENBLATT

Introduction to Symmetry Analysis
BRIAN J. CANTWELL

Hydrodynamic Instabilities
FRANÇOIS CHARRU

A First Course in Continuum Mechanics
OSCAR GONZALEZ AND ANDREW M. STUART

Theory of Vortex Sound
M. S. HOWE

Applied Solid Mechanics
PETER HOWELL, GREGORY KOZYREFF, AND JOHN OCKENDON

Practical Applied Mathematics: Modelling, Analysis, Approximation
SAM HOWISON

A First Course in the Numerical Analysis of Differential Equations (2nd Edition)
ARIEH ISERLES

A First Course in Combinatorial Optimization
JON LEE

An Introduction to Parallel and Vector Scientific Computation
RONALD W. SHONKWILER AND LEW LEFTON

AN INTRODUCTION TO STOCHASTIC DYNAMICS

JINQIAO DUAN
Illinois Institute of Technology

Shaftesbury Road, Cambridge CB2 8EA, United Kingdom

One Liberty Plaza, 20th Floor, New York, NY 10006, USA

477 Williamstown Road, Port Melbourne, VIC 3207, Australia

314–321, 3rd Floor, Plot 3, Splendor Forum, Jasola District Centre, New Delhi – 110025, India

103 Penang Road, #05–06/07, Visioncrest Commercial, Singapore 238467

Cambridge University Press is part of Cambridge University Press & Assessment, a department of the University of Cambridge.

We share the University's mission to contribute to society through the pursuit of education, learning and research at the highest international levels of excellence.

www.cambridge.org
Information on this title: www.cambridge.org/9781107428201

First published 2015

A catalogue record for this publication is available from the British Library

Library of Congress Cataloging-in-Publication data
Duan, Jinqiao.
An introduction to stochastic dynamics / Jinqiao Duan, Illinois Institute of Technology.
pages cm. – (Cambridge texts in applied mathematics)
Includes bibliographical references and index.
ISBN 978-1-107-07539-9 (hardback) – ISBN 978-1-107-42820-1 (pbk.)
1. Stochastic processes. 2. Probabilities. I. Title.
QC20.7.S8D83 2015
519.2′3 – dc23 2014046699

ISBN 978-1-107-07539-9 Hardback
ISBN 978-1-107-42820-1 Paperback

Dedicated to the memory of
my grandparents, Duan Chongxiang and Ye Youxiang
and
my parents, Duan Jianhua and Chen Yuying

Contents

Preface

Dynamical systems are often subject to random influences, such as external fluctuations, internal agitation, fluctuating initial conditions, and uncertain parameters. In building mathematical models for these systems, some less-known, less-well-understood, or less-well-observed processes (e.g., highly fluctuating fast- or small-scale processes) are ignored because of a lack of knowledge or limitations in our analytical skills and computational capability. This ignorance also contributes to uncertainty in mathematical models of complex dynamical systems. However, uncertainty or randomness may have a delicate and profound impact on the overall evolution of complex dynamical systems. Indeed, there is clear recognition of the importance of taking randomness into account when modeling complex phenomena in biological, chemical, physical, and other systems.

Stochastic differential equations are usually appropriate models for randomly influenced systems. Although the theoretical foundation for stochastic differential equations has been provided by stochastic calculus, better understanding the dynamical behaviors of these equations is desirable.

Who Is This Book For?

There is growing interest in stochastic dynamics in the applied mathematics community. This book is written primarily for applied mathematicians who may not have the necessary background to go directly to advanced reference books or the research literature in stochastic dynamics. My goal is to provide an introduction to basic techniques for understanding solutions of stochastic differential equations, from analytical, deterministic, computational, and structural perspectives. In deterministic dynamical systems, invariant manifolds and other invariant structures provide global information for dynamical evolution. For stochastic dynamical systems, in addition to these invariant structures, certain computable quantities, such as the mean exit time and escape probability (reminiscent of quantities like "eigenvalues" and

"Poincaré index" in deterministic dynamics and "entropy" in statistical physics), also offer insights into global dynamics under uncertainty. The mean exit time and escape probability are computed by solving deterministic, local or nonlocal, partial differential equations. Thus, I treat them as deterministic tools for understanding stochastic dynamics. It is my hope that this book will help the reader in accessing advanced monographs and the research literature in stochastic dynamics.

What Does This Book Do?

A large part of the material in this book is based on my lecture notes for the graduate course Stochastic Dynamics that I have taught many times since 1997. Among the students who have taken this course, about two-thirds are from applied mathematics, and the remaining one-third are from departments such as physics, computer science, bioengineering, mechanical engineering, electrical engineering, and chemical engineering. I would like to thank those graduate students for helpful feedback and for solutions to some exercises. For this group of graduate students, selection of topics and choice of presentation style are necessary. Thus, some interesting topics are not included. The choice of topics is personal but is influenced by my teaching these graduate students, who have basic knowledge in differential equations, dynamical systems, probability, and numerical analysis. Some materials are adopted from my recent research with collaborators; these include the most probable phase portraits in Chapter 5 and random invariant manifolds in Chapter 6, together with mean exit time, escape probability, and nonlocal Fokker-Planck equations for systems with non-Gaussian Lévy noise in Chapter 7.

I have tried to strike a balance between mathematical precision and accessibility for the readers of this book. For example, some proofs are presented, whereas some are outlined and others direct to the references. Some definitions are presented in separate paragraphs starting with "Definition," but many others are introduced less formally as they occur in the body of the text. As far as possible, I have tried to make connections between new concepts in stochastic dynamics and old concepts in deterministic dynamics.

After some motivating examples (Chapter 1), background in analysis and probability (Chapter 2), a mathematical model for white noise (Chapter 3), and a crash course in stochastic differential equations (Chapter 4), I focus on three topics:

- **Quantities that carry stochastic dynamical information (Chapter 5):** This includes moments, probability densities, most probable phase portraits, mean exit time, and escape probability.
- **Structures that build stochastic dynamics (Chapter 6):** This includes the multiplicative ergodic theorem and Hartman-Grobman theorem for linearized stochastic systems and invariant manifolds for nonlinear stochastic systems.

- **Non-Gaussian stochastic dynamics (Chapter 7):** This is an introduction to systems driven by non-Gaussian, α-stable Lévy motion.

This book is full of examples, together with many figures. There are separate Matlab simulation sections in Chapters 2–4, whereas in Chapters 5 and 7, numerical simulations are included in various sections. Although Chapter 6 contains no numerical simulations for its nature, it has examples and problems that require detailed derivations or calculations by hand. At the end of each chapter are homework problems, including some numerical simulation problems; Matlab is sufficient for this purpose. Most of these problems have been tested in the classroom. Hints or solutions to most problems are provided at the end of the book.

A section with an asterisk may be skipped on a first reading.

Some additional references are provided in the "Further Readings" section, for more advanced readers.

What Prerequisites Are Assumed?

For the reader, it is desirable to have basic knowledge of dynamical systems, such as the material contained in

- Chapters 1–2 of *Nonlinear Oscillations, Dynamical Systems, and Bifurcations of Vector Fields* by J. Guckenheimer and P. Holmes
- Chapters 1–2 of *Introduction to Applied Nonlinear Dynamical Systems and Chaos* by S. Wiggins
- Chapters 1–2 of *Differential Equations and Dynamical Systems* by L. Perko
- Chapters 1–3 of *Nonlinear Dynamics and Chaos* by S. H. Strogatz

Ideally, it is also desirable to have elementary knowledge of stochastic differential equations, such as

- Chapters 1–6 of *Stochastic Differential Equations* by L. Arnold
- Chapters 1–5 of *An Introduction to Stochastic Differential Equations* by L. C. Evans
- Chapters 1–5 of *Stochastic Differential Equations* by B. Oksendal
- Chapters 1–3 of *Stochastic Methods* by C. Gardiner

Realizing that some readers may not be familiar with stochastic differential equations, I review this topic in Chapter 4.

Acknowledgments

I would like to thank Philip Holmes for suggesting that I write this book back in 2004, when we were taking an academic tour in China. Steve Wiggins has also encouraged me to publish this book. I am especially grateful to Ludwig Arnold, who has always inspired and encouraged my learning and research in stochastic dynamics. I appreciate Bernt Oksendal's encouragement and comments. I have benefited from many years of productive collaboration and interaction with many colleagues, especially Peter Bates, Peter Baxendale, Dirk Blomker, Tomas Caraballo, Michael Cranston, Hans Crauel, Manfred Denker, David Elworthy, Franco Flandoli, Hongjun Gao, Martin Hairer, Peter Imkeller, Peter E. Kloeden, Kening Lu, Navaratnam Sri Namachchivaya, Anthony Roberts, Michael Scheutzow, Bjorn Schmalfuss, Richard Sowers, Xu Sun, Yong Xu, and Huaizhong Zhao. My interest in non-Gaussian stochastic dynamics started with a joint paper with D. Schertzer, M. Larcheveque, V. V. Yanovsky, and S. Lovejoy in 2000 and was further inspired and enhanced by Peter Imkeller and Ilya Pavlyukevich during my sabbatical leave at Humboldt University in Berlin in 2006.

Ludwig Arnold proofread this entire book and provided invaluable comments and suggestions. Han Crauel, Peter Imkeller, Peter Kloeden, Jicheng Liu, Guangying Lu, Mark Lytell, Bjorn Schmalfuss, Renming Song, Xiangjun Wang, and Jiang-Lun Wu proofread parts of this book, and their comments and corrections helped improve the book in various ways.

I thank Jia-an Yan for helpful discussions about topics in Chapter 4. I am very grateful to Zhen-Qing Chen, Xiaofan Li, Huijie Qiao, Renming Song, Xiangjun Wang, and Jiang-Lun Wu for helpful discussions about topics in Chapter 7. My former and current graduate students, especially Xiaopeng Chen, Hongbo Fu, Ting Gao, Zhongkai Guo, Tao Jiang, Xingye Kan, Mark Lytell, Jian Ren, Jiarui Yang, and Yayun Zheng, have helped with generating figures, proofreading some chapters, and providing solutions to some problems.

I would also like to acknowledge the National Science Foundation for its many years of generous support of my research. A part of this book was written while I was at the Institute for Pure and Applied Mathematics (IPAM), Los Angeles, during 2011–13. Diana Gillooly at Cambridge University Press has provided valuable professional help for the completion of this book.

My wife, Yan Xiong, and my children, Victor and Jessica, are constant sources of inspiration and happiness. Their love and understanding made this book possible.

Jinqiao Duan
Chicago, April 2014

Notation

$\triangleq$ is defined to be

$|x|$ absolute value of $x \in \mathbb{R}^1$

$\|x\|$ Euclidean norm of $x \in \mathbb{R}^n$

$a \wedge b \triangleq \min\{a, b\}$

$a \vee b \triangleq \max\{a, b\}$

$a^+ \triangleq \max\{a, 0\}$

$a^- \triangleq \max\{-a, 0\}$

B_t Brownian motion

$\mathcal{B}(\mathbb{R}^n)$ Borel σ-field of $\mathbb{R}^n$

$\mathcal{B}(S)$ Borel σ-field of state space S

$C(\mathbb{R}^n)$ space of continuous functions on $\mathbb{R}^n$

$C_0(\mathbb{R}^n)$ space of continuous functions on $\mathbb{R}^n$ that have compact support

$C^k(\mathbb{R}^n)$ space of continuous functions on $\mathbb{R}^n$ that have up to kth-order continuous derivatives

$C_0^k(\mathbb{R}^n)$ space of continuous functions on $\mathbb{R}^n$ that (1) have up to kth-order continuous derivatives and (2) have compact support

$C^\infty(\mathbb{R}^n)$ space of continuous functions on $\mathbb{R}^n$ that have derivatives of all orders

$C_0^\infty(\mathbb{R}^n)$ space of continuous functions on $\mathbb{R}^n$ that (1) have derivatives of all orders and (2) have compact support

$C^\alpha(D)$ space of functions that are locally Hölder continuous in D with exponent α

$C^\alpha(\bar{D})$ space of functions that are uniformly Hölder continuous in D with exponent α

$C^{k,\alpha}(D)$ space of continuous functions in D whose kth-order derivatives are locally Hölder continuous in D with exponent α

$\delta(\xi)$ Dirac delta function

$\mathbb{E}$ expectation

$\mathbb{E}^x$ expectation with respect to the probability measure $\mathbb{P}^x$ induced by a solution process starting at x

$F_X(x)$ distribution function of the random variable X

$\mathcal{F}^X$ or $\sigma(X)$ σ-field generated by the random variable X

$\mathcal{F}^{X_t} = \sigma(X_s, s \in \mathbb{R})$ σ-field generated by a stochastic process X_t; it is the smallest σ-field with which X_t is measurable for every t

$\mathcal{F}^\xi$ σ-field generated by the stochastic process ξ_t

$\mathcal{F}_t^B \triangleq \sigma(B_s : s \le t)$ filtration generated by Brownian motion B_t

$\mathcal{F}_\infty \triangleq \sigma(\bigcup_{t\ge 0} \mathcal{F}_t)$

$\mathcal{F}_{t+} \triangleq \bigcap_{\varepsilon>0} \mathcal{F}_{t+\varepsilon}$

$\mathcal{F}_{t-} \triangleq \sigma(\bigcup_{s<t} \mathcal{F}_s)$

$\mathcal{F}_t^X \triangleq \sigma(X_s : 0 \le s \le t)$ filtration generated by a stochastic process X_t

$\mathcal{F}_{-\infty}^t \triangleq \sigma(\cup_{s\le t} \mathcal{F}_s^t)$ also denoted as $\bigvee_{s\le t} \mathcal{F}_s^t$

$\mathcal{F}_s^\infty \triangleq \sigma(\cup_{t\ge s} \mathcal{F}_s^t)$ also denoted as $\bigvee_{t\ge s} \mathcal{F}_s^t$

$H(f)$ Hessian matrix of a scalar function $f : \mathbb{R}^n \to \mathbb{R}^1$

$H(\xi)$ Heaviside function

$H^k(D)$ Sobolev space

$H_0^k(D)$ Sobolev space of functions with compact support

$\|\cdot\|_k$ Sobolev norm in $H^k(D)$ or $H_0^k(D)$

lim in m.s. convergence in mean square, i.e., convergence in $L^2(\Omega)$

l^p space of infinite sequences $\{x_i\}_{i=1}^\infty$ such that $\sum_{i=1}^\infty |x_i|^p < \infty$

$L^2(\mathbb{R}^n)$ space of square-integrable functions defined on $\mathbb{R}^n$

$L^p(\mathbb{R}^n)$ space of p-integrable functions defined on $\mathbb{R}^n$, with $p \ge 1$

$L^p(D)$ space of p-integrable functions defined on a domain $D \subset \mathbb{R}^n$, with $p \ge 1$

$L^2(\Omega)$ or $L^2(\Omega, \mathbb{R}^n)$ space of random variables, taking values in Euclidean space $\mathbb{R}^n$, with finite variance

$L^2(\Omega)$ or $L^2(\Omega, H)$ space of random variables, taking values in Hilbert space H, with finite variance

$L^p(\Omega)$ or $L^p(\Omega, \mathbb{R}^n)$ $\{X : \mathbb{E}|X|^p < \infty\}$ for $p \ge 1$

$L_t(\omega)$ Lévy motion

$L_t^{\alpha}(\omega)$ α-stable Lévy motion

$\mathbb{N}$ set of the natural numbers

$\mathcal{N}(\mu, \sigma^2)$ normal (or Gaussian) distribution with mean μ and variance σ^2

$\nu(dy)$ Lévy jump measure

$\mathbb{P}$ probability measure

$\mathbb{P}(A)$ or $\mathbb{P}\{A\}$ probability of an event A

P^X distribution measure induced by the random variable X

$\mathbb{P}^x$ probability measure induced by a solution process starting at x

$P(\lambda)$ Poisson distribution with parameter $\lambda > 0$

$\mathbb{R}$ two-sided time axis

$\mathbb{R}^+$ one-sided time set $\{t : t \geq 0\}$

$\mathbb{R}^1$ one-dimensional Euclidean space

$\mathbb{R}^n$ n-dimensional Euclidean space

$\sigma(X)$ or $\mathcal{F}^X$ σ-field generated by the random variable X; it is the smallest σ-field with which X is measurable

$\mathrm{Supp}(f) \triangleq$ closure of $\{x \in \mathbb{R}^n : f(x) \neq 0\}$ support of function f

$\mathrm{Tr}(A)$ trace of A

$U(a, b)$ uniform distribution on the interval $[a, b]$

$\bigvee_{s \leq t} \mathcal{F}_s^t \triangleq \sigma(\cup_{s \leq t} \mathcal{F}_s^t)$ also denoted as $\mathcal{F}_{-\infty}^t$

$\bigvee_{t \geq s} \mathcal{F}_s^t \triangleq \sigma(\cup_{t \geq s} \mathcal{F}_s^t)$ also denoted as $\mathcal{F}_s^{\infty}$

1
Introduction

Noisy fluctuations are abundant in complex systems. In some cases, noise is not negligible, whereas in some other situations, noise could even be beneficial. It is desirable to have a better understanding of the impact of noise on dynamical evolution of complex systems. In other words, it becomes crucial to take randomness into account in mathematical modeling of complex phenomena under uncertainty.

In 1908, Langevin devised a stochastic differential equation for the motion of Brownian particles in a fluid, under random impacts of surrounding fluid molecules. This stochastic differential equation, although important for understanding Brownian motion, went largely unnoticed in the mathematical community until after stochastic calculus emerged in the late 1940s. Introductory books on stochastic differential equations (SDEs) include [8, 88, 213].

The goal for this book is to examine and present select dynamical systems concepts, tools, and methods for understanding solutions of SDEs. To this end, we also need basic information about deterministic dynamical systems modeled by ordinary differential equations (ODEs), as presented in the first couple of chapters in one of the references [110, 290].

In this introductory chapter, we present a few examples of deterministic and stochastic dynamical systems, then briefly outline the contents of this book.

1.1 Examples of Deterministic Dynamical Systems

We recall a few examples of deterministic dynamical systems, where short time-scale forcing and nonlinearity can affect dynamics in a profound way.

Example 1.1 *A double-well system.* Consider a one-dimensional dynamical system $\dot{x} = x - x^3$. It has three equilibrium states, -1, 0, and 1, at which the vector field

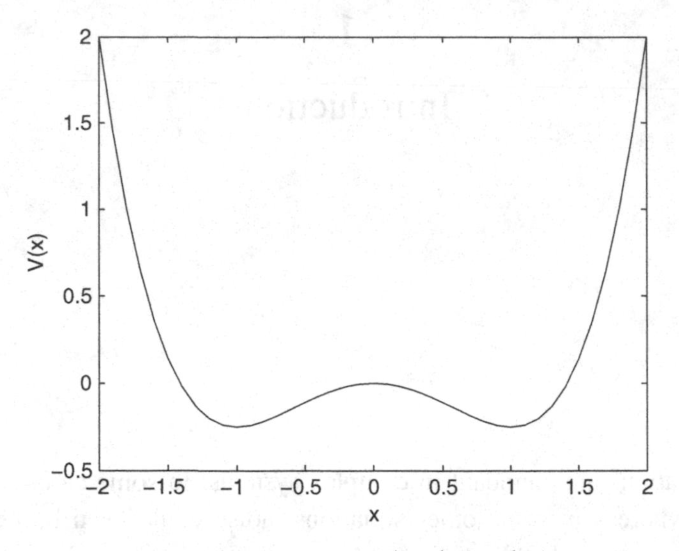

Figure 1.1 Plot of $V(x) \triangleq -\frac{1}{2}x^2 + \frac{1}{4}x^4$.

$x - x^3$ is zero. Observe that

$$\dot{x} = x - x^3 = x(1 - x^2) = \begin{cases} < 0, & -1 < x < 0 \quad \text{or} \quad 1 < x < \infty, \\ = 0, & x = -1, 0, 1, \\ > 0, & -\infty < x < -1 \quad \text{or} \quad 0 < x < 1. \end{cases}$$

Note that $\dot{x} = x - x^3 \triangleq -\frac{dV(x)}{dx}$, where the potential function $V(x) \triangleq -\frac{1}{2}x^2 + \frac{1}{4}x^4$ has two minimal values (sometimes called "wells"); see Figure 1.1.

A solution curve, or orbit, or trajectory, starting with $x(0) = x_0$ in $(-1, 0)$, decreases in time (because $\dot{x} < 0$ on this interval) and approaches the equilibrium state -1 as $t \to +\infty$, whereas an orbit starting with $x(0) = x_0$ in $(-\infty, -1)$ increases in time (because $\dot{x} > 0$ on this interval) and approaches the equilibrium state -1 as $t \to +\infty$. Thus the equilibrium point $\{-1\}$ is a stable equilibrium state and is an *attractor*, that is, it attracts nearby orbits. Likewise $\{1\}$ is also an attractor. But the equilibrium state $\{0\}$ is unstable and is called an *repeller*. See Figure 1.2 for a few representative solutions curves.

An orbit starting near one equilibrium state $\{-1\}$ cannot go anywhere near the other equilibrium state $\{1\}$, and vice versa. There is no transition between these two stable states.

If we only look at the solution curves in the state space, $\mathbb{R}^1$, where state x lives, we get a state portrait, or as it is often called, a phase portrait.

Figure 1.3 shows the phase portrait for this double-well system.

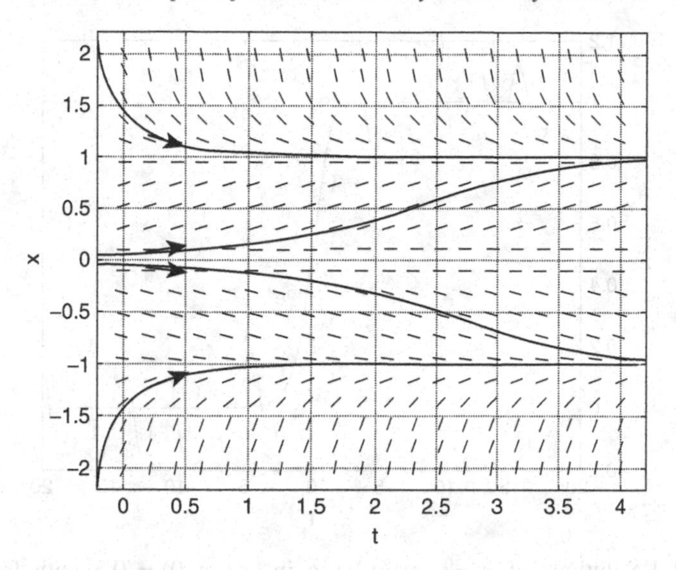

Figure 1.2 Solution curves for $\dot{x} = x - x^3$.

Example 1.2 *High-frequency (or short time-scale) forcing.* Consider a simple one-dimensional nonlinear system with time-periodic forcing with frequency ω:

$$\dot{x} = -x + x^3 + \varepsilon \sin(\omega t), \ x(0) = 0.5. \tag{1.1}$$

Solution curves with frequency $\omega = 2$ and $\omega = 10$ are shown in Figures 1.4 and 1.5, respectively. The difference between low- and high-frequency forcing is visible.

Example 1.3 *Small nonlinearity leads to fundamental change in dynamics.* Consider a harmonic oscillator (a spring-mass system) of mass m and spring constant k, under damping that is proportional to the cubic of velocity: $m\ddot{x} = -kx - \varepsilon \dot{x}^3$, where ε is a positive constant. For simplicity, we take m, k both equal to 1. This can also be achieved by rescaling the time. Thus,

$$\ddot{x} = -x - \varepsilon \dot{x}^3 \tag{1.2}$$

or equivalently,

$$\dot{x} = y, \tag{1.3}$$

$$\dot{y} = -x - \varepsilon y^3, \tag{1.4}$$

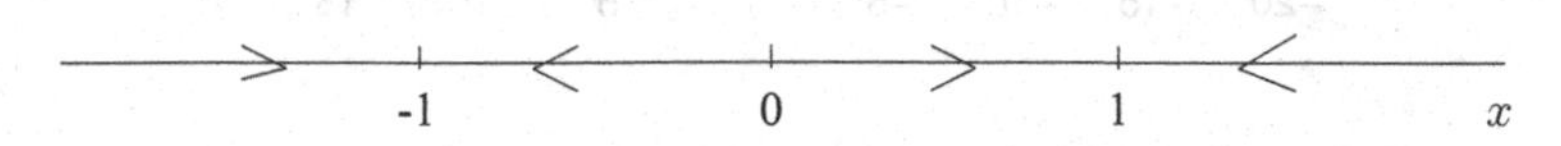

Figure 1.3 Phase portrait for $\dot{x} = x - x^3$.

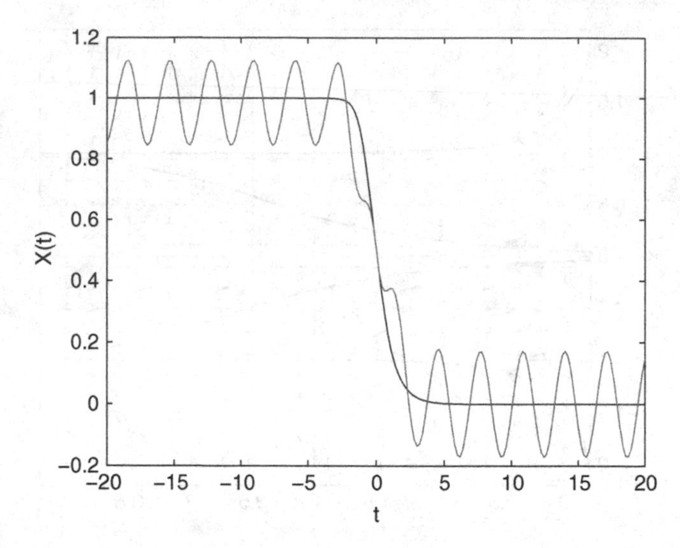

Figure 1.4 Solutions of $\dot{x} = -x + x^3 + \varepsilon \sin(\omega t)$, $x(0) = 0.5$ with frequency $\omega = 2$: $\varepsilon = 0$ (no "oscillations" or black line) and $\varepsilon = 0.35$ (with "oscillations" or gray line).

where x is the displacement and y is the velocity of the oscillator. The equilibrium state is $(0, 0)$.

Without damping ($\varepsilon = 0$), the model equations become

$$\dot{x} = y, \tag{1.5}$$

$$\dot{y} = -x. \tag{1.6}$$

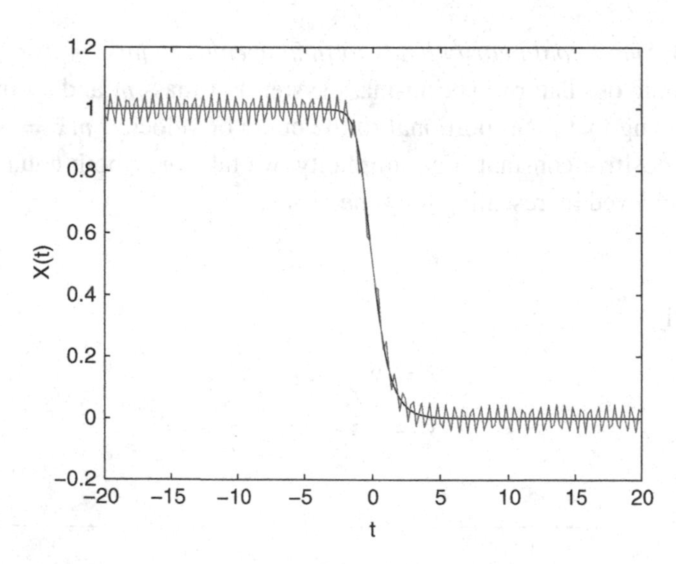

Figure 1.5 Solutions of $\dot{x} = -x + x^3 + \varepsilon \sin(\omega t)$, $x(0) = 0.5$ with frequency $\omega = 10$: $\varepsilon = 0$ (no "oscillations" or black line) and $\varepsilon = 0.35$ (with "oscillations" or gray line).

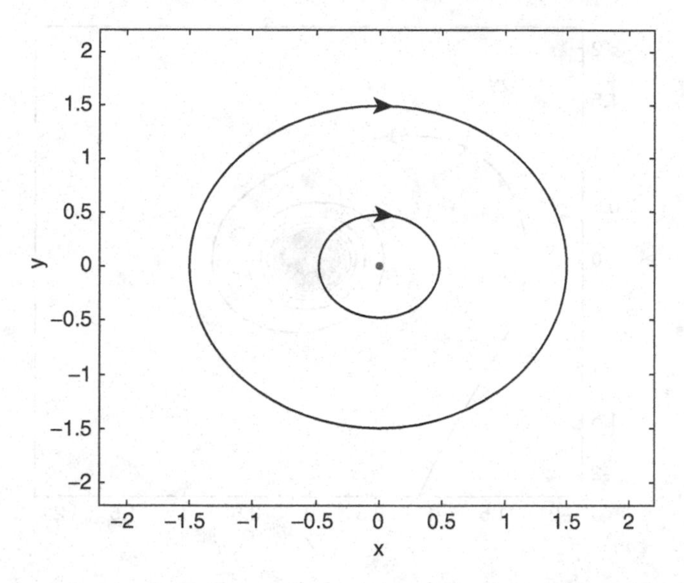

Figure 1.6 Phase portrait for harmonic oscillator $\dot{x} = y$, $\dot{y} = -x - \varepsilon y^3$: $\varepsilon = 0$ (no damping).

Dividing these two equations, we obtain

$$\frac{dy}{dx} = -\frac{x}{y}$$

or

$$xdx + ydy = 0.$$

Integrating this equation, we see that the solution curves $(x(t), y(t))$ satisfy the conservation of energy

$$x^2 + y^2 = c \tag{1.7}$$

for an arbitrary (nonnegative) constant of integration, c. Thus, the solution curves are circles; see Figure 1.6.

In the case of damping, that is, when $\varepsilon > 0$, the energy is not conserved,

$$\frac{d}{dt}(x^2 + y^2) = 2x\dot{x} + 2y\dot{y} = 2xy + 2y(-x - \varepsilon y^3) = -2\varepsilon y^4 < 0,$$

at all points, except the equilibrium point (0, 0). Thus, all orbits approach the equilibrium point (0, 0) as $t \to \infty$, no matter how small the damping coefficient ε is, as shown in Figure 1.7. Comparing Figure 1.6 and Figure 1.7, we see that the dynamics, with or without damping, are drastically different.

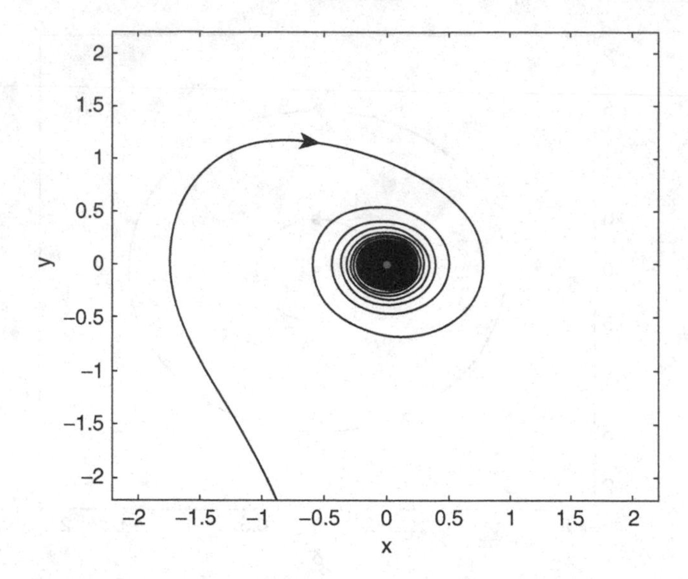

Figure 1.7 Phase portrait for damped harmonic oscillator $\dot{x} = y$, $\dot{y} = -x - \varepsilon y^3$: $\varepsilon = 0.5$.

Example 1.4 *Simple pendulum.* Consider a simple pendulum of mass m and length l: $\ddot{x} = -\frac{g}{l}\sin x$ where x is the angular displacement from the vertical downward (equilibrium) position. Note that m does not appear in the equation. Introducing a time change $\tau \triangleq \sqrt{\frac{g}{l}}\, t$, but still denoting $\frac{dx}{d\tau}$ by $\dot{x}$, we have

$$\ddot{x} = -\sin x \tag{1.8}$$

or, equivalently,

$$\dot{x} = y, \tag{1.9}$$

$$\dot{y} = -\sin x, \tag{1.10}$$

where y is the angular velocity of the pendulum. The equilibrium states are $(0, 0)$ and $(\pm n\pi, 0)$, for $n = 1, 2, \ldots$.

Dividing these two equations, we obtain

$$\frac{dy}{dx} = -\frac{\sin x}{y}$$

or, equivalently,

$$\sin x dx + y dy = 0.$$

This leads to the conservation of energy

$$\frac{1}{2}y^2 - \cos x = c$$

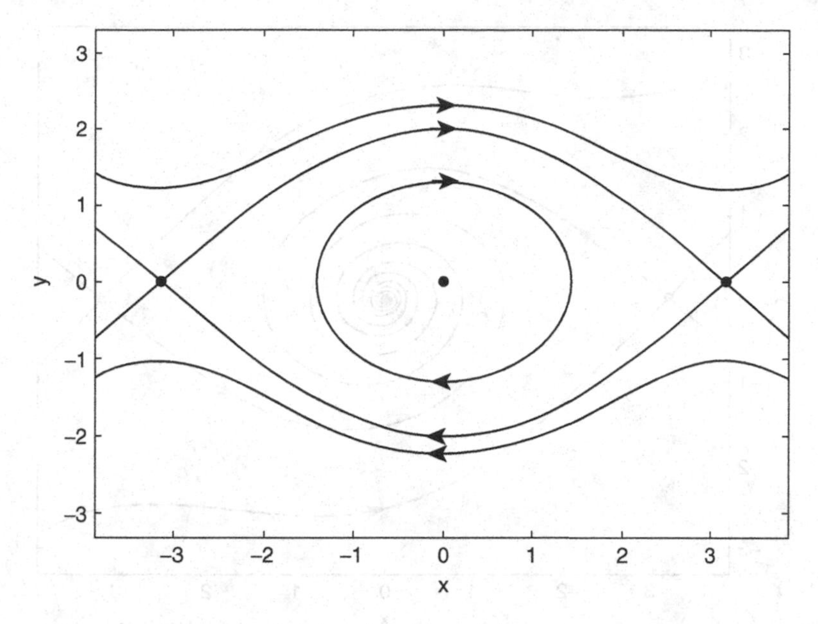

Figure 1.8 Phase portrait for simple pendulum: $\dot{x} = y,\ \dot{y} = -\sin x$.

for an arbitrary constant c. See Figure 1.8 for the phase portrait of this simple pendulum system.

Now consider the simple pendulum under damping, $\ddot{x} = -\sin x - \varepsilon\dot{x}$, with $\varepsilon > 0$. This equation is rewritten as

$$\dot{x} = y, \tag{1.11}$$

$$\dot{y} = -\sin x - \varepsilon y. \tag{1.12}$$

In this case, the energy is not conserved,

$$\frac{d}{dt}\left(\frac{1}{2}y^2 - \cos x\right) = y\dot{y} + (\sin x)\dot{x} = y(-\sin x - \varepsilon y) + (\sin x)y = -\varepsilon y^2 < 0,$$

for all (x, y), except the equilibrium points. As was the case of Example 1.3, the dynamics with damping is very different, no matter how small the parameter ε is. See Figure 1.9 for the phase portrait of this damped simple pendulum.

Example 1.5 *A coupled system: Small change in one part affects the others dramatically.* Consider a coupled system

$$\dot{x} = 0.001x - xy, \tag{1.13}$$

$$\dot{y} = -6y + \varepsilon x^2, \tag{1.14}$$

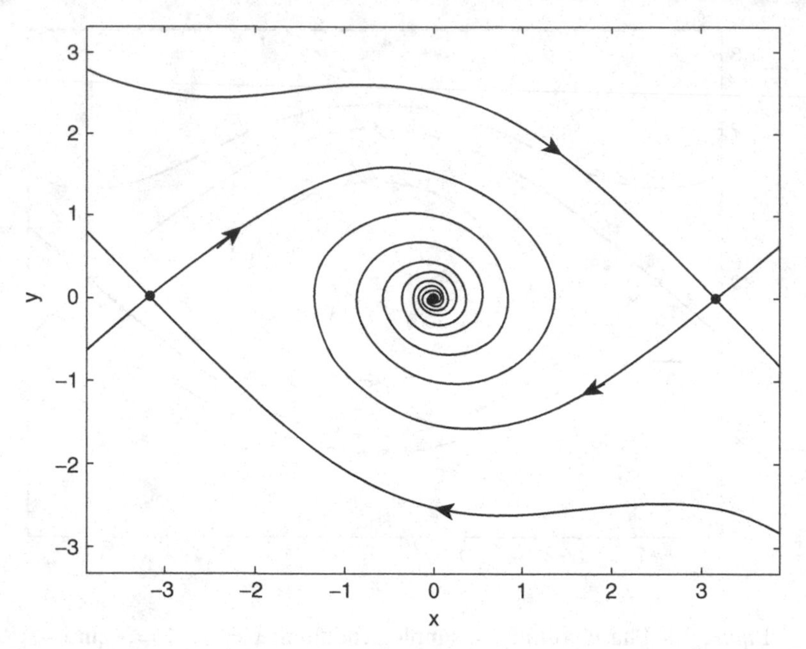

Figure 1.9 Phase portrait for simple pendulum with damping: $\dot{x} = y$, $\dot{y} = -\sin x - \varepsilon y$ with $\varepsilon = 0.25$.

where ε is a small real parameter. A small change in the y-part affects the dynamics dramatically: when $\varepsilon = 0$, the system has an unbounded attracting set (i.e., the whole x-axis), whereas for $\varepsilon = 0.01$, the system has a bounded attracting set (in fact, an inertial manifold). Compare Figures 1.10 and 1.11.

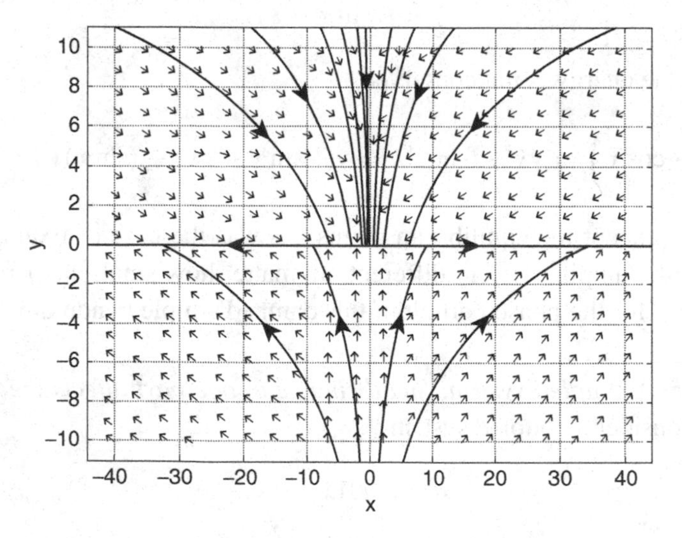

Figure 1.10 Phase portrait for $\dot{x} = 0.001x - xy$, $\dot{y} = -6y$.

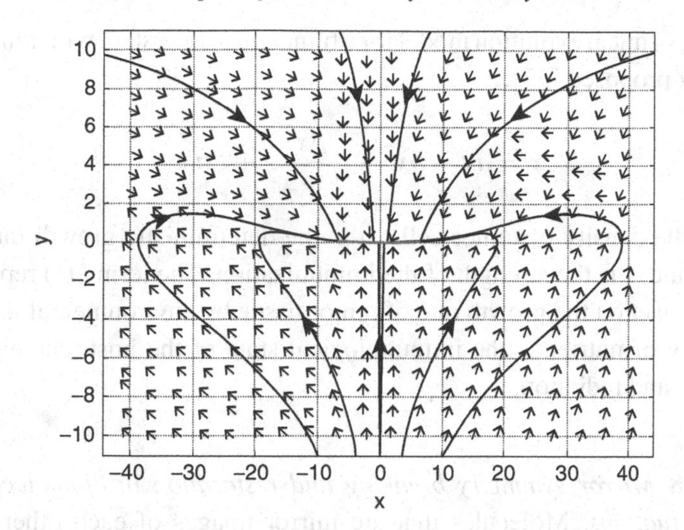

Figure 1.11 Phase portrait for $\dot{x} = 0.001x - xy$, $\dot{y} = -6y + 0.01x^2$: the global attractor is clearly seen (gray curve within $-7 < x < 7$ and near the x-axis).

1.2 Examples of Stochastic Dynamical Systems

Let us look at a few examples of stochastic differential equations arising in the modeling of complex phenomena from various disciplines.

Example 1.6 *Motion of "particles" subject to a random force, in physics, biophysics, and geophysics.* A particle moving in one dimension x, subject to a driving force $K(x)$, a friction force $-\gamma\dot{x}$ with a nonnegative parameter γ, and a random force $\xi(t)$, is described by [280, 179]:

$$\ddot{x} + \gamma\dot{x} = K(x) + \xi(t). \tag{1.15}$$

This includes the model equation of motion for a simple pendulum under a random force,

$$\ddot{x} = \sin(x) + \xi(t), \tag{1.16}$$

where x is now the angular displacement.

A similar stochastic differential equation, still in one dimension, arises to model a mechanism (socalled ergodic pumping) that drives biomolecular conformation changes [189, 190]. This equation in two or three dimensions is a model for the motion of geophysical fluid particles under random influences [40, 229, 108].

Example 1.7 *A tumor growth system with immunization.* In the absence of an immune reaction, tumor evolution is thought to follow a growth law that can be approximated by a logistic function. With the tumor-immune interactions taken into account, a dynamical evolution model for a tumor growth system with immunization was recently proposed [37]:

$$\dot{x} = ax - bx^2 - \frac{\beta x^2}{1+x^2} + \xi(t), \tag{1.17}$$

where x is the density of tumor cells, a is the deterministic growth rate, b is the decay rate, and β is the strength of the immunization. The term $\xi(t)$ represents the uncertain impact on the growth rate of tumor tissue by environmental factors, such as the supply of nutrients, the immunological state of the host, chemical agents, temperature, and radiation.

Example 1.8 *Mirror symmetry breaking and restoration in biomolecules under random fluctuations.* Molecules that are mirror images of each other are called enantiomers. However, this mirror or chiral symmetry is broken in all biological systems. When this symmetry breaking occurred and whether it may be restored are unanswered questions. The chiral structures include proteins, almost always as the left-handed enantiomers (L), and DNA, RNA polymers, and sugars with chiral building blocks composed by right-handed (D) monocarbohydrates. To study this chiral symmetry breaking and restoration, a stochastic model is introduced [124] to quantify the dynamical evolution of the enantiomeric excess $x = ([L] - [D])/([L] + [D])$, which is the order parameter for mirror symmetry breaking, and the net chiral matter $y = [L] + [D]$,

$$\dot{x} = -2k_1 Ax/y + \frac{1}{2}(k_3 - k_{-2})xy(1 - x^2) + \xi(t), \tag{1.18}$$

$$\dot{y} = 2k_1 A - y^2 \left[k_{-2} + \frac{1}{2}(k_3 - k_{-2})(1 - x^2) \right] + (k_2 A - k_{-1})y, \tag{1.19}$$

where $\xi(t)$ is a noise process due to internal and external fluctuations of the chemical system, A is the achiral reactant (kept at constant concentration), and $k_{\pm 1}$, $k_{\pm 2}$ together with k_3 are various reaction, amplification, or inhibition rates.

Example 1.9 *Dynamics of two inhibitory neural units.* Brown and Holmes [42] considered a stochastic dynamical model for two mutually inhibiting, leaky, neural units characterized by state variables x_j, subject to external stimuli ρ_j, additive noises modeled by independent Brownian motions (to be defined in Chapter 3) $B_{j(t)}$, and "priming biases" $i_0 + b_j$, including an overall level i_0 and separate unit biases b_j. Each unit inhibits the other via an activation function $f(x; g, b) = 1/[1 + \exp(-g(x - b))]$ with gain g that achieves half level at $x = b$.

This model is a two-dimensional stochastic dynamical system:

$$dx_1 = (-kx_1 - \beta f(x_2) + i_0 + b_1 + \rho_1)\,dt + dB_1(t), \tag{1.20}$$

$$dx_2 = (-kx_2 - \beta f(x_1) + i_0 + b_2 + \rho_2)\,dt + dB_2(t), \tag{1.21}$$

where k is the leak and β is the inhibition level.

Example 1.10 *An epidemic model.* Consider a population system of S susceptible individuals and I infected individuals, with the constant nonnegative contact rate β and recovery rate γ. Taking environmental fluctuations into account, this system is described by the following coupled SDEs [5]:

$$dS = -\frac{\beta SI}{S+I}dt - \sqrt{\frac{\beta SI}{S+I}}\,dB_1(t), \tag{1.22}$$

$$dI = \left(\frac{\beta SI}{S+I} - \gamma I\right)dt + \sqrt{\frac{\beta SI}{S+I}}\,dB_1(t) - \sqrt{\gamma I}\,dB_2(t), \tag{1.23}$$

where B_1, B_2 are two independent scalar Brownian motions, to be defined in Chapter 3.

1.3 Mathematical Modeling with Stochastic Differential Equations

In the previous section, we saw that SDEs are mathematical models for physical, biological, and medical systems. Indeed, SDEs arise naturally as mathematical models for dynamical systems under uncertainty [203, 188, 4]. These include SDE models for dilute fluid particle suspension [45], climate evolution [117], stochastic parameterization of geophysical processes [216], geophysical fluid particle dynamics [40, 108], a laser system [31], data assimilation [113, 195], multiscale modeling [222], noise-induced phenomena [125, 276], stochastic micro-macroscopic modeling [157], stochastic resonance [130, 120], nonequilibrium statistical physics [308], stochastic dynamics and control [311], biological noise attenuation [287], tumor growth [37], genetic transitions triggered by uncertain events [278, 150], and nonlinear expectation and risk modeling [225, 220, 224].

1.4 Outline of This Book

We consider the following stochastic system defined by stochastic differential equations (SDEs) in $\mathbb{R}^n$:

$$dX_t = b(X_t)dt + \sigma(X_t)d\xi_t, \tag{1.24}$$

where $b \in \mathbb{R}^n$ and $\sigma \in \mathbb{R}^{n\times m}$, respectively, and $\xi_t \in \mathbb{R}^m$ is a stochastic process that could be a Brownian motion B_t (a Gaussian process) or a Lévy motion L_t (a non-Gaussian process). Note that the natural numbers n and m may be equal or different. We treat X, X_t, $X(t)$, or $X_t(\omega)$ as the same random quantity. The b term is often called a drift or a vector field, and the noise term $\sigma d\xi_t$ is a model for uncertainty. This model uncertainty could be caused by external random influences or by fluctuating coefficients and parameters in a mathematical model. When the stochastic process ξ_t is a Brownian motion B_t, σ is often called a diffusion coefficient (although it may be better called the noise intensity).

This book is organized as follows.

Chapter 2: Background in Analysis and Probability

We introduce basic facts and definitions in analysis and probability.

Chapter 3: Noise

We recall basic definitions and facts about Brownian motion. The discussion on white noise follows [8].

Chapter 4: A Crash Course in Stochastic Differential Equations

We review basic facts about stochastic integration and stochastic differential equations with Brownian motion. See [213] for more details.

Chapter 5: Deterministic Quantities for Stochastic Dynamics

We consider how to estimate moments of solutions of stochastic differential equations via Itô's formula. Then, we quantify some aspects of stochastic dynamics by mean exit time and escape probability [196, 253, 234].

Chapter 6: Invariant Structures for Stochastic Dynamics

The discussions on deterministic invariant manifolds follow [59, 226]. Definitions on cocycles and random invariance follow [9, Chapters 1, 2, 3]. The random invariant manifolds are finite-dimensional versions of our earlier work on stochastic partial differential equations [47, 79].

Chapter 7: Dynamical Systems Driven by Non-Gaussian Lévy Motion

This chapter is an overview on stochastic systems with α-stable Lévy motion. Basic information about these systems is from [7]. Numerical approaches for computing mean exit time are from [53, 97]. The discussions on escape probability follow [234].

To help readers who may not be familiar with deterministic dynamical systems, I have incorporated some deterministic dynamical systems concepts in a few places.

1.5 Problems

1.1 A one-dimensional system

(a) Consider the so-called logistic equation $\dot{x} = x(1 - x)$. What are equilibrium states? Are they stable or unstable? Sketch the phase portrait.
(b) Do the same for the one-dimensional dynamical system $\dot{x} = -x + x^2$.

1.2 A two-dimensional system

Consider a spring-mass system $\ddot{x} = -kx$, with k a positive constant. It can be converted into a two-dimensional dynamical system $\dot{x} = y, \ \dot{y} = -kx$. What is the equilibrium state? Is the energy $\frac{1}{2}kx^2 + \frac{1}{2}y^2$ conserved. Sketch the phase portrait.

1.3 A saddle system

Consider a two-dimensional linear dynamical system $\dot{x} = -x, \ \dot{y} = y$. What is the equilibrium state? Write this in matrix form $\dot{X} = AX$ for $X = (x, y)^T$, where ^{T} denotes the transpose of a vector or a matrix. What are the eigenvalues of this linear system (i.e., eigenvalues of A) and the corresponding eigenspaces? Sketch the phase portrait.

2

Background in Analysis and Probability

In this chapter we briefly review some basic facts, concepts, and tools in analysis and probability. For basic references in analysis, see Apostol [6] and Kreyszig [158], whereas for more information in probability, we refer to Ash [13], Jacod and Protter [138], and Durrett [84].

2.1 Euclidean Space

Euclidean space $\mathbb{R}^n$ is equipped with the usual scalar product $x \cdot y = \sum_{j=1}^n x_j y_j$, which induces the Euclidean norm (or length) $\|x\| = \sqrt{\sum_{j=1}^n x_j^2} = \sqrt{\langle x, x\rangle}$ and distance $d(x, y) = \sqrt{\sum_{j=1}^n (x_j - y_j)^2} = \sqrt{\langle x - y, x - y\rangle}$. A point x in $\mathbb{R}^n$ is also called a vector, which is usually written in a column but is sometimes represented as a row vector. The scalar product $x \cdot y$ is also denoted by $\langle x, y\rangle$ or $x^{\mathrm{T}} y$. With this norm or distance, we can define an open ball centered at x_0 with radius r: $\{x : d(x, x_0) < r\}$. A set M is called open if, for every $x \in M$, there is an open ball centered at x that is contained in M. An open connected set is called a domain.

A σ-field $\mathcal{F}$ in $\mathbb{R}^n$ is a collection of subsets which satisfy that

(i) the empty set $\emptyset$ is in $\mathcal{F}$;
(ii) if $A \in \mathcal{F}$, then so is its complement $A^c = \mathbb{R}^n \setminus A$;
(iii) if $A_1, A_2, \ldots$ are in $\mathcal{F}$, then so is the union $\cup A_i$; thus $\mathbb{R}^n$ is also in $\mathcal{F}$.

The Borel σ-field of $\mathbb{R}^n$, that is, $\mathcal{B}(\mathbb{R}^n)$, is generated (via unions, intersections, and complements) by all open balls in $\mathbb{R}^n$. Every element in $\mathcal{B}(\mathbb{R}^n)$ is called a Borel set in $\mathbb{R}^n$. Moreover, $\mathcal{B}(\mathbb{R}^n)$ is the smallest σ-field containing all open sets of $\mathbb{R}^n$.

A function $f : \mathbb{R}^n \to \mathbb{R}^1$ is Borel measurable if, for every Borel set A in $\mathbb{R}^1$, the preimage $f^{-1}(A)$ is a Borel set in $\mathbb{R}^n$.

Let M be a subset of Euclidean space $\mathbb{R}^n$. A point x_0 of X (which may or may not be a point of M) is called a limit point of M if every neighborhood of x_0 contains at least one point y of M distinct from x_0. The set consisting of all points of M and its limit points is called the closure of M, and it is denoted by $\bar{M}$. In fact, $x \in \bar{M}$ if and only if there is a sequence $\{x_n\}$ in M such that $x_n \to x$. A set M is called closed if itself and its closure coincide. A set M in $\mathbb{R}^n$ is called compact if it is closed and bounded.

A set S is called dense in $\mathbb{R}^n$ if $S \subset \mathbb{R}^n$ and if the closure of S is the whole $\mathbb{R}^n$, that is, $\bar{S} = \mathbb{R}^n$.

In this book, $\mathbb{R}^1$ denotes the real line (i.e., one-dimensional Euclidean space), whereas $\mathbb{R}$ is for the whole time set (including past, current and future time instants).

2.2 Hilbert, Banach, and Metric Spaces

A vector space (or linear space) X is a set with two operations $+$ (addition) and $\cdot$ (scalar multiplication) which satisfy the usual properties.

A scalar product $\langle \cdot, \cdot \rangle$ in a vector space X satisfies the following conditions for $x, y, z \in X$ and any scalar α:

(i) $\langle x + y, z \rangle = \langle x, z \rangle + \langle y, z \rangle$;
(ii) $\langle \alpha x, y \rangle = \alpha \langle x, y \rangle$;
(iii) $\langle x, y \rangle = \overline{\langle y, x \rangle}$;
(iv) $\langle x, x \rangle \geq 0$ and $\langle x, x \rangle = 0$ if and only if $x = 0$.

A scalar product naturally induces a norm (and thus also a metric) $\|x\| = \sqrt{\langle x, x \rangle}$. A space $(X, \langle \cdot, \cdot \rangle)$ is called a Hilbert space if every Cauchy sequence in X converges in X.

A norm $\|x\|$ in a vector space X offers a concept of magnitude or length for the vector or point x. It satisfies the following conditions for $x, y, z \in X$ and any scalar α:

(i) $\|x\| \geq 0$;
(ii) $\|x\| = 0$ if and only if $x = 0$;
(iii) $\|\alpha x\| = |\alpha| \|x\|$;
(iv) $\|x + y\| \leq \|x\| + \|y\|$.

Property (iv) is called the triangle inequality. A norm naturally induces a metric $d(x, y) = \|x - y\|$ and, thus, endows a concept of convergence. A sequence $\{x_n\}$ in X converges to x, under the norm $\| \cdot \|$, if $\|x_n - x\| \to 0$ as $n \to \infty$. A space $(X, \| \cdot \|)$ is called a Banach space if every Cauchy sequence in X converges in X.

Two norms, $\| \cdot \|_1$ and $\| \cdot \|_2$, are called equivalent if there exist two positive constants, c_1 and c_2, such that

$$\|x\|_1 \leq c_2 \|x\|_2 \text{ and } \|x\|_2 \leq c_1 \|x\|_1$$

for every $x \in X$. For two equivalent norms, the convergence of a sequence in one equivalent norm implies convergence in another, and vice versa.

Note that all possible norms in Euclidean space $\mathbb{R}^n$ are equivalent. So, we often only need to consider the usual norm $\|x\| = \sqrt{\sum_{j=1}^{n} x_j^2}$, that is, the Euclidean norm.

A metric $d(x, y)$ in a set X provides a concept of distance between points x and y. It satisfies the following conditions for $x, y, z \in X$:

(i) $d(x, y) \geq 0$;
(ii) $d(x, y) = 0$ if and only if $x = y$;
(iii) symmetry $d(x, y) = d(y, x)$;
(iv) $d(x, y) \leq d(x, z) + d(z, y)$.

Property (iv) is called the triangle inequality. A metric space (X, d) is called complete if every Cauchy sequence in X converges in X.

Euclidean space, Hilbert space, and Banach space are all examples of complete metric spaces.

A metric space (X, d) is called compact if every sequence X_n has a convergent subsequence X_{n_k}. In fact, in a metric space, other notions of compactness are equivalent to this sequential compactness [204, Theorem 28.2].

2.3 Taylor Expansions

We recall Taylor expansions [6, Chapter 12] in Euclidean space $\mathbb{R}^n$.

Taylor expansion for $f: \mathbb{R}^1 \to \mathbb{R}^1$

$$f(x) = f(a) + f'(a)(x - a) + o((x - a)^2).$$

Taylor expansion for $f: \mathbb{R}^2 \to \mathbb{R}^1$

$$\begin{aligned} f(x, y) = {} & f(a, b) + f_x(a, b)(x - a) + f_y(a, b)(y - b) \\ & + \frac{1}{2}[f_{xx}(a, b)(x - a)^2 + 2f_{xy}(a, b)(x - a)(y - b) + f_{yy}(a, b)(y - b)^2] \\ & + o((x - a)^2 + (y - b)^2). \end{aligned}$$

Taylor expansion for $f: \mathbb{R}^n \to \mathbb{R}^1$

$$f(x) = f(a) + \nabla f(a) \cdot (x - a) + \frac{1}{2}(x - a)^{\mathrm{T}} H(f(a))(x - a) + o(\|x - a\|^2),$$

where ∇f is the gradient of f (which is also often denoted as Df and called the Jacobian of f), $H(f) = (f_{x_i x_j})$ is the $n \times n$ Hessian matrix, and $\cdot$ denotes the usual scalar product in $\mathbb{R}^n$.

Introducing the following notations for $f: \mathbb{R}^n \to \mathbb{R}^1$, $h \in \mathbb{R}^n$

$$f'(x)h \triangleq \sum_{i=1}^{n} f_{x_i} h_i = \nabla f \cdot h,$$

$$f''(x)h^2 \triangleq \sum_{i=1}^{n} f_{x_i,x_j} h_i h_j,$$

$$f'''(x)h^3 \triangleq \sum_{i=1}^{n} f_{x_i,x_j,x_k} h_i h_j h_k,$$

and similarly for higher-order derivatives. Each of these derivatives is a multilinear operator. For example,

$$f'(x) : \mathbb{R}^n \to \mathbb{R}^1,$$
$$f''(x) : \mathbb{R}^n \times \mathbb{R}^n \to \mathbb{R}^1,$$
$$f'''(x) : \mathbb{R}^n \times \mathbb{R}^n \times \mathbb{R}^n \to \mathbb{R}^1,$$

and so on.

Now, we have the more general Taylor expansion

$$f(x+h) = f(x) + f'(x)h + \frac{1}{2!} f''(x)h^2 + \cdots + \frac{1}{m!} f^{(m)}(x)h^m + \frac{1}{(m+1)!} f^{(m+1)}(z)h^{m+1},$$

where z is some point in the line segment connecting x and $x+h$.

For $f: \mathbb{R}^n \to \mathbb{R}^n$, with given $x_0 \in \mathbb{R}^n$ and $h \in \mathbb{R}^n$, the Jacobian matrix is defined as

$$Df(x_0) = \left(\frac{\partial f_i}{\partial x_j}(x_0) \right).$$

2.4 Improper Integrals and Cauchy Principal Values

In this book, we use various integrals, such as the Riemann integral $\int_a^b f(t)dt$, the Riemann-Stieltjes integral $\int_a^b f(t)dg(t)$, and integrals with respect to a measure μ, including the Lebesgue measure and a probability measure. In particular, we will need the integral with respect to a probability measure $\mathbb{P}$ (see Section 2.7), for example, $\int X d\mathbb{P}$.

Let us also recall the improper Riemann integral [6, Section 10.13].

Definition 2.1 (Improper integral)

(i) If f is Riemann integrable on $[a, b]$ for every $b > a$, then the improper Riemann integral $\int_a^{+\infty} f(t)dt$ is defined as

$$\int_a^{+\infty} f(t)dt \triangleq \lim_{b \to +\infty} \int_a^b f(t)dt, \tag{2.1}$$

whenever the limit on the right hand side exists.
Similarly,

$$\int_{-\infty}^{+\infty} f(t)dt \triangleq \lim_{a \to -\infty} \int_a^p f(t)dt + \lim_{b \to +\infty} \int_p^b f(t)dt, \tag{2.2}$$

whenever both limits on the right hand side exist, for some fixed real p.

(ii) If f is unbounded at point a, and is Riemann integrable on $[c, b]$ for every $c \in (a, b)$, then the improper Riemann integral $\int_a^b f(t)dt$ is defined as

$$\int_a^b f(t)dt \triangleq \lim_{c \to a+} \int_c^b f(t)dt, \tag{2.3}$$

whenever the limit on the right-hand side exists. Similarly, if f is unbounded at $r \in (a, b)$ but Riemann integrable on $[a, c_1]$ for every $c_1 \in (a, r)$, and Riemann integrable on $[c_2, b]$ for every $c_2 \in (r, b)$, then the improper Riemann integral $\int_a^b f(t)dt$ is defined as

$$\int_a^b f(t)dt \triangleq \lim_{c_1 \to r-} \int_a^{c_1} f(t)dt + \lim_{c_2 \to r+} \int_{c_2}^b f(t)dt, \tag{2.4}$$

whenever both limits on the right-hand side exist.

When an improper Riemann integral of f is well defined (i.e., it has finite value), we say f is an improper Riemann integrable function.

If f is an improper Riemann integrable function on $[a, +\infty)$, and additionally, $\int_a^b f(t)dt$ is bounded for $b \geq a$, then f is also Lebesgue integrable, and both integrals have the same value [6, p. 276].

We also recall the definition for Cauchy principal values for improper Riemann integrals.

Definition 2.2 (Cauchy principal values)

(i) If f is Riemann integrable on $[-b, b]$ for every $b > 0$, then the Cauchy principal value of the improper Riemann integral $\int_{-\infty}^{+\infty} f(t)dt$ is defined and denoted as

$$\text{P. V.} \int_{-\infty}^{+\infty} f(t)dt \triangleq \lim_{b \to +\infty} \int_{-b}^b f(t)dt \tag{2.5}$$

whenever the limit on the right-hand side exists.

(ii) If f is unbounded at $r \in (a, b)$ but Riemann integrable on $[a, c_1]$ for every $c_1 \in (a, r)$, and Riemann integrable on $[c_2, b]$ for every $c_2 \in (r, b)$, then the Cauchy principal value for the improper Riemann integral $\int_a^b f(t)dt$ is defined and denoted as

$$\text{P. V.} \int_a^b f(t)dt \triangleq \lim_{\varepsilon \to 0+} \left[\int_a^{r-\varepsilon} f(t)dt + \int_{r+\varepsilon}^b f(t)dt \right] \tag{2.6}$$

whenever the limit on the right-hand side exists.

The key point in the definition of the Cauchy principal values is the symmetric limit. Note that the Cauchy principal value might exist even when the corresponding improper integral does not exist.

Example 2.3 The improper integral $\int_{-1}^1 \frac{1}{t} dt$ does not exist, since

$$\int_{-1}^1 \frac{1}{t} dt = \lim_{c_1 \to 0-} \int_{-1}^{c_1} \frac{1}{t} dt + \lim_{c_2 \to 0+} \int_{c_2}^1 \frac{1}{t} dt$$

is not well defined (which gives $-\infty + \infty$). However, the Cauchy principal value exists:

$$\text{P. V.} \int_{-1}^1 \frac{1}{t} dt = \lim_{\varepsilon \to 0+} \left[\int_{-1}^{-\varepsilon} \frac{1}{t} dt + \int_{\varepsilon}^1 \frac{1}{t} dt \right] = 0.$$

2.5 Some Useful Inequalities

We recall some useful inequalities [303, 158, 306] for estimating solution orbits of stochastic differential equations. These inequalities hold only when each term in them is well defined or meaningful. Some inequalities are stated for scalar random variables or scalar functions, although they hold also for random vectors or vector functions.

2.5.1 Young's Inequality

Young's inequality has been used widely in estimating products of two quantities (e.g., a product of two solutions).

Let a and b be two nonnegative real numbers. Then,

$$ab \le \frac{a^p}{p} + \frac{b^q}{q},$$

where $1 < p, q < \infty$, and

$$\frac{1}{p} + \frac{1}{q} = 1.$$

A special case is $ab \leq \frac{1}{2}a^2 + \frac{1}{2}b^2$. Another special case is the following:

$$ab = \sqrt{\epsilon}a \, \frac{b}{\sqrt{\epsilon}} \leq \frac{\epsilon}{2}a^2 + \frac{1}{2\epsilon}b^2 \quad \text{for } a, b \geq 0 \text{ and } \epsilon > 0.$$

2.5.2 Gronwall Inequality

This inequality has differential and integral forms.

Gronwall Inequality: Differential Form

If $y(t) \geq 0$, $g(t)$ and $h(t)$ are integrable, and $\frac{dy}{dt} \leq g(t)y + h(t)$ for $t \geq t_0$, then

$$y(t) \leq y(t_0)e^{\int_{t_0}^t g(\tau)d\tau} + \int_{t_0}^t h(s)\left[e^{\int_s^t g(\tau)d\tau}\right]ds, \quad t \geq t_0.$$

In particular, if $\frac{dy}{dt} \leq gy + h$ for $t \geq t_0$ with g, h being constants and $t_0 = 0$, we have

$$y(t) \leq y(0)e^{gt} - \frac{h}{g}(1 - e^{gt}), \quad t \geq 0.$$

Note that when constant $g < 0$, we conclude that $\lim_{t\to\infty} y(t) = -\frac{h}{g}$.

Gronwall Inequality: Integral Form [110, 64]

If $u(t)$, $v(t)$ and $c(t)$ are all nonnegative, $c(t)$ is differentiable, and $v(t) \leq c(t) + \int_0^t u(s)v(s)ds$ for $t \geq t_0$, then

$$v(t) \leq v(t_0)e^{\int_{t_0}^t u(\tau)d\tau} + \int_{t_0}^t c'(s)\left[e^{\int_s^t u(\tau)d\tau}\right]ds, \quad t \geq t_0.$$

In particular, if $y(t)$ is nonnegative and continuous and $y(t) \leq C + K\int_0^t y(s)ds$, with C, K being positive constants, for $t > 0$, then

$$y(t) \leq Ce^{Kt}, \quad t \geq 0.$$

2.5.3 Cauchy-Schwarz Inequality

The Cauchy-Schwarz inequality holds in any Hilbert space. We review it in various familiar Hilbert spaces.

The Cauchy-Schwarz inequality in $\mathbb{R}^n$ is

$$|x \cdot y| \leq \|x\| \, \|y\|,$$

where $\cdot$ denotes the usual scalar product in $\mathbb{R}^n$. Namely,

$$\left|\sum_{i=1}^{n} x_i y_i\right| \leq \sqrt{\sum_{i=1}^{n} x_i^2}\sqrt{\sum_{i=1}^{n} y_i^2}.$$

The Cauchy-Schwarz inequality in the space of square-summable infinite sequences l^2 is

$$\left|\sum_{i=1}^{\infty} x_i y_i\right| \leq \sqrt{\sum_{i=1}^{\infty} x_i^2}\sqrt{\sum_{i=1}^{\infty} y_i^2}.$$

The Cauchy-Schwarz inequality in the Lebesgue space $L^2(D)$ of square-integrable functions defined on a domain $D \subset \mathbb{R}^n$ is

$$\left|\int_D f(x)g(x)dx\right| \leq \sqrt{\int_D f^2(x)dx}\sqrt{\int_D g^2(x)dx}.$$

The Cauchy-Schwarz inequality in more general Hilbert space H is

$$|\langle x, y\rangle| \leq \|x\| \, \|y\|.$$

2.5.4 Hölder Inequality

The Hölder inequality may be regarded as a generalization of the Cauchy-Schwarz inequality. It holds in various Banach spaces as follows:

Let $1 < p, q < \infty$ with $\frac{1}{p} + \frac{1}{q} = 1$.

The Hölder inequality in $\mathbb{R}^n$, for $x = (x_1, \ldots, x_n)$, $y = (y_1, \ldots, y_n)$ in $\mathbb{R}^n$, is

$$\left|\sum_{i=1}^{n} x_i y_i\right| \leq \left(\sum_{i=1}^{n} x_i^p\right)^{\frac{1}{p}} \left(\sum_{i=1}^{n} y_i^q\right)^{\frac{1}{q}}.$$

The Hölder inequality in the space of infinite sequences, for $x = (x_1, x_2, \ldots) \in l^p$ and $y = (y_1, y_2, \ldots) \in l^q$, is

$$\left|\sum_{i=1}^{\infty} x_i y_i\right| \leq \left(\sum_{i=1}^{\infty} x_i^p\right)^{\frac{1}{p}} \left(\sum_{i=1}^{\infty} y_i^q\right)^{\frac{1}{q}},$$

where $l^r = \{x = (x_1, x_2, \ldots) \colon \sum_{i=1}^{\infty} |x_i|^r < \infty\}$ for $r > 0$.

The Hölder inequality in the Lebesgue space, for $X \in L^p(D)$ and $Y \in L^q(D)$, is

$$\left| \int_D f(x)g(x)dx \right| \leq \left(\int_D |f(x)|^p dx \right)^{\frac{1}{p}} \left(\int_D |g(x)|^q dx \right)^{\frac{1}{q}},$$

where $L^p(D) \triangleq \{f\colon \int_D |f(x)|^p dx < \infty\}$ for a domain $D \subset \mathbb{R}^n$.

When $p = q = 2$, Hölder inequalities become the Cauchy-Schwarz inequalities, as stated earlier.

2.5.5 *Minkowski Inequality*

Let $1 \leq p < \infty$.

The Minkowski inequality in $\mathbb{R}^n$ is

$$\left(\sum_{i=1}^{n} |x_i \pm y_i|^p \right)^{\frac{1}{p}} \leq \left(\sum_{i=1}^{n} |x_i|^p \right)^{\frac{1}{p}} + \left(\sum_{i=1}^{n} |y_i|^p \right)^{\frac{1}{p}}.$$

The Minkowski inequality in l^p is

$$\left(\sum_{i=1}^{\infty} |x_i \pm y_i|^p \right)^{\frac{1}{p}} \leq \left(\sum_{i=1}^{\infty} |x_i|^p \right)^{\frac{1}{p}} + \left(\sum_{i=1}^{\infty} |y_i|^p \right)^{\frac{1}{p}}.$$

The Minkowski inequality for random variables in $L^p(\Omega)$ is

$$(\mathbb{E}|X(\omega) \pm Y(\omega)|^p)^{\frac{1}{p}} \leq (\mathbb{E}|X|^p)^{\frac{1}{p}} + (\mathbb{E}|Y|^p)^{\frac{1}{p}}.$$

The Minkowski inequality in the Lebesgue space $L^p(D)$ is

$$\left(\int_D |f(x) \pm g(x)|^p dx \right)^{\frac{1}{p}} \leq \left(\int_D |f(x)|^p dx \right)^{\frac{1}{p}} + \left(\int_D |g(x)|^p dx \right)^{\frac{1}{p}}.$$

2.6 Hölder Spaces, Sobolev Spaces, and Related Inequalities

Let D be a domain in $\mathbb{R}^n$, with boundary ∂D. Let $C(D)$ or $C(\bar{D})$ be the set of continuous functions on D (or $\bar{D}$) with the usual sup-norm. They are also denoted as $C^0(D)$ or $C^0(\bar{D})$. Similarly, for a positive integer k, let $C^k(D)$ or $C^k(\bar{D})$ be the set of functions having all derivatives of order less than or equal to k continuous in D or in $\bar{D}$. The support of a function is the closure of the set on which the function does not vanish. We denote by $C_0^k(D)$ the set of functions in $C^k(D)$ with compact support in D. Moreover, $C^\infty(D)$ is defined to be the set of functions having derivatives of all orders and $C_0^\infty(D)$ to be the set of functions in $C^\infty(D)$ with compact support in D. When $D = \mathbb{R}^n$, we have $C^k(\mathbb{R}^n)$, $C_0^k(\mathbb{R}^n)$, $C^\infty(\mathbb{R}^n)$, and $C_0^\infty(\mathbb{R}^n)$.

Now, we consider Hölder spaces. Let $\alpha \in (0, 1]$. A function f is called Hölder continuous with exponent α at x_0 in D if the quantity

$$[f]_{\alpha,x_0} \triangleq \sup_D \frac{|f(x) - f(x_0)|}{|x - x_0|^\alpha} \tag{2.7}$$

is finite. When $\alpha = 1$, f is said to be Lipschitz continuous at x_0.

A function f is called uniformly Hölder continuous with exponent α in D if the quantity

$$[f]_\alpha \triangleq \sup_{x,y\in D, x\neq y} \frac{|f(x) - f(y)|}{|x - y|^\alpha} \tag{2.8}$$

is finite. We say f is locally Hölder continuous with exponent α in D if it is uniformly Hölder continuous with exponent α on compact subsets of D. The Hölder space $C^{k,\alpha}(D)$ (or $C^{k,\alpha}(\bar{D})$) is the subspace of $C^k(D)$ (or $C^k(\bar{D})$) consisting of functions whose kth-order derivatives are locally (or uniformly) Hölder continuous with exponent α in D.

Furthermore, $C_0^{k,\alpha}(D)$ is the space of functions in $C^{k,\alpha}(D)$ having compact support in D. We also denote $C^{0,\alpha}(D) = C^\alpha(D)$, $C^{0,\alpha}(\bar{D}) = C^\alpha(\bar{D})$, $C^{k,0}(D) = C^k(D)$, and $C^{k,0}(\bar{D}) = C^k(\bar{D})$.

Definition 2.4 A bounded domain D in $\mathbb{R}^n$ is called a $C^{k,\alpha}$ domain if each point of its boundary ∂D has a neighborhood in which ∂D is the graph of a $C^{k,\alpha}$ function. We also say that D has a $C^{k,\alpha}$ boundary. In this case, D also satisfies the exterior sphere condition, that is, for each point x in ∂D, there exists a ball B so that $\bar{B} \bigcap \bar{D} = \{x\}$.

We recall some inequalities for weakly differentiable functions ([89, chapter 5] or [105, chapter 7]). The Lebesgue spaces $L^p = L^p(D)$, $p \geq 1$, are the spaces of measurable functions that are pth-order Lebesgue integrable on a domain D in $\mathbb{R}^n$. The norm of f in L^p is defined by

$$\|f\|_{L^p} \triangleq \left(\int_D |f(x)|^p dx \right)^{\frac{1}{p}}.$$

In particular, $L^2(D)$ is a Hilbert space with the following scalar product $\langle \cdot, \cdot \rangle$ and norm $\| \cdot \|$:

$$\langle f, g \rangle \triangleq \int_D f\bar{g} dx, \qquad \|f\| \triangleq \sqrt{\langle f, f \rangle} = \sqrt{\int_D |f(x)|^2 dx},$$

for f, g in $L^2(D)$.

Now, we introduce some common Sobolev spaces. For $k = 1, 2, \ldots$, define

$$H^k(D) \triangleq \{f : \partial^\alpha f \in L^2(D), |\alpha| \leq k\}.$$

Here, $\alpha = (\alpha_1, \ldots, \alpha_n)$, with α_is being nonnegative integers, $|\alpha| = \alpha_1 + \cdots + \alpha_n$, and $\partial^\alpha \triangleq \partial_{x_1}^{\alpha_1} \cdots \partial_{x_n}^{\alpha_n}$. Each of these is a Hilbert space with scalar product

$$\langle u, v \rangle_k = \int_D \sum_{|\alpha| \le k} \partial^\alpha u \, \partial^\alpha \bar{v} \, dx$$

and the norm

$$\|u\|_k = \sqrt{\langle u, u \rangle_k} = \sqrt{\int_D \sum_{|\alpha| \le k} |\partial^\alpha u|^2 dx}.$$

For $k = 1, 2, \ldots$ and $p \ge 1$, we further define another class of Sobolev spaces

$$W^{k,p}(D) \triangleq \{u \colon \partial^\alpha u \in L^p(D), |\alpha| \le k\},$$

with norm

$$\|u\|_{k,p} = \left(\sum_{|\alpha| \le k} \|\partial^\alpha u\|_{L^p}^p \right)^{\frac{1}{p}}.$$

Recall that $C_0^\infty(D)$ is the space of infinitely differentiable functions with compact support in the domain D. Then $H_0^k(D)$ is the closure of $C_0^\infty(D)$ in Hilbert space $H^k(D)$ (under the norm $\| \cdot \|_k$). It is a Hilbert space inside $H^k(D)$. Similarly, $W_0^{k,p}(D)$ is the closure of $C_0^\infty(D)$ in Banach space $W^{k,p}(D)$ (under the norm $\| \cdot \|_{k,p}$), and it is a Banach space inside $W^{k,p}(D)$.

When $D = \mathbb{R}^n$, we similarly have $H^k(\mathbb{R}^n)$, $H_0^k(\mathbb{R}^n)$, $W^{k,p}(\mathbb{R}^n)$, and $W_0^{k,p}(\mathbb{R}^n)$.

Standard abbreviations $L^2 = L^2(D)$, $H_0^k = H_0^k(D)$, $k = 1, 2, \ldots$, are used for these common Sobolev spaces.

Let us list several common inequalities involving function spaces.

2.6.1 *Cauchy-Schwarz Inequality*

For $f, g \in L^2(D)$,

$$\left| \int_D f(x) g(x) dx \right| \le \sqrt{\int_D |f(x)|^2 dx} \sqrt{\int_D |g(x)|^2 dx}.$$

2.6.2 *Hölder Inequality*

For $f \in L^p(D)$ and $g \in L^q(D)$ with $\frac{1}{p} + \frac{1}{q} = 1$, $p > 1, q > 1$,

$$\left| \int_D f(x) g(x) dx \right| \le \left(\int_D |f(x)|^p dx \right)^{\frac{1}{p}} \left(\int_D |g(x)|^q dx \right)^{\frac{1}{q}}.$$

2.6.3 Minkowski Inequality

For $f, g \in L^p(D)$,

$$\left(\int_D |f(x) \pm g(x)|^p dx\right)^{\frac{1}{p}} \leq \left(\int_D |f(x)|^p dx\right)^{\frac{1}{p}} + \left(\int_D |g(x)|^p dx\right)^{\frac{1}{p}}.$$

2.6.4 Poincaré Inequality in H_0^1

For $g \in H_0^1(D)$,

$$\|g\|^2 = \int_D |g(x)|^2\, dx \leq \left(\frac{|D|}{\omega_n}\right)^{\frac{1}{n}} \int_D |\nabla g|^2\, dx,$$

where $|D|$ is the Lebesgue measure of the domain D and ω_n is the volume of the unit ball in $\mathbb{R}^n$. In terms of the Gamma function Γ (see Remark 2.5),

$$\omega_n = \frac{\pi^{\frac{n}{2}}}{\Gamma(\frac{n}{2}+1)}.$$

For example, $\omega_1 = 2$, $\omega_2 = \pi$, and $\omega_3 = \frac{4}{3}\pi$.

Remark 2.5 The Gamma function Γ is defined by

$$\Gamma(z) \triangleq \int_0^{+\infty} t^{z-1} e^{-t} dt, \tag{2.9}$$

for $z > 0$. As an improper integral, it actually converges for complex numbers z with a positive real part. The Gamma function is a generalization of the factorial function since

$$\Gamma(n) = (n-1)!.$$

In particular, $\Gamma(1) = \Gamma(2) = 1$. Moreover,

$$\Gamma(z+1) = z\Gamma(z),$$

$$\Gamma(1-z)\Gamma(z) = \frac{\pi}{\sin(\pi z)},$$

$$\Gamma(z)\Gamma(z+\frac{1}{2}) = 2^{1-2z}\sqrt{\pi}\,\Gamma(2z).$$

In particular, $\Gamma(\frac{1}{2}) = \sqrt{\pi}$, and $\Gamma(n+\frac{1}{2}) = \frac{(2n)!\sqrt{\pi}}{n!2^{2n}}$ for $n = 1, 2, \ldots$.

The surface area of the sphere of radius r in $\mathbb{R}^n$ is [265, Appendix A.5] represented via Γ function

$$A_n = \frac{2\pi^{n/2} r^{n-1}}{\Gamma(n/2)}.$$

For example, $A_2 = 2\pi r$ and $A_3 = 4\pi r^2$.

The Gamma function Γ and the Riemann zeta function $\zeta(s) \triangleq \sum_{n=1}^{\infty} \frac{1}{n^s}$, $s > 1$, are related in the following way:

$$\Gamma(s)\zeta(s) = \int_0^\infty \frac{x^{s-1}}{e^x - 1} dx. \tag{2.10}$$

2.6.5 *Poincaré Inequality in* $W^{1,p}$

For $u \in W_0^{1,p}(D)$, $1 \leq p < \infty$, and $D \subset \mathbb{R}^n$ a bounded domain,

$$\|u\|_p \leq \left(\frac{|D|}{\omega_n}\right)^{\frac{1}{n}} \|\nabla u\|_p.$$

Let $u \in W^{1,p}(D)$, $1 \leq p < \infty$, and $D \subset \mathbb{R}^n$ be a bounded convex domain. Let $S \subset D$ be any measurable subset, and define the spatial average of u over S by $u_S = \frac{1}{|S|} \int_S u dx$ (with $|S|$ being the Lebesgue measure of S). Then,

$$\|u - u_S\|_p \leq \left(\frac{\omega_n}{|D|}\right)^{1-\frac{1}{n}} d^n \|\nabla u\|_p,$$

where d is the diameter of D.

2.6.6 *Agmon Inequality*

Let $D \subset \mathbb{R}^n$ be an open domain with piecewise smooth boundary. There exists a constant $C > 0$, depending only on domain D, such that

$$\|u\|_{L^\infty(D)} \leq C \|u\|^{\frac{1}{2}}_{H^{\frac{n-1}{2}}(D)} \|u\|^{\frac{1}{2}}_{H^{\frac{n+1}{2}}(D)}, \quad \text{for } n \text{ odd},$$

$$\|u\|_{L^\infty(D)} \leq C \|u\|^{\frac{1}{2}}_{H^{\frac{n-2}{2}}(D)} \|u\|^{\frac{1}{2}}_{H^{\frac{n+2}{2}}(D)}, \quad \text{for } n \text{ even}.$$

In particular, for $n = 1$ and $u \in H^1(0, l)$,

$$\|u\|_{L^\infty(0,l)} \leq C \|u\|^{\frac{1}{2}}_{L^2(0,l)} \|u\|^{\frac{1}{2}}_{H^1(0,l)}.$$

Moreover, for $n = 1$ and $u \in H_0^1(0, l)$,

$$\|u\|_{L^\infty(0,l)} \leq C \|u\|^{\frac{1}{2}}_{L^2(0,l)} \|u_x\|^{\frac{1}{2}}_{L^2(0,l)}.$$

For more information on analysis, see Zeidler [305, 306] or Yosida [303].

2.7 Probability Spaces

We recall some basic concepts in probability [13, 84]. A sample space Ω is the collection of all outcomes (also called samples) of a random experiment. While flipping a coin, there are two outcomes, head (H) or tail (T), and the natural sample space is $\Omega = \{H, T\}$. For tossing a cubic dice, there are six possibilities, and we can label the six faces by 1, 2, 3, 4, 5, 6. The sample space is thus $\Omega = \{1, 2, 3, 4, 5, 6\}$. In a deterministic experiment, we can think of tossing a ball. There is only one sample (i.e., we can only write one number on the surface of a ball, without getting confused). Thus the corresponding sample space is $\Omega = \{1\}$. With only one sample, it is not necessary to indicate it in a deterministic experiment.

A σ-field $\mathcal{F}$ is a family of subsets of the sample space, and these subsets satisfy that

(i) the empty set $\emptyset$ is in $\mathcal{F}$;
(ii) if $A \in \mathcal{F}$, then so is its complement $A^c = \Omega \setminus A$;
(iii) if $A_1, A_2, \ldots$ are in $\mathcal{F}$, then so is the union $\cup A_i$.

Thus Ω is also in $\mathcal{F}$. By the well-known De Morgan laws, $(\cup A_i)^c = \cap A_i^c$ and $(\cap A_i)^c = \cup A_i^c$, we see that the intersection $\cap A_i$ is also in $\mathcal{F}$. Each element of $\mathcal{F}$ is a subset of Ω and is called an event. Two events A and B are called disjoint if their intersection is empty, that is, $A \cap B = \emptyset$.

A probability measure is a mapping $\mathbb{P}\colon \mathcal{F} \to [0, 1]$ satisfying (i) $\mathbb{P}(\Omega) = 1$; (ii) $\mathbb{P}(\cup A_i) = \sum_i \mathbb{P}(A_i)$ for a family of pairwise disjoint events $A_1, A_2, \ldots$. If $A \subset B$, then $B = A \cup (B \setminus A)$. Thus, $\mathbb{P}(B) = \mathbb{P}(A) + \mathbb{P}(B \setminus A)$, or $\mathbb{P}(B \setminus A) = \mathbb{P}(B) - \mathbb{P}(A)$. In particular, $\mathbb{P}(\emptyset) = \mathbb{P}(\Omega) - \mathbb{P}(\Omega) = 0$. From the fact $A \cup A^c = \Omega$, now we have $\mathbb{P}(A^c) = 1 - \mathbb{P}(A)$. In this book, we will not distinguish $\mathbb{P}(A)$ from $\mathbb{P}\{A\}$; both mean the probability of the event A.

Definition 2.6 A probability space $(\Omega, \mathcal{F}, \mathbb{P})$ is composed of a sample space Ω, a σ-field $\mathcal{F}$, and a probability (or probability measure) $\mathbb{P}$.

A probability space, or its σ-field $\mathcal{F}$, is called complete if $\mathcal{F}$ contains all subsets of sets of probability 0. A set of probability 0 is called a null set or zero probability event. It is possible to make a probability space complete. This is called the completion of a probability space or its σ-field $\mathcal{F}$, which requires at least adding all sets of probability 0 and their complements into $\mathcal{F}$.

2.7.1 Scalar Random Variables

In the real line $\mathbb{R}^1$, a natural σ-field is the Borel σ-field $\mathcal{B}(\mathbb{R}^1)$. Recall that it is the smallest σ-field containing all open intervals, and thus it is called the σ-field generated by open intervals. In this book, $\mathbb{R}^1$ is always endowed with its Borel σ-field. Each set in this Borel σ-field is called a Borel set.

For two σ-fields $\mathcal{F}_1$ and $\mathcal{F}_2$, from two different probability spaces Ω_1 and Ω_2, their product σ-field $\mathcal{F}_1 \times \mathcal{F}_2$ is the smallest σ-field containing $\{A_1 \times A_2 \colon A_1 \in \mathcal{F}_1,\ A_2 \in \mathcal{F}_2\}$. Similarly, we can define the product σ-field of σ-fields $\mathcal{F}_1, \mathcal{F}_2, \ldots, \mathcal{F}_k$.

A random variable X taking real values is a measurable function $X \colon \Omega \to \mathbb{R}^1$. That is, for every Borel set A in $\mathcal{B}(\mathbb{R}^1)$, the preimage $X^{-1}(A)$ is in the σ-field $\mathcal{F}$.

The mean, or mathematical expectation, of a scalar real-valued random variable X is a real value defined as [46, 13, 30]

$$\mathbb{E}(X) = \int_\Omega X(\omega)\, d\mathbb{P}(\omega),$$

and the variance of X is a real value defined by

$$\mathrm{Var}(X) = \mathbb{E}((X - E(X))^2.$$

Here and hereafter, the mean and variance make sense when the right-hand sides in the preceding two equations are finite.

The covariance of two scalar random variables X and Y is defined by

$$\mathrm{Cov}(X, Y) = \mathbb{E}[(X - E(X))(Y - E(Y))]\,.$$

The distribution function $F_X(x)$ of X is defined as

$$F_X(x) = \mathbb{P}\{\omega \in \Omega \colon X(\omega) \le x\}\,, \tag{2.11}$$

which is a monotonically increasing function such that $F_X(-\infty) = 0$ and $F_X(\infty) = 1$. For example, $F_X(100)$ quantifies the likelihood that $X \le 100$.

Theorem 2.7 *If X is a scalar-continuous random variable, then $\mathbb{P}(X = a) = 0$ for every real value a, that is, the probability for X exactly assuming a given value is zero.*

Proof For any $h > 0$,

$$\mathbb{P}(X = a) \le \mathbb{P}(a - h < X \le a + h) = F(a + h) - F(a - h) \to 0,$$

as $h \to 0+$. Thus $\mathbb{P}(X = a) = 0$. □

Remark 2.8 By this theorem, we see that for any continuous random variable X, $\mathbb{P}(X < a) = \mathbb{P}(X \le a) = F_X(a)$.

Recall the fundamental theorem of calculus: if F is differentiable on $[a, b]$ and $F' = f$ is Riemann integrable on $[a, b]$, then

$$F(b) - F(a) = \int_a^b f(x)dx. \tag{2.12}$$

In particular, $F(x) = \int_{-\infty}^x f(x)dx$.

If there is a function $f : \mathbb{R}^1 \to \mathbb{R}^1$ such that

$$F_X(x) = \int_{-\infty}^x f(s)ds\,, \tag{2.13}$$

then f is called the probability density function of the random variable X. Note that $\frac{d}{dx}F_X(x) = f(x)$, at every continuity point x for f. Being the derivative of a monotonically increasing function, the probability density function is nonnegative: $f(x) \geq 0$. Moreover, $\int_{-\infty}^{\infty} f(x)dx = 1$ since $F_X(\infty) = 1$. In this case, the mean and variance of X can be calculated via the probability density function f as follows:

$$\mathbb{E}X = \int_{\mathbb{R}^1} xf(x)dx, \tag{2.14}$$

$$\mathrm{Var}(X) = \int_{\mathbb{R}^1} (x - \mathbb{E}X)^2 f(x)dx. \tag{2.15}$$

2.7.2 Random Vectors

Similarly, with the Borel σ-field $\mathcal{B}(\mathbb{R}^n)$, we define a random variable taking values in Euclidean space $\mathbb{R}^n$.

The mean or mathematical expectation of a $\mathbb{R}^n$-valued random variable X is a vector in $\mathbb{R}^n$ defined as [46, 13, 30]

$$\mathbb{E}(X) = \int_\Omega X(\omega)\, d\mathbb{P}(\omega),$$

whenever the integral in the right-hand side is well defined componentwise. The covariance of X is the $n \times n$ matrix defined by

$$\mathrm{Cov}(X, X) = \mathbb{E}[(X - \mathbb{E}(X))(X - \mathbb{E}(X))^{\mathrm{T}}],$$

where $(\cdot)^{\mathrm{T}}$ denotes the transpose of a vector or a matrix. The covariance of $\mathbb{R}^n$-valued random variables X and Y is the $n \times n$ matrix defined by

$$\mathrm{Cov}(X, Y) = \mathbb{E}[(X - \mathbb{E}(X))(Y - \mathbb{E}(Y))^{\mathrm{T}}],$$

whenever the right-hand side makes sense.

For $p \geq 1$, denote

$$L^p(\Omega) \triangleq L^p(\Omega, \mathbb{R}^n) = \{Y : Y \text{ taking values in } \mathbb{R}^n,\ \mathbb{E}\|Y\|^p < \infty\}.$$

In particular, $L^2(\Omega)$ is a Hilbert space with the scalar product $\langle X, Y\rangle = \mathbb{E}(XY)$. The sequence $X_n \to X$ in L^p if $\lim_{n\to\infty} \mathbb{E}|X_n - X|^p = 0$. When $p = 2$, this convergence is called convergence in mean square.

The following useful inequalities are for random variables in $L^p(\Omega) \triangleq L^p(\Omega, \mathbb{R}^n)$.

2.7.3 Chebyshev Inequality

For $c > 0$, $p > 0$, and $X \in L^p(\Omega)$,

$$\mathbb{P}(\omega\colon \|X(\omega)\| \geq c) \leq \frac{\mathbb{E}\|X\|^p}{c^p}.$$

2.7.4 Jensen Inequality

Let f be a convex function and let both X and $f(X)$ be in $L^1(\Omega)$. Then,

$$f(\mathbb{E}X) \leq \mathbb{E}f(X).$$

2.7.5 Lyapunov Inequality

For $0 < q < p$ and $X \in L^p(\Omega)$,

$$(\mathbb{E}\|X\|^q)^{\frac{1}{q}} \leq (\mathbb{E}\|X\|^p)^{\frac{1}{p}}.$$

This says $L^p(\Omega) \subset L^q(\Omega)$ for $p > q > 0$. It also implies that an even-order moment is an upper bound for all its lower-order moments. In fact, $\mathbb{E}(X \cdot X)^k = \mathbb{E}\|X\|^{2k} \geq (\mathbb{E}\|X\|^q)^{\frac{2k}{q}}$ for positive q, k with $q < 2k$ and $k = 1, 2, \ldots$.

2.7.6 Cauchy-Schwarz Inequality

The Cauchy-Schwarz inequality in the space of square-integrable random variables $L^2(\Omega)$, is, for $X, Y \in L^2(\Omega)$,

$$|\mathbb{E}(X(\omega) \cdot Y(\omega))| \leq \sqrt{\mathbb{E}\|X\|^2}\,\sqrt{\mathbb{E}\|Y\|^2}.$$

2.7.7 Hölder Inequality

The Hölder inequality in the space of random variables, for $X \in L^p(\Omega)$ and $Y \in L^q(\Omega)$, is

$$|\mathbb{E}(X(\omega) \cdot Y(\omega))| \leq (\mathbb{E}\|X\|^p)^{\frac{1}{p}}\,(\mathbb{E}\|Y\|^q)^{\frac{1}{q}}.$$

2.7.8 Gaussian Random Variables

A scalar normal or Gaussian random variable $X\colon \Omega \to \mathbb{R}^1$ has distribution $\mathcal{N}(\mu, \sigma^2)$, with probability density function

$$f(x) = \frac{1}{\sqrt{2\pi}\,\sigma} \exp\left[\frac{-(x-\mu)^2}{2\sigma^2}\right],$$

where μ, σ are real constants and $\sigma \neq 0$.

It is known that $\mathbb{E}(X) = \mu$ and $\mathrm{Var}(X) = \sigma^2$. We often write $X \sim \mathcal{N}(\mu, \sigma^2)$. In fact, it can be further shown that the odd central moments $\mathbb{E}(X-\mu)^{2k+1} = 0$ for $k = 0, 1, 2, \ldots$ and the even central moments $\mathbb{E}(X-\mu)^{2k} = 1 \cdot 3 \cdot 5 \cdots (2k-1)\sigma^{2k}$ for $k = 1, 2, \ldots$. This implies that all moment of a Gaussian random variable can be expressed in terms of the first two moments or in terms of its mean and variance.

The characteristic function for a scalar Gaussian random variable, $X \sim \mathcal{N}(\mu, \sigma^2)$, is

$$\Phi(u) = \mathbb{E}e^{iuX} = e^{i\mu u - \frac{1}{2}\sigma^2 u^2}, \tag{2.16}$$

and its moment-generating function is

$$M(u) = \mathbb{E}e^{uX} = e^{\mu u + \frac{1}{2}\sigma^2 u^2}. \tag{2.17}$$

The kth moment of X can be expressed in terms of the kth derivative of the moment-generating function as follows: $\mathbb{E}(X^k) = m^{(k)}(0)$.

For a Gaussian random variable $X \sim \mathcal{N}(0, \sigma^2)$, let us calculate its tail distribution: for $a > 0$,

$$\begin{aligned}
\mathbb{P}(X > a) &= \frac{1}{\sqrt{2\pi}\,\sigma} \int_a^\infty \exp\left[\frac{-x^2}{2\sigma^2}\right] dx \\
&\leq \frac{1}{a\sqrt{2\pi}\,\sigma} \int_a^\infty x \exp\left[\frac{-x^2}{2\sigma^2}\right] dx \\
&= \frac{\sigma}{a\sqrt{2\pi}} \exp\left[\frac{-a^2}{2\sigma^2}\right].
\end{aligned} \tag{2.18}$$

So, this tail is "light" as it decays exponentially fast in a.

A random variable, taking values in $\mathbb{R}^n$,

$$X\colon \Omega \to \mathbb{R}^n,$$

is called Gaussian if, for any $a = (a_1, \ldots, a_n)^{\mathrm{T}} \in \mathbb{R}^n$, $X \cdot a = a_1 X_1 + \cdots + a_n X_n$ is a scalar Gaussian random variable. A Gaussian random variable in $\mathbb{R}^n$ is denoted as $X \sim \mathcal{N}(m, Q)$, with mean vector $\mathbb{E}(X) = m$ and covariance matrix $\mathrm{Cov}(X, X) = Q$. The covariance matrix Q, defined by

$$Q = (Q_{ij}) = (\mathbb{E}[(X_i - m_i)(X_j - m_j)]),$$

is symmetric and nonnegative definite (i.e., eigenvalue $\lambda_j \geq 0,\ j = 1, \ldots, n$). The trace of Q is written as $\mathrm{Tr}(Q) = \lambda_1 + \cdots + \lambda_n$. The probability density function for this Gaussian random variable X in $\mathbb{R}^n$ is

$$\begin{aligned} f(x) = f(x_1, \ldots, x_n) &= \frac{\sqrt{\det(A)}}{(2\pi)^{n/2}} e^{-\frac{1}{2}\sum_{j,k=1}^{n}(x_j - m_j)a_{jk}(x_k - m_k)} \\ &= \frac{\sqrt{\det(A)}}{(2\pi)^{n/2}} e^{-\frac{1}{2}(x-m)^{\mathrm{T}}A(x-m)} \\ &= \frac{1}{(2\pi)^{n/2}\sqrt{\det(Q)}} \exp\left[-\frac{1}{2}(x - m)^{\mathrm{T}} Q^{-1}(x - m)\right], \qquad (2.19) \end{aligned}$$

where $A = Q^{-1} = (a_{jk})$.

Here are some observations. For $a, b \in \mathbb{R}^n$,

$$\mathbb{E}\langle X, a\rangle = \mathbb{E}\sum_{i=1}^{n} a_i X_i = \sum_{i=1}^{n} a_i \mathbb{E}(X_i) = \sum_{i=1}^{n} a_i m_i = \langle m, a\rangle, \qquad (2.20)$$

and

$$\begin{aligned} \mathbb{E}(\langle X - m, a\rangle\langle X - m, b\rangle) &= \mathbb{E}\left(\sum_i a_i(X_i - m_i) \sum_j b_j(X_j - m_j)\right) \\ &= \sum_{i,j} a_i b_j \mathbb{E}[(X_i - m_i)(X_j - m_j)] \\ &= \sum_{i,j} a_i b_j Q_{ij} = \langle Qa, b\rangle. \qquad (2.21) \end{aligned}$$

In particular, $\langle Qa, a\rangle = \mathbb{E}\langle X - m, a\rangle^2 \geq 0$, which confirms that Q is nonnegative definite. Also, $\langle Qa, b\rangle = \langle a, Qb\rangle$, which implies that Q is symmetric.

The characteristic function for a Gaussian random variable in $\mathbb{R}^n$, $X \sim \mathcal{N}(m, Q)$ with mean vector m and covariance matrix Q, is

$$\Phi(u) = \mathbb{E}e^{i\langle u, X\rangle} = e^{i\langle m, u\rangle - \frac{1}{2}\langle u, Qu\rangle}, \qquad (2.22)$$

and its moment-generating function is

$$M(u) = \mathbb{E}e^{\langle u, X\rangle} = e^{\langle m, u\rangle + \frac{1}{2}\langle u, Qu\rangle}. \qquad (2.23)$$

Here m is its mean vector and Q is the symmetric nonnegative definite covariance matrix.

2.7.9 Non-Gaussian Random Variables

In Chapter 7, we also consider non-Gaussian random variables, especially α-stable random variables.

2.7.10 More Probability Concepts

We recall more probability concepts for random variables taking values in $\mathbb{R}^1$. These concepts may be similarly introduced for random variables taking values in $\mathbb{R}^n$.

Note that for a random variable $X: \Omega \to \mathbb{R}^1$, preimages of the form $X^{-1}(A) = \{\omega \in \Omega: X(\omega) \in A\}$, for a Borel set in $\mathbb{R}^1$, carry information about X. For example, for $A = (-\infty, a]$, $X^{-1}(A) = F_X(a)$. Therefore, it is worthwhile to quantify or describe these preimages.

The probability distribution measure P^X, induced by a scalar random variable X, is defined as

$$P^X(A) = \mathbb{P}(X(\omega) \in A)$$

for every Borel set $A \in \mathcal{B}(\mathbb{R}^1)$. When $A = (-\infty, x]$, $P^X((-\infty, x]) = \mathbb{P}(X \leq x) = F_X(x)$. Therefore, the distribution measure P^X, for a random variable X, is a generalization of the usual distribution function F_X. This probability measure is also called the law of X and is sometimes denoted by $\mathcal{L}_X$.

Theorem 2.9 *The distribution measure $\mathbb{P}^X$ is indeed a probability measure on the Euclidean space $\mathbb{R}^1$, and thus $(\mathbb{R}^1, \mathcal{B}(\mathbb{R}^1), P^X)$ is a probability space.*

Proof Check Definition 2.6. In particular, $P^X(\mathbb{R}^1) = \mathbb{P}(X \in \mathbb{R}^1) = \mathbb{P}(\Omega) = 1$. □

Because preimages like $X^{-1}(A)$ carry information for X, it is also worthwhile to examine them together as a special σ-field. The σ-field, $\mathcal{F}^X$, generated by a scalar random variable X is the smallest σ-field containing all preimages of the form $\{\omega: a < X(\omega) < b\}$, for all $a, b \in \mathbb{R}^1$. We also use the notation $\mathcal{F}^X = \sigma(X)$. This σ-field holds all information about X and is the smallest σ-field on which X is measurable.

Now it is natural to discuss the independence of two scalar random variables X and Y. Recall that two events, A and B, in $(\Omega, \mathcal{F}, \mathbb{P})$ are independent if $\mathbb{P}(A \cap B) = \mathbb{P}(A)\mathbb{P}(B)$. Thus, it becomes clear that random variables X and Y are independent if $\mathbb{P}(A \cap B) = \mathbb{P}(A)\mathbb{P}(B)$ for every $A \in \mathcal{F}^X$ and every $B \in \mathcal{F}^Y$, that is, $\mathbb{P}(X^{-1}(E) \cap Y^{-1}(F)) = \mathbb{P}(X^{-1}(E))\mathbb{P}(Y^{-1}(F))$ for all Borel sets, E and F, in $\mathbb{R}^1$. In other words, X, Y are independent if

$$\mathbb{P}(X \in E, Y \in F) = \mathbb{P}(X \in E)\mathbb{P}(Y \in F) \tag{2.24}$$

for all Borel sets E and F. In this case, we also say the two σ-fields, $\mathcal{F}^X$ and $\mathcal{F}^Y$, are independent.

Similarly, events $A_1, A_2, \ldots, A_k$ are called independent if

$$\mathbb{P}(A_1 \cap A_2 \ldots \cap A_k) = \mathbb{P}(A_1)\mathbb{P}(A_2) \ldots \mathbb{P}(A_k).$$

The random variables $X_1, X_2, \ldots, X_k$ are independent if the σ-fields generated by them, $\mathcal{F}^{X_1}, \mathcal{F}^{X_2}, \ldots, \mathcal{F}^{X_k}$, are independent, that is, $A_1, A_2, \ldots, A_k$ are independent for every $A_1 \in \mathcal{F}^{X_1}, A_2 \in \mathcal{F}^{X_2}, \ldots, A_k \in \mathcal{F}^{X_k}$. This concept will be needed in Chapter 3, when we introduce a Brownian motion, which has independent increments.

Theorem 2.10 *If X, Y are two independent random variables, and $f, g : \mathbb{R}^1 \to \mathbb{R}^1$ are two Borel measurable functions, then $f(X), g(Y)$ are independent.*

Proof For Borel sets E, F in $\mathbb{R}^1$,

$$\begin{aligned}\mathbb{P}(f(X) \in E, g(Y) \in F) &= \mathbb{P}(X \in f^{-1}(E), Y \in g^{-1}(F)) \\ &= \mathbb{P}(X \in f^{-1}(E))\mathbb{P}(Y \in g^{-1}(F)) \\ &= \mathbb{P}(f(X) \in E)\mathbb{P}(g(Y) \in F).\end{aligned}$$

Therefore, $f(X)$ and $g(Y)$ are independent.

If X, Y have probability density functions $f_X(x)$ and $f_Y(y)$, respectively, together with the joint probability density function $f(x, y)$, then they are independent if and only if

$$f(x, y) = f_X(x) f_Y(y). \tag{2.25}$$

The following theorem also characterizes the independence of random variables [46].

Theorem 2.11 *The random variables X, Y are independent if and only if*

$$\mathbb{E}(f(X)g(Y)) = \mathbb{E} f(X)\, \mathbb{E} g(Y) \tag{2.26}$$

for all Borel measurable bounded functions f and g.

Recall that the covariance and correlation, for scalar random variables X and Y, are defined as

$$\operatorname{Cov}(X, Y) \triangleq \mathbb{E}[(X - \mathbb{E}X)(Y - \mathbb{E}Y)], \tag{2.27}$$

$$\operatorname{Corr}(X, Y) \triangleq \frac{\mathbb{E}[(X - \mathbb{E}X)(Y - \mathbb{E}Y)]}{\sqrt{\mathbb{E}(X - \mathbb{E}X)^2}\sqrt{\mathbb{E}(Y - \mathbb{E}Y)^2}}, \tag{2.28}$$

whenever the right-hand sides in these two equations make sense. When $\mathrm{Corr}(X, Y) = 0$, we say X and Y are uncorrelated.

Remark 2.12 Theorem 2.11 implies that X and Y are independent if and only if $f(X)$ and $g(Y)$ are uncorrelated, for all Borel measurable bounded functions f and g.

In particular, if X and Y are independent, then they are uncorrelated, that is, $\mathrm{Corr}(X, Y) = 0$, as shown in [46, Proposition 5.7, p. 139]. But the converse is not generally true. However, if (X, Y) is a two-dimensional Gaussian random variable, then X and Y are independent if and only if they are uncorrelated [151, Theorem 2.19, p. 41].

The space where a random variable takes its values is called a state space, usually denoted as S, and it may be $\mathbb{R}^1$, $\mathbb{R}^n$, a Hilbert space, such as $L^2(\mathbb{R}^n)$, or a Banach space, such as $C(\mathbb{R}^n)$. The state space S is often equipped with a Borel σ-field $\mathcal{B}(S)$.

2.8 Stochastic Processes

A stochastic process X_t is a family of random variables taking values in a state space S, on a probability space $(\Omega, \mathcal{F}, \mathbb{P})$, parameterized by time $t \in \mathbb{R}$ (a special case is one-sided time, $t \in \mathbb{R}^+$). In this book, unless stated otherwise, we consider stochastic processes with two-sided time $\mathbb{R}$. We also write $X(t)$ as X_t, $X_t(\omega)$ or $X(t, \omega)$.

The σ-field generated by a stochastic process X_t is $\mathcal{F}^X = \sigma(X_s, s \in \mathbb{R})$, which is the smallest σ-field containing all preimages $X_s^{-1}(A)$, for $A \in \mathcal{B}(S)$. It is the smallest σ-field on which X_t is measurable for every t.

Two stochastic processes X_t and Y_t are independent if their generated σ-fields, $\sigma(X_s, s \in \mathbb{R})$ and $\sigma(Y_s, s \in \mathbb{R})$, are independent, that is, $\mathbb{P}(A \cap B) = \mathbb{P}(A)\mathbb{P}(B)$ for every $A \in \sigma(X_s, s \in \mathbb{R})$ and every $B \in \sigma(Y_s, s \in \mathbb{R})$. We can also talk about the independence of a random variable ξ and a stochastic process X_t, when σ-fields $\sigma(\xi)$ and $\sigma(X_s, s \in \mathbb{R})$ are independent.

There are three ways to consider a real-valued stochastic process $X_t(\omega)$, $t \in \mathbb{R}$.

The first way is to think of it as a family of random variables on $(\Omega, \mathcal{F}, \mathbb{P})$. Given a particular t, it is a random variable.

The second way is to think of $X_t(\omega) = X(t, \omega)$ as a function on the product space $\mathbb{R}^1 \times \Omega$.

The third way is to treat it as a random variable taking values in an infinite dimensional function space $(\mathbb{R}^1)^{\mathbb{R}^1}$, the space of all functions $f: \mathbb{R}^1 \to \mathbb{R}^1$. This space is endowed with the product σ-field $\mathcal{B}((\mathbb{R}^1)^{\mathbb{R}^1})$. Then, the stochastic process $X_t(\omega)$ is a random variable $X: \Omega \to (\mathbb{R}^1)^{\mathbb{R}^1}$. This random variable induces a probability measure $\mathbb{P}^X$ on $(\mathbb{R}^1)^{\mathbb{R}^1}$ defined as

$$\mathbb{P}^X(B) \triangleq \mathbb{P}(X^{-1}(B)) = \mathbb{P}(\omega: X(\omega) \in B), \quad \text{for } B \in \mathcal{B}((\mathbb{R}^1)^{\mathbb{R}^1}). \quad (2.29)$$

Definition 2.13 A stochastic process X_t is continuous at t_0 if $\lim_{t \to t_0} X_t = X_{t_0}$, a.s. A stochastic process X_t is continuous in probability at t_0 (or stochastically continuous at t_0) if $X_t \to X_{t_0}$ in probability, that is, $\lim_{t \to t_0} \mathbb{P}(|X_t - X_{t_0}| > \epsilon) = 0$ for all $\epsilon > 0$.

Because convergence a.s. implies convergence in probability, this stochastic continuity is weaker than the usual continuity.

Let X_t and Y_t be two stochastic processes. We say that Y_t is a modification or a version of X_t if $\mathbb{P}(X_t = Y_t) = 1$ for all t. A modification Y_t is called indistinguishable if $\mathbb{P}(X_t = Y_t, \text{ for all } t) = 1$, that is, X_t and Y_t almost surely (i.e., with probability one) have the same sample paths.

The following stochastic Fubini theorem is often useful [46, Theorem 6.13]. It tells us when we may exchange the order of integrals if the integrand is a stochastic process. This is used whenever we take mathematical expectation in examining stochastic differential equations. For other versions of the Fubini theorem, see [29, 281, 151].

Consider a stochastic process $X(t), t \in (0, T)$. Let $\mathcal{B}(0, T)$ be the Borel σ-field of $(0, T)$. The product σ-field $\mathcal{F} \times \mathcal{B}(0, T)$ is the smallest σ-field containing all events like $A_1 \times A_2$, for $A_1 \in \mathcal{F}$ and $A_2 \in \mathcal{B}(0, T)$. We say that $X: \Omega \times (0, T) \to \mathbb{R}^1$ is measurable if it is measurable with respect to this product σ-field. If X is integrable with respect to this product σ-field, that is,

$$\int_{\Omega \times (0,T)} |X(t, \omega)| d\mathbb{P}(\omega) dt < \infty,$$

then we denote $X \in L^1(\Omega \times (0, T))$, and we often say that X is integrable with respect to ω and t.

Theorem 2.14 (Fubini theorem)
If $X \in L^1(\Omega \times (0, T))$, then

$$\int_0^T \mathbb{E}X(t, \omega) dt = \mathbb{E} \int_0^T X(t, \omega) dt.$$

For more information on stochastic processes, see [5, 43, 169].

2.8.1 Filtration

Let $(\Omega, \mathcal{F}, \mathbb{P})$ be a complete probability space.

To keep track of information for a stochastic process, we need a concept called filtration. The filtration helps us to have a sense of past, present, and future, so that we can ask how much an observer of the process knows about it at present, as compared to how much he knew at some instant in the past or will know at some time in the future [147]. A filtration $\mathcal{F}_t$ in a probability space $(\Omega, \mathcal{F}, \mathbb{P})$ is a family of sub$-\sigma$-fields of $\mathcal{F}$ such that (i) $\mathcal{F}_s \subset \mathcal{F}_t$ if $s \leq t$; (ii) $\bigcap_{\varepsilon>0} \mathcal{F}_{t+\varepsilon} = \mathcal{F}_t$; and (iii) each $\mathcal{F}_t$ contains all null sets (i.e., zero probability events) of $\mathcal{F}$. Sometimes $(\Omega, \mathcal{F}, \mathcal{F}_t, \mathbb{P})$ is called a filtered probability space. A filtration describes an increasing flow of information as time goes on.

We define $\mathcal{F}_{t^+} \triangleq \bigcap_{\varepsilon>0} \mathcal{F}_{t+\varepsilon}$ to be the σ-field of events immediately after $t \geq 0$, and $\mathcal{F}_{t^-} \triangleq \sigma(\bigcup_{s<t} \mathcal{F}_s)$ to be the σ-field of events strictly prior to $t > 0$. Moreover, we denote $\mathcal{F}_\infty \triangleq \sigma(\bigcup_{t\geq 0} \mathcal{F}_t)$. The filtration is called right or left continuous if $\mathcal{F}_t = \mathcal{F}_{t^+}$ or $\mathcal{F}_t = \mathcal{F}_{t^-}$ for all t. It is said to be continuous if it is both right and left continuous.

The X_t is called adapted (or nonanticipating) to a filtration $\mathcal{F}_t$ (or $\mathcal{F}_t$-adapted) if, for any t, X_t is $\mathcal{F}_t$-measurable, that is, $\mathcal{F}_t$ contains all information about X_t.

The filtration generated by a stochastic process $X_t, t \geq 0$, is defined as

$$\mathcal{F}_t^X \triangleq \sigma(X_s : 0 \leq s \leq t), \tag{2.30}$$

which is the smallest σ-field with respect to which X_s is measurable for every $s \in [0, t]$. Every stochastic process is adapted to the filtration generated by itself. Most often we consider the filtration generated by the Brownian motion

$$\mathcal{F}_t^B \triangleq \sigma(B_s : 0 \leq s \leq t). \tag{2.31}$$

2.8.2 Stopping Times

Now we introduce a useful concept for stochastic processes.

Definition 2.15 (Stopping time)
A nonnegative random variable τ is called a stopping time (with respect to a filtration $\mathcal{F}_t$) if, for each t, the event $\{\tau \leq t\}$ is in $\mathcal{F}_t$.

In the context of stochastic differential equations with Brownian motion B_t (or Lévy motion L_t), a stopping time is usually defined with respect to the filtration $\mathcal{F}_t^B$ (or $\mathcal{F}_t^L$) generated by B_t (or L_t). Note that a filtration generated by the solution process is contained in the filtration $\mathcal{F}_t^B$ (or $\mathcal{F}_t^L$), because the solution process is measurable with respect to $\mathcal{F}_t^B$ (or $\mathcal{F}_t^L$).

2.9 Convergence Concepts

We only consider convergence concepts for scalar random variables, as these concepts for random vectors are similarly defined.

Consider a sequence of random variables X_n in a probability space $(\Omega, \mathcal{F}, \mathbb{P})$. Strictly speaking, we better write this sequence X_n as $\{X_n\}$, but we will not distinguish these two notations.

The sequence $X_n \to X$ almost surely (a.s.) if this convergence holds for almost all samples ω. The sequence $X_n \to X$ in probability if $\lim_{n\to\infty} \mathbb{P}(|X_n - X| > \epsilon) = 0$ for any $\epsilon > 0$. If $X_n \to X$ in probability, then there exists a subsequence X_{n_k} that converges almost surely. The sequence X_n converges to X in mean square if $\mathbb{E}|X_n - X|^2 \to 0$ as $n \to \infty$. This is also called the convergence in $L^2(\Omega)$. The convergence in $L^p(\Omega)$ is similarly defined for $p > 0$.

Theorem 2.16 *If a sequence X_n converges to X in mean square, then it also converges to X in probability.*

Proof By the Chebyshev inequality,

$$\mathbb{P}(|X_n - X| > \epsilon) \leq \frac{\mathbb{E}|X_n - X|^2}{\epsilon^2}$$

for every $\epsilon > 0$. Thus, if $\mathbb{E}|X_n - X|^2 \to 0$, then $\mathbb{P}(|X_n - X| > \epsilon) \to 0$ as $n \to \infty$, for every $\epsilon > 0$. □

Note that $X_n \to X$ in probability if and only if $\mathbb{E}\left(\frac{|X_n - X|}{1+|X_n - X|}\right) \to 0$ as $n \to \infty$. Thus, convergence in probability is metrized by the metric $d(X, Y) \triangleq \mathbb{E}\left(\frac{|X-Y|}{1+|X-Y|}\right)$. In fact, it is also metrized by the Ky Fan metric [83].

The sequence $X_n \to X$ in distribution (or convergence in law) if the distribution measures P^{X_n} converges to P^X weakly. That is, $\int f(x)P^{X_n}(dx)$ converges to $\int f(x)P^X(dx)$ for every continuous and bounded function f on $\mathbb{R}^d$. Therefore, $X_n \to X$ in distribution if and only if

$$\lim_{n\to\infty} \mathbb{E}f(X_n) = \mathbb{E}f(X) \tag{2.32}$$

for every continuous and bounded functions f. Let $F_n(x) \triangleq \mathbb{P}(X_n \leq x)$ and $F(x) \triangleq \mathbb{P}(X \leq x)$ be the distribution functions of X_n and X, respectively. Then, $X_n \to X$ in distribution if and only if $F_n(x) \to F(x)$ at all continuity points of F in $\mathbb{R}^1$.

Both convergence in L^p and convergence a.s. imply convergence in probability. Convergence in probability further implies convergence in distribution.

If $X_n \to X$ and $Y_n \to Y$, both in probability, then $X_nY_n \to XY$ in probability.

If X_n and X are Gaussian random variables, and $X_n \to X$ in distribution, then the limit X is also Gaussian. This can be shown using the characteristic function and weak convergence [139, p. 171]. Moreover, this implies that if a Gaussian random sequence $X_n \to X$ in probability or in mean square, then the limit X is also Gaussian.

Here are some more properties for various convergence concepts [13]. Let $g: \mathbb{R}^1 \to \mathbb{R}^1$ be a continuous function. Then,

if $X_n \to X$ a.s., then $g(X_n) \to g(X)$ a.s.;
if $X_n \to X$ in probability, then $g(X_n) \to g(X)$ in probability;
if $X_n \to X$ in distribution, then $g(X_n) \to g(X)$ in distribution.

For more information on probability and measure theory, see [13, 83, 300, 127].

2.10 Simulation

A building block for stochastic simulation is the generation of random numbers [242, 152]. Nowadays, random numbers are generated numerically by computers, and they are often called pseudorandom numbers. In Matlab [116], there are two built-in random number generators. The first one is the random number generator `rand`, which utilizes a scheme like the linear congruential pseudorandom number generation algorithm, for the standard uniform distribution $U(0, 1)$, on the interval $[0, 1]$, with probability density function

$$f(x) = \begin{cases} 1, & 0 \le x \le 1, \\ 0, & \text{otherwise}, \end{cases}$$

with the distribution function

$$F(x) = \begin{cases} 0, & x < 0, \\ x, & 0 \le x \le 1, \\ 1, & x > 1. \end{cases}$$

Thus, whenever x is in the range of values for the standard uniform random variable, $F(x) = x$.

The second one is the random number generator `randn` for the standard normal (or Gaussian) distribution, which utilizes a scheme like the Box-Muller algorithm, for the distribution $\mathcal{N}(0, 1)$ with probability density function

$$f(x) = \frac{1}{\sqrt{2\pi}} e^{\frac{-x^2}{2}}, \quad x \in \mathbb{R}^1.$$

The corresponding distribution function is $F(x) = \int_{-\infty}^{x} f(\xi)d\xi$, and the distribution measure is $P(A) = \int_A f(x)dx$, for every Borel set in $\mathbb{R}^1$.

The command `rand(N)` returns an N $\times$ N matrix containing pseudorandom values drawn from the standard uniform distribution on the open interval (0, 1).

The command `rand(M,N)` returns an M × N matrix. By default, `rand` generates a scalar random value, that is, it is the same as `rand(1, 1)`. For example, `rand(1, 10)` generates a sequence of 10 random numbers from the distribution $U(0, 1)$. To generate a sequence of 10 random numbers from the uniform distribution $U(a, b)$ on the interval $[a, b]$, we use the following command: $Y = a + (b - a). * \text{rand}(1, 10)$.

Similarly, `randn(N)` returns an N × N matrix containing pseudorandom values drawn from the standard normal distribution, while `randn(M,N)` returns an M × N matrix. By default, `randn` generates a scalar random value, that is, it is the same as `randn(1, 1)`. For example, `randn(1, 10)` generates a sequence of 10 random numbers from the standard normal distribution $\mathcal{N}(0, 1)$. To generate a sequence of 10 random numbers from a normal distribution, $\mathcal{N}(1, 4)$, with mean 1 and standard deviation 2, we use the command $Y = 1 + 2. * \texttt{randn}(1, 10)$.

In fact, we can generate random numbers from any continuous distributions by the so-called inverse transform algorithm [242].

Theorem 2.17 (Inverse transform algorithm) *Let F be a given continuous distribution function and Y be a random variable with the standard uniform distribution* $U(0, 1)$. *Then, the random variable defined by*

$$X = F^{-1}(Y)$$

has distribution F. Here, $F^{-1}(y)$ *means the inverse image of a value y, that is,* $F(x) = y$.

In other words, in Matlab, $X(\omega) = F^{-1}(\texttt{rand}(1, N))$ generates a sequence of N random numbers from the distribution F.

Proof Recall that F, being a distribution function, is monotonically increasing. For the standard uniform random variable Y, $F_Y(y) = y$ for $y \in [0, 1]$. Let us calculate the distribution function $F_X(x)$ for $X = F^{-1}(Y)$:

$$\begin{aligned} F_X(x) &= \mathbb{P}(X \leq x) \\ &= \mathbb{P}(F^{-1}(Y) \leq x) \\ &= \mathbb{P}(F\ F^{-1}(Y) \leq F(x)) \\ &= \mathbb{P}(Y \leq F(x)) \\ &= F(x). \end{aligned}$$

This completes the proof. □

We have proposed a new algorithm [72] for generating pseudo-random (pseudo-generic) numbers of conformal measures of a continuous map acting on a compact space and for a Hölder continuous potential.

2.11 Problems

2.1 Random variables

Let $X\colon \Omega \to \mathbb{R}^1$ be a real-valued random variable and $f\colon \mathbb{R}^1 \to \mathbb{R}^1$ be a Borel measurable function. Show that $f(X)$ is also a real-valued random variable.

2.2 Gaussian random variables

A normal/Gaussian random variable $X\colon \Omega \to \mathbb{R}^1$ has the distribution function $\mathcal{N}(\mu, \sigma^2)$; that is, the density function, with μ, σ real constants ($\sigma > 0$), is

$$f(x) = \frac{1}{\sqrt{2\pi}\sigma} \exp \frac{-(x-\mu)^2}{2\sigma^2}, \quad x \in \mathbb{R}^1.$$

Show the following:

(a) The mean and variance are $\mathbb{E}(X) = \mu$ and $\mathrm{Var}(X) = \sigma^2$, using the definition and direct computation.

(b) The odd central moments $\mathbb{E}(X-\mu)^{2k-1} = 0$, while the even central moments $\mathbb{E}(X-\mu)^{2k} = 1 \cdot 3 \cdot 5 \ldots (2k-1)\sigma^{2k}$, for $k = 1, 2, \ldots$. This implies that all moments of a Gaussian random variable can be expressed in terms of the first two moments (mean and variance).

(c) For two arbitrary scalar Gaussian random variables $X \sim \mathcal{N}(\mu_1, \sigma_1^2)$ and $Y \sim \mathcal{N}(\mu_2, \sigma_2^2)$, are $X + Y$ and XY still scalar Gaussian random variables? Is $Z = (X, Y)$ a Gaussian random variable in $\mathbb{R}^2$? How about when X and Y are independent?

2.3 Gaussian random vectors

Consider a Gaussian random variable in $\mathbb{R}^3$: $X \sim \mathcal{N}(m, Q)$, with the mean vector $m = (1, 2, 0)^{\mathrm{T}}$ and covariance matrix

$$Q = \begin{pmatrix} 1 & 1 & 0 \\ 1 & 4 & 1 \\ 0 & 1 & 1 \end{pmatrix}.$$

What is its probability density function $p(x_1, x_2, x_3)$, and what is the trace of the covariance matrix, $\mathrm{Tr}(Q)$? What is its moment-generating function $M_X(u)$ and characteristic function $\Phi_X(u)$?

2.4 Random vectors

Let $X\colon \Omega \to \mathbb{R}^n$ be a random variable in $\mathbb{R}^n$. For a given $h \in \mathbb{R}^n$, is $z(\omega) := \langle X, h \rangle = h_1 X_1(\omega) + \cdots + h_n X_n(\omega)$ a scalar random variable in $\mathbb{R}^1$? If X is a Gaussian random variable in $\mathbb{R}^n$, is z a scalar Gaussian random variable?

2.5 Probability distribution

Let $X\colon \Omega \to \mathbb{R}^1$ be a random variable with the probability distribution function $F(x)$. Explain how to find the probability distribution function for new random variables X^2 and $aX + b$, with a and b being nonzero constants. Show the details for a special case when X is the standard Gaussian random variable with distribution $\mathcal{N}(0, 1)$.

2.6 Joint probability distribution

If random variables X and Y have joint density function $f(x, y)$, what is the density function for the new random variable $Z = X + Y$? Show the details for the case of independent Gaussian random variables $X \sim \mathcal{N}(0, 4)$ and $Y \sim \mathcal{N}(0, 5)$.

2.7 Probability distribution measure or law

Let $X: \Omega \to \mathbb{R}^1$ be a random variable. What is the probability distribution measure $\mathcal{L}_X$ induced by X on the state space $(\mathbb{R}^1, \mathcal{B}(\mathbb{R}^1))$, that is, the law of X? If X has the probability density function $p(x)$, what is $\mathcal{L}_X$ for open interval (or, in general, Borel set) $A \in \mathcal{B}(\mathbb{R}^1)$? Show the details for $X \sim \mathcal{N}(0, 1)$.

2.8 Random vector

Consider a random vector $(X, Y): \Omega \to \mathbb{R}^2$ with the joint probability density function

$$p(x, y) = \begin{cases} \frac{1}{50}(x^2 + y^2), & \text{if } 0 < x < 2 \text{ and } 1 < y < 4; \\ 0, & \text{otherwise.} \end{cases}$$

Calculate the likelihood that X and Y have sums larger than 4, that is, $\mathbb{P}(\omega\colon X(\omega) + Y(\omega) > 4)$.

2.9 Chebyshev inequality

Verify the Chebyshev inequality $\mathbb{P}(|X| \geq c) \leq \frac{\mathbb{E}X^2}{c^2}$ for a Gaussian random variable $X \sim \mathcal{N}(\mu, \sigma^2)$.

2.10 Independence and uncorrelation

Let X and Y be in $L^2(\Omega)$.

(i) Show that independent random variables X, Y must be uncorrelated.
(ii) Are uncorrelated random variables X, Y also independent, in general? Explain, disprove, or prove.
(iii) Are uncorrelated Gaussian random variables X, Y also independent? Explain, disprove, or prove.

2.11 Space of random variables of finite variance

Consider the Hilbert space $L^2(\Omega)$ with the usual scalar product $\langle X, Y \rangle = \int_\Omega X(\omega)Y(\omega)d\mathbb{P}(\omega)$. Explain how to estimate the mean of the product, $\mathbb{E}(XY)$, via the Cauchy-Schwarz inequality. Show the details for $X \sim \mathcal{N}(\mu_1, \sigma_1^2)$ and $Y \sim \mathcal{N}(\mu_2, \sigma_2^2)$.

2.12 Mean square convergence

Show that if X_n and Y_n converge to X and Y, respectively, in $L^2(\Omega)$, as $n \to \infty$, then $X_n + Y_n$ converges to $X + Y$ in $L^2(\Omega)$ as $n \to \infty$.

2.13 Convergence in probability

If X_n, Y_n converge to X, Y in probability as $n \to \infty$, respectively, does $X_n Y_n$ converge to XY in probability? Why?

2.14 Uniform random number generators

Assume that a and b are real numbers, $a < b$, and that the random variable X has the standard uniform distribution $U(0, 1)$. Find the distribution function $F_Y(y)$ and probability density function for the random variable $Y = a + (b - a)X$, to verify that Y, indeed, has the uniform distribution $U(a, b)$.

2.15 Gaussian random number generators

In Matlab, describe how to generate a sequence of K random numbers from a normal distribution $\mathcal{N}(\mu, \sigma^2)$ with mean μ and standard deviation $\sigma > 0$.

2.16 Inverse transform algorithm

Discuss how to generate a sequence of random numbers from an exponential distribution with the probability density function

$$f(x) = \begin{cases} \frac{1}{2}e^{-\frac{x}{2}}, & x > 0, \\ 0, & \text{otherwise}. \end{cases}$$

Write Matlab code (i.e., a few commands in a sequential order) to generate a sequence of 20 random numbers from this exponential distribution.

3
Noise

Complex systems in science and engineering are often under the influence of fluctuations or noise. For example, fluctuations appear in the winds in the atmosphere, in electric circuits [293], in movements of tiny grains of pollen suspended in liquid, and in vacuum tubes [223]. They also appear in superconductors, disordered conductors, quantum ballistic systems, electromagnetic fields, and electric currents [154]. In statistical mechanics, thermal fluctuations or agitations are random deviations of a system from its equilibrium. In fact, thermal fluctuations are a source of noise in various mechanical and electronic systems. Moreover, molecules, subcellular organelles, and cells, immersed in a liquid environment, are all subject to thermal fluctuations [24]. Fluctuations also arise in gene expressions, protein concentrations [150], and biochemical reactions [23].

In this chapter, we consider noise defined in terms of a special stochastic process, that is, Brownian motion. A stochastic process X_t is a family of random variables, in a probability space $(\Omega, \mathcal{F}, \mathbb{P})$, parameterized by time $t \in \mathbb{R}$. Sometimes $t \in [0, +\infty)$ or $t \in [0, T]$. A stochastic process X_t is adapted to a filtration $\mathcal{F}_t$ if, for every t, X_t is $\mathcal{F}_t$-measurable, that is, $\mathcal{F}_t$ contains all information about X_t.

3.1 Brownian Motion

The physical phenomenon *Brownian motion*, which owes its name to its discovery by the English botanist Robert Brown in 1827, is due to the incessant hitting of pollen by the much smaller molecules of the liquid. The hits occur a large number of times in any small time internal, independently of each other, and the effect of a particular hit is small compared to the total effect. In 1900, Bachelier [14] discussed the use of Brownian motion to model stock price evolution. The physical theory of this motion, set up by Albert Einstein in 1905, suggests that the motion is random and has the following properties:

1. The motion is continuous.
2. It has independent increments.
3. The increments are Gaussian random variables.

Intuitively speaking, property 1 says that the sample path of the Brownian motion is continuous. Property 2 means that the displacements of a pollen particle over disjoint time intervals are independent random variables. Property 3 is natural considering the *central limit theorem.*

We now describe Brownian motion in mathematical language; that is, we introduce the definition of Brownian motion [13, 43].

3.1.1 Brownian Motion in $\mathbb{R}^1$

We first look at a scalar Brownian motion (also called Wiener process).

We adopt the following definition, from [13, p. 401] and [200, p. 33].

Definition 3.1 A stochastic process $\{B_t(\omega): t \geq 0\}$ defined on a probability space $(\Omega, \mathcal{F}, P)$ is called a *Brownian motion* or a *Wiener process* if the following conditions hold:

(i) $B_0 = 0$, a.s.
(ii) The paths $t \to B_t(\omega)$ are continuous, a.s.
(iii) B_t has independent increments, that is, if $0 \leq t_1 < t_2 < \cdots < t_n$, then the random variables $B_{t_2} - B_{t_1}, \ldots, B_{t_n} - B_{t_{n-1}}$ are independent.
(iv) B_t has stationary increments that are Gaussian distributed, that is, $B_t(\omega) - B_s(\omega)$ has the normal distribution with mean 0 and variance $t - s$. Namely, $B_t(\omega) - B_s(\omega) \sim \mathcal{N}(0, t - s)$ for any $0 \leq s < t$.

Remark 3.2 The item (iv) implies that $\mathbb{E}B_t = 0$ and $\mathbb{E}(B_t - B_s)^2 = t - s$. It also says that $B_t - B_s$ and B_{t-s} have the same distribution $\mathcal{N}(0, t - s)$, for $t > s > 0$. However, this does not mean that $B_t - B_s$ equals B_{t-s} pathwisely. In fact, $B_t - B_s \neq B_{t-s}$, a.s.

An equivalent definition is provided by the following theorem.

Theorem 3.3 *A stochastic process B_t is a Brownian motion, or Wiener process, if and only if*

(i) $B_0 = 0$ a.s.
(ii) the paths $t \to B_t(\omega)$ are continuous, a.s.
(iii) for every $n \geq 2$ and $0 \leq t_1 < t_2 < \cdots < t_n$, the random variable $B_{t_n} - B_{t_{n-1}}$ is independent of the random variables $B_{t_1}, B_{t_2}, \ldots, B_{t_{n-1}}$.

(iv) $B_t - B_s$ *has the normal distribution with mean* 0 *and variance* $t - s$ *for every* s, t *with* $t > s \geq 0$, *namely,* $B_t(\omega) - B_s(\omega) \sim \mathcal{N}(0, t - s)$ *for every* s, t *with* $t > s \geq 0$.

Proof See [160, p. 33]. □

Theorem 3.4 *A Brownian motion has the following basic properties:*

(i) *A Brownian motion* B_t *has distribution* $\mathcal{N}(0, t)$, *that is, its probability density function is* $\frac{1}{\sqrt{2\pi t}} e^{-\frac{x^2}{2t}}$ *for* $t > 0$.
(ii) $\mathbb{E}(B_s B_t) = \min\{s, t\}$.
(iii) *For given* $c > 0$, *the process* $B_{t+c} - B_c$ *is a Brownian motion. Also, for any* $c \neq 0$, *the process* $cB_{\frac{t}{c^2}}$ *is a Brownian motion.*
(iv) *The process* $-B_t$ *is a Brownian motion.*
(v) *The process* $\tilde{B}_t$ *defined as*

$$\tilde{B}_t = \begin{cases} 0, & t = 0 \\ tB_{\frac{1}{t}}, & t > 0 \end{cases}$$

is a Brownian motion.
(vi) *For every fixed* s *with* $0 \leq s < t$, $B_t - B_s$ *is independent of the* σ*-field* $\mathcal{F}_s^B = \sigma(B_u, 0 \leq u \leq s)$.

Remark 3.5 The preceding properties (ii)–(v) imply that although, statistically speaking, there is only one Brownian motion, it can have different versions that all have the same statistical properties as listed in the definition.

Proof The property (i) is directly from the definition. The property (ii) is proved as follows. Without loss of generality, let us assume that $s < t$. Owing to the independence of increments, $\mathbb{E}(B_s B_t) = \mathbb{E}[(B_t - B_s)(B_s - B_0) + B_s^2] = \mathbb{E}(B_t - B_s)\, \mathbb{E}(B_s - B_0) + \mathbb{E}B_s^2 = s = \min\{s, t\}$. Properties (iii)–(v) can be verified by definition of B_t. A proof of (vi) is in [147, p. 49]. □

Recall the definition of "finite variation." Consider all possible partitions of the time interval $[0, T]$ into small subintervals: $0 = t_0 < t_1 < \cdots < t_i < t_{t+1} < \cdots < t_n = T$. If there exists a positive number M such that

$$\sum_{i=0}^{n-1} |g(t_{i+1}) - g(t_i)| \leq M$$

for all partitions of $[0, T]$, then g is said to be of finite variation (or bounded variation); otherwise it is of infinite variation.

Theorem 3.6 *A Brownian motion has the following advanced properties:*

(i) For every positive number $\alpha < \frac{1}{2}$, almost all paths of a Brownian motion B_t are Hölder continuous, with exponent α, on every bounded interval. So almost all paths of Brownian motion are Hölder continuous with exponents less than one-half; that is, they can not be Lipschitz continuous.

(ii) First variation: Almost every path of Brownian motion has infinite variation on every finite interval.

(iii) Quadratic variation: Let $\{t_0^{(n)}, \ldots, t_k^{(n)}\}$ be a sequence of partitions of an time interval $[a, b]$, for $n = 1, 2, \ldots$. Then

$$\sum_i \left|B\left(t_{i+1}^{(n)}\right) - B\left(t_i^{(n)}\right)\right|^2 \to b - a \ \ in\ L^2(\Omega) \ \ as\ \max_i \left|t_{i+1}^{(n)} - t_i^{(n)}\right| \to 0. \quad (3.1)$$

(iv) The paths of Brownian motion are almost surely nowhere differentiable.

Remark 3.7 The property (i) says that a path of Brownian motion cannot be Lipschitz continuous. Recall that a Lipschitz continuous function must be differentiable almost everywhere.

For (ii), let us look at a case where the subintervals are of the same length Δt (i.e., there are $\frac{b-a}{\Delta t}$ subintervals) and consider $\sum_i |B(t_{i+1}) - B(t_i)|$. Taking the mean and noticing that $\mathbb{E}|B_t| = \sqrt{\frac{2t}{\pi}}$ (see Problem 3.5), we conclude that

$$\begin{aligned}
\mathbb{E}\sum_i |B(t_{i+1}) - B(t_i)| &= \sum_i \mathbb{E}|B(t_{i+1}) - B(t_i)| \\
&= \sum_i \mathbb{E}|B(\Delta t)| \\
&= \sum_i \sqrt{\frac{2\Delta t}{\pi}} \\
&= \sqrt{\frac{2\Delta t}{\pi}}\frac{b-a}{\Delta t} \\
&= \sqrt{\frac{2}{\pi}}\frac{b-a}{\sqrt{\Delta t}}.
\end{aligned}$$

This is unbounded when Δt becomes small. Thus, intuitively, or in the sense of expectation, the quantity $\sum_i |B(t_{i+1}) - B(t_i)|$ is unbounded. This indicates that B_t might not be of finite variation.

For (iii), note that $\mathbb{E}\sum_i |B(t_{i+1}^{(n)}) - B(t_i^{(n)})|^2 = \sum_i \mathbb{E}|B(t_{i+1}^{(n)}) - B(t_i^{(n)})|^2 = \sum_i [t_{i+1}^{(n)} - t_i^{(n)}] \equiv b - a$. This says that the mean of the sequence $X_n = \sum_i |B(t_{i+1}^{(n)}) - B(t_i^{(n)})|^2$ is always the same as the time interval length (on which B_t

is defined), but it does not imply that this sequence converges to $b - a$ in a mean square sense.

The properties (ii) and (iii) are reminiscent of the fact that $\sum_{i=1}^{\infty} \frac{1}{i} = \infty$, but $\sum_{i=1}^{\infty} \left(\frac{1}{i}\right)^2 = \frac{\pi^2}{6} < \infty$.

For (iv), let us consider whether $\frac{B_{t+\Delta t} - B_t}{\Delta t}$ has a limit as $\Delta t \to 0$. The mean of this quotient is zero, but the variance (i.e., the second moment in this case) is

$$\mathbb{E}\left(\frac{B_{t+\Delta t} - B_t}{\Delta t}\right)^2 = \frac{\mathbb{E}(B_{\Delta t})^2}{(\Delta t)^2} = \frac{\Delta t}{(\Delta t)^2} = \frac{1}{\Delta t},$$

which has no limit as $\Delta t \to 0$. This indicates intuitively that B_t might not be differentiable.

Proof A proof of (i) is in [93, p. 38, Vol. 1]. A proof of (ii) is in [200, p. 189] and also in [13, p. 409], and a proof of (iii) can be found in [13, p. 410]. A weaker version of (iii), that is, convergence in probability instead of convergence in mean square, is proved in [160, p. 37]. A proof of (iv) is in [200, p. 188] and also in [13, p. 408]. □

There are other ways to quantify the Brownian motion. For example, $p(t) = \mathbb{P}(B_t(\omega) \in [1, 4])$ specifies the likelihood that B_t lies between 1 and 4 at time t.

Moreover, Brownian motion B_t satisfies the following two laws: the law of large numbers and the law of the iterated logarithms.

Theorem 3.8 *Law of large numbers*

$$\lim_{t\to\infty} \frac{B_t}{t} = 0, \text{ a.s.} \tag{3.2}$$

Note that $\mathbb{E}(\frac{B_t}{t})^2 = \frac{t}{t^2} = \frac{1}{t} \to 0$ as $t \to \infty$. So $\lim_{t\to\infty} \frac{B_t}{t} = 0$ in mean square, which further implies that $\frac{B_t}{t}$ converges to 0 in probability. But this does not imply the convergence almost surely.

Theorem 3.9 *Law of the iterated logarithms*

$$\limsup_{t\to\infty} \frac{B_t}{\sqrt{2t \ln \ln t}} = +1, \text{ a.s.} \tag{3.3}$$

$$\liminf_{t\to\infty} \frac{B_t}{\sqrt{2t \ln \ln t}} = -1, \text{ a.s.} \tag{3.4}$$

$$\limsup_{t\to 0} \frac{B_t}{\sqrt{2t \ln \ln 1/t}} = +1, \text{ a.s.} \tag{3.5}$$

$$\liminf_{t\to 0} \frac{B_t}{\sqrt{2t \ln \ln 1/t}} = -1, \text{ a.s.} \tag{3.6}$$

For proofs of these two laws, see, for example, [13, 271, 121].

Two-sided Brownian Motion

In stochastic dynamics, we consider a two-sided Brownian motion B_t, $t \in \mathbb{R}$. It is defined as (see [9, Appendix A.3]) a stochastic process with $B_0 = 0$ and stationary independent increments satisfying $B_t - B_s \sim \mathcal{N}(0, |t-s|I)$, for any t, s in $\mathbb{R}$. Alternatively, we may define a two-sided Brownian motion B_t, by means of two independent usual Brownian motions B_t^1 and B_t^2 $(t \geq 0)$:

$$B_t = \begin{cases} B_t^1, & t \geq 0, \\ B_{-t}^2, & t < 0. \end{cases} \tag{3.7}$$

For a two-sided Brownian motion, we have $\lim_{t \to \pm\infty} \frac{B_t}{t} = 0$, a.s.

3.1.2 Brownian Motion in $\mathbb{R}^n$

The Brownian motion B_t, taking values in $\mathbb{R}^n$, is a Gaussian stochastic process on an underlying probability space $(\Omega, \mathcal{F}, \mathbb{P})$. Being a Gaussian process, B_t is characterized by its mean vector (taken to be the zero vector) and its covariance matrix (taken to be the identity matrix). More specifically, B_t satisfies the following conditions [13, 147, 207, 202]:

(a) $B_0 = 0$, a.s.
(b) B_t has continuous paths, a.s.
(c) B_t has independent increments.
(d) B_t has stationary increments, and $B_t - B_s \sim \mathcal{N}(0, (t-s)I)$, for $t > s \geq 0$, where I is the $n \times n$ identity matrix.

By this definition, we have the following conclusions:

(i) The covariance matrix for the Brownian motion $B(t)$ in $\mathbb{R}^n$ is $t\,I$, with I the $n \times n$ identity matrix, and its trace is $\mathrm{Tr}(t\,I) = nt$. For convenience, we just call I the covariance matrix for B_t.
(ii) Because the covariance matrix is I, the components of $B(t)$ are pairwise uncorrelated. The Gaussianity further implies that they are pairwise independent.
(iii) $B(t) \sim \mathcal{N}(0, tI)$, that is, $B(t)$ has probability density function $p_t(x) = \frac{1}{(2\pi t)^{\frac{n}{2}}} e^{-\frac{x_1^2 + \cdots + x_n^2}{2t}}$. This joint probability density function is the product of the probability density functions for the scalar components of $B(t)$. Thus, the components of $B(t)$ are independent scalar Brownian motions (not just pairwise independent).

Remark 3.10 The Brownian motion so defined is called the standard Brownian motion, as the covariance matrix is the identity matrix. We may revise the preceding definition to allow the covariance matrix to be a general positive definite, symmetric matrix Q.

Definition 3.11 (Brownian motion with covariance matrix Q) An n-dimensional Brownian motion with covariance matrix Q is defined by

$$B_t^Q = \sigma B_t,$$

where σ is an $n \times m$ real nonzero matrix and B_t is an m-dimensional standard Brownion motion, such that $Q = \sigma\sigma^{\mathrm{T}}$.

This terminology is used in Chapter 7. Note that $Q = \sigma\sigma^{\mathrm{T}}$ is symmetric and positive definite, as $x^{\mathrm{T}} Q x = x^{\mathrm{T}} \sigma\sigma^{\mathrm{T}} x = (\sigma^{\mathrm{T}} x)^{\mathrm{T}} (\sigma^{\mathrm{T}} x) = \|\sigma^{\mathrm{T}} x\|^2 > 0$ for $x \neq 0$. Thus, Q has positive real eigenvalues.

A two-sided Brownian motion in $\mathbb{R}^n$ is similarly defined as in (3.7).

3.2 What Is Gaussian White Noise?

Although the random fluctuations in a complex system arise from the specific situation, such fluctuations appear to have common features. They are generally regarded as stationary stochastic processes, with zero mean and with special correlations at different time instants.

We assume that stochastic processes are defined for all real-time $t \in (-\infty, \infty)$. Let $X_t(\omega)$ be a scalar stochastic process defined in a probability space $(\Omega, \mathcal{F}, \mathbb{P})$.

Definition 3.12 A stochastic process X_t is called stationary if, for any natural number k and any real numbers $t_1 < t_2 < \cdots < t_k$, the distribution of $(X_{t_1+t}, X_{t_2+t}, \ldots, X_{t_k+t})$ does not depend on t, that is,

$$\mathbb{P}(\{(\omega\colon X_{t_1+t}(\omega), \ldots, X_{t_k+t}(\omega)) \in A\}) = \mathbb{P}(\{\omega\colon (X_{t_1}(\omega), \ldots, X_{t_k}(\omega)) \in A\})$$

for every Borel set A in $\mathbb{R}^1$ and for all real numbers t.

Such a stationary process X_t is also called a strongly stationary process; see [219].

A stochastic process $X(t)$ is called a weakly stationary process if the mean $\mathbb{E}X(t)$ is a constant and the autocorrelation $\mathbb{E}(X(t_1)X(t_2))$ depends only on the time lag $t_2 - t_1$.

Moreover, we say that X_t has stationary increments if, for any integer k and any real numbers $t_0 < t_1 < \cdots < t_k$, the distribution of $X_{t_j} - X_{t_{j-1}}$ depends on t_j

and t_{j-1} only through the difference $t_j - t_{j-1}$ where $j = 1, \ldots, k$. It means that if $t_j - t_{j-1} = t_i - t_{i-1}$ for some $i, j \in \{1, \ldots, k\}$, then $(X_{t_j} - X_{t_{j-1}}) \stackrel{d}{=} (X_{t_i} - X_{t_{i-1}})$, that is, the both sides have the same distributions.

A (Gaussian) noise is a special stationary stochastic process $\eta_t(\omega)$, with mean $\mathbb{E}\eta_t = 0$ and covariance $\mathbb{E}(\eta_t \eta_s) = K\ c(t-s)$ for all t and s, for a positive constant K and a function $c(\cdot)$. When $c(t-s)$ is the Dirac delta function $\delta(t-s)$, the noise η_t is called a white noise; otherwise, it is a colored noise.

Gaussian white noise may be modeled in terms of the "time derivative" of Brownian motion. Let us first discuss this formally [159, 219]. Recall that a scalar Brownian motion B_t is a Gaussian process with stationary (and also independent) increments, together with mean $\mathbb{E}B_t = 0$ and covariance $\mathbb{E}(B_t B_s) = t \wedge s = \min\{t, s\}$. By the formal formula $\mathbb{E}(\dot{X}_t \dot{X}_s) = \partial^2 \mathbb{E}(X_t X_s)/\partial t \partial s$, we see that [159]

$$\mathbb{E}(\dot{B}_t \dot{B}_s) = \frac{\partial^2}{\partial t \partial s}\mathbb{E}(B_t B_s) = \frac{\partial^2}{\partial t \partial s}(t \wedge s)$$

$$= \frac{\partial}{\partial t}\frac{\partial}{\partial s}\begin{cases} t, & t-s<0 \\ s, & t-s \geq 0 \end{cases} = \frac{\partial}{\partial t}\begin{cases} 0, & t-s<0 \\ 1, & t-s \geq 0 \end{cases} = \delta(t-s).$$

Here in the final step, we have used a fact for the Heaviside function $H(\xi)$:

$$H(\xi) = \begin{cases} 1 & \text{if } \xi \geq 0, \\ 0 & \text{otherwise.} \end{cases} \tag{3.8}$$

Namely, $\frac{d}{d\xi}H(\xi) = \delta(\xi)$. See Section 3.3.1 for the Heaviside function and the delta function. This says that $\dot{B}_t$ is uncorrelated at different time instants.

So the spectral density function for $\dot{B}_t$, that is, the Fourier transform $\mathbb{F}$ for its covariance function $\mathbb{E}(\dot{B}_t \dot{B}_s)$, has constant absolute value

$$|\mathbb{F}(\mathbb{E}(\dot{B}_t \dot{B}_s))| = |\mathbb{F}(\delta(t-s))| = \left|\frac{1}{\sqrt{2\pi}}\, e^{-iks}\right| = \frac{1}{\sqrt{2\pi}}.$$

Recall that the Fourier transform of g is $\mathbb{F}(g) = \frac{1}{\sqrt{2\pi}}\int_{\mathbb{R}^1} e^{-ikt} g(t)dt$. Moreover, the increments like $B_{t+\Delta t} - B_t \approx \frac{1}{\Delta t}\, \dot{B}_t$ are stationary, and formally, $\mathbb{E}\dot{B}_t \approx \mathbb{E}\frac{B_{t+\Delta t} - B_t}{\Delta t} = \frac{0}{\Delta t} = 0$. Thus $\eta_t = \dot{B}_t$ is taken as a mathematical model for white noise. Note that Brownian motion does not have an usual time derivative. It is necessary to interpret $\dot{B}_t$ as a generalized time derivative and make the argument rigorous.

3.3 *A Mathematical Model for Gaussian White Noise

This section may be omitted in the first reading. To make the discussions in the previous section more rigorous, we discuss generalized time derivatives for stochastic processes in Section 3.3.1, and then consider white noise in Section 3.3.2.

3.3.1 Generalized Derivatives

To examine noise more rigorously, we consider the Heaviside function

$$H(t) = \begin{cases} 1 & \text{if } t \geq 0, \\ 0 & \text{otherwise.} \end{cases} \tag{3.9}$$

This function is certainly not $C^1(\mathbb{R}^1)$, but if it had a (generalized) derivative $H'(t)$, then by formal integration by parts, $H'(t)$ should satisfy

$$\int_{-\infty}^{+\infty} H'(t)v(t)dt = -\int_{-\infty}^{+\infty} H(t)v'(t)dt \tag{3.10}$$

$$= -\int_{0}^{+\infty} v'(t)dx$$

$$= v(0) - v(+\infty) = v(0) \tag{3.11}$$

for every test function $v \in C_0^1(\mathbb{R}^1)$ (i.e., smooth functions with compact support on $\mathbb{R}^1$).

Because $H'(t) \equiv 0$ for $t \neq 0$, this suggests that "$H'(0) = \infty$" in such a way that $\int_{-\infty}^{+\infty} H'(t)v(t)dt = v(0)$. Of course, no true function behaves like this, so we call $H'(t)$ a generalized function. The preceding computation of the derivatives is via formal integration by parts against a function in $C_0^\infty(\mathbb{R}^1)$. This is called generalized differentiation, and the functions in $C_0^\infty(\mathbb{R}^1)$ are called test functions.

The classical derivative is defined via a pointwise limit, revealing that it is a "local" concept. But the generalized derivative (sometimes called the weak derivative) is defined via integration by parts, indicating that it is a "nonlocal" concept.

We define the delta function $\delta(t)$ such that

$$\int_{-\infty}^{+\infty} \delta(t)v(t)dt = v(0) \tag{3.12}$$

for every test function $v \in C_0^\infty(\mathbb{R}^1)$. Hence we find that

$$H'(t) = \delta(t). \tag{3.13}$$

Note that we also similarly define multivariate delta function $\delta(x)$ for $x \in \mathbb{R}^n$.

In fact, we can also view this in the context of generalized functions, which are linear functionals on the test space $C_0^\infty(\mathbb{R}^1)$. For example, $\Phi_f(v) = \int_{\mathbb{R}^1} v(t)f(t)dt$, with $f(t)$ given and $v \in C_0^\infty(\mathbb{R}^1)$ is a generalized function. When there is no confusion, we may omit subscript f. Thus, (3.13) may be reformulated as

$$\dot{\Phi}_H(v) = -\Phi_\delta(v'), \quad v \in C_0^\infty(\mathbb{R}^1). \tag{3.14}$$

This is another way to say that the generalized derivative of the Heaviside function H is the Dirac δ function.

In the next subsection, we consider a mathematical model for Gaussian white noise.

3.3.2 Gaussian White Noise

This subsection follows [8, Chapter 3]. In engineering, white noise is generally understood as a stationary process ξ_t, with zero mean $\mathbb{E}\xi_t = 0$ and constant spectral density. Let $\mathbb{E}\xi_s\xi_{t+s} = C(t)$ be the covariance function of ξ_t. Thus, for a white noise, "the Fourier transform of the covariance function $C(t)$" (i.e., the spectral density) needs to be a constant; that is,

$$\frac{1}{\sqrt{2\pi}} \int_{-\infty}^{+\infty} e^{-ikt} C(t) dt = \frac{c_0}{\sqrt{2\pi}},$$

where c_0 is a constant. This relation holds for a stochastic process ξ_t with covariance function $C(t) = \delta(t)$, the Dirac delta function. This says that white noise is a stationary stochastic process that has zero mean and is uncorrelated at different time instants. From the intuitive discussions in the previous subsection, we know that $\xi_t = \dot{B}_t$ is such a stochastic process and thus may be regarded as "white noise." See [159, 293] for discussions of white noise in engineering and see [76, 77] for more applications.

If white noise ξ_ts covariance $\mathrm{Cov}(\xi_t, \xi_s)$ and $\dot{B}_t$s covariance $\mathrm{Cov}(\dot{B}_t, \dot{B}_s)$ are the same, then we can take $\dot{B}_t$ as a mathematical model for white noise ξ_t. This is indeed the case, but we have to verify this in the context of generalized functions, because $\frac{dB_t}{dt}$ has no meaning in the sense of ordinary functions.

In fact, white noise was correctly described in connection with the theory of generalized functions [8]. From the previous subsection, we know that

$$\Phi_f(\varphi) = \int_{-\infty}^{+\infty} \varphi(t) f(t) dt \tag{3.15}$$

defines a generalized function for a given f. The function Φ_f depends linearly and continuously on test functions φ. It is the generalized function corresponding to f. With this representation, we regard f as a generalized function. In fact, we may identify f with this linear functional Φ_f.

In particular, the generalized function defined by

$$\Phi(\varphi(t)) = \varphi(t_0)$$

with a fixed t_0, for $\varphi \in C_0^\infty(\mathbb{R})$, is called the Dirac delta function and is also symbolically denoted by $\delta(t - t_0)$ or $\delta_{t_0}(t)$, a shifted δ function. In contrast with classical functions, generalized functions always have derivatives of every order, which again are generalized functions. By the derivative $\dot{\Phi}$ of Φ, we mean the generalized function defined by

$$\dot{\Phi}(\varphi) = -\Phi(\dot{\varphi}). \tag{3.16}$$

A generalized stochastic process is now simply a random generalized function in the following sense: for every test function φ, a random variable $\Phi(\varphi)$ is assigned such that the functional Φ is linear and continuous.

Definition 3.13 A generalized stochastic process Φ is called a Gaussian process if, for every set of linearly independent functions $\varphi_1, \dots \varphi_n \in K$, the random variable $(\Phi(\varphi_1), \dots \Phi(\varphi_n))$ is normally distributed. Just as in the classical case, a generalized Gaussian process is uniquely defined by a continuous linear mean functional

$$\mathbb{E}\Phi(\varphi) = m(\varphi)$$

and a continuous bilinear positive-definite covariance functional

$$\mathbb{E}[\Phi(\varphi) - m(\varphi))(\Phi(\psi) - m(\psi)] = C(\varphi, \psi).$$

One of the important advantages of a generalized stochastic process Φ is the fact that its derivative $\dot{\Phi}$ always exists and is itself a generalized stochastic process. In fact, the derivative $\dot{\Phi}$ of Φ is the process defined by setting

$$\dot{\Phi}(\varphi) = -\Phi(\dot{\varphi}).$$

The derivative of a generalized Gaussian process with mean $m(\varphi)$ and covariance $C(\varphi, \psi)$ is again a generalized Gaussian process, and it has mean $\dot{m}(\varphi) = -m(\dot{\varphi})$ and covariance $C(\dot{\varphi}, \dot{\psi})$.

Now let us look at Brownian motion B_t and its generalized derivative. As we know, its classical derivative does not exist.

Theorem 3.14 *The generalized derivative of Brownian motion B_t is a white noise.*

Proof The generalized stochastic process corresponding to B_t is the following linear functional:

$$\Phi(\varphi) = \int_{-\infty}^{+\infty} \varphi(t) B_t dt \tag{3.17}$$

for $\varphi \in C_0^\infty(\mathbb{R})$. We better denote this linear functional by Φ_B, but for simplicity, we still call it by Φ. With this representation, we regard B_t as a generalized stochastic process. In fact, we may identify B_t with this linear functional Φ. We conclude that the mean functional

$$m(\varphi) \equiv 0$$

and the covariance functional

$$C(\varphi, \psi) = \int_0^\infty \int_0^\infty \min\{t, s\}\ \varphi(t)\psi(s)dtds.$$

After some elementary manipulations and integration by parts, we get

$$C(\varphi, \psi) = \int_0^\infty (\hat{\varphi}(t) - \hat{\varphi}(\infty))(\hat{\psi}(t) - \hat{\psi}(\infty))dt,$$

where

$$\hat{\varphi}(t) = \int_0^t \varphi(s)ds \text{ and } \hat{\psi}(t) = \int_0^\infty \int_0^t \psi(s)ds.$$

The derivative of B_t, that is, the derivative of $\Phi(\varphi) = \int_{-\infty}^{+\infty} \varphi(t)B_t dt$, is also a generalized Gaussian process with mean

$$\dot{m}(\varphi) = -m(\dot{\varphi}) = 0 \tag{3.18}$$

and covariance

$$\dot{C}(\varphi, \psi) = C(\dot{\varphi}, \dot{\psi}) = \int_0^\infty \varphi(t)\psi(s)dt.$$

This covariance formula can be rewritten as

$$\dot{C}(\varphi, \psi) = \int_0^\infty \int_0^\infty \delta(t-s)\varphi(t)\psi(s)dtds.$$

Therefore, the covariance of the derivative of Brownian motion B_t is the generalized stochastic process with mean zero and covariance

$$\dot{C}(s, t) = \delta(t - s). \tag{3.19}$$

This is precisely the covariance for white noise ξ_t. Thus, $\dot{B}_t$ is a mathematical model for white noise ξ_t. □

With the preceding theorem, we see the rationality of the notation

$$\xi_t = \dot{B}_t \tag{3.20}$$

frequently used in engineering literature and occasionally in stochastic differential equations.

3.4 Simulation

In Matlab, the standard Gaussian random number generator `randn` generates random numbers from the distribution $\mathcal{N}(0, 1)$.

Now let us generate sample paths for a scalar Brownian motion B_t, $t \in [0, T]$, for some given $T > 0$. Partition the time interval $[0, T]$ into small subintervals: $t_0 = 0 < t_1 < t_2 < \cdots < t_k < \cdots < t_N = T$. To get an approximate sample path for B_t, we generate $B(t_0), \ldots, B(t_N)$ and then connect these points on the (t, B_t) plane by interpolation (note that the sample paths for B_t are continuous in time t, almost surely). First we generate independent standard normal random numbers (using `randn` in Matlab) $Z_1, Z_2, \ldots, Z_N$ and compute the independent increments:

$$B(t_i) - B(t_{i-1}) = \sqrt{t_i - t_{i-1}}\, Z_i, \; i = 1, \ldots, N. \tag{3.21}$$

Note that $B(t_i) - B(t_{i-1}) \sim \mathcal{N}(0, t_i - t_{i-1})$, which is the same as $\sqrt{t_i - t_{i-1}}\mathcal{N}(0, 1)$. Then, to simulate the values $B(t_1), \ldots, B(t_N)$, we use the recursive algorithm

$$B(t_i) = B(t_{i-1}) + (B(t_i) - B(t_{i-1})) = B(t_{i-1}) + \sqrt{t_i - t_{i-1}}\, Z_i, \tag{3.22}$$

for $i = 1, \ldots, N$. Very often, we take equal partition of the interval $[0, T]$ with the stepsize $\Delta t > 0$. That is, $N = \frac{T}{\Delta t}$ (taking its integer part if needed) and

$$t_0 = 0, t_1 = \Delta t, \ldots, t_i = t_{i-1} + \Delta t, \ldots, t_N = T, \tag{3.23}$$

and

$$\begin{aligned} &B(t_0) = 0, \\ &B(t_i) = B(t_{i-1}) + \sqrt{\Delta t}\, Z_i, \; i = 1, \ldots, N. \end{aligned} \tag{3.24}$$

A Matlab code can then be developed to plot sample paths for B_t. A sample path of Brownian motion is shown in Figure 3.1.

We can also generate sample paths for two-sided Brownian motion B_t, $t \in [-T, T]$ for some given $T > 0$ and sample paths for higher-dimensional Brownian motions.

Forward in time (just as for the usual one-sided Brownian motion), for a partition of $[0, T]$, $0 = t_0 < t_1 < \cdots < t_i < t_{i+1} < \cdots < t_N = T$ of equal subinterval length (or stepsize) Δt and

$$t_0 = 0, t_1 = \Delta t, \ldots, t_i = t_{i-1} + \Delta t, \ldots, t_N = T,$$

and

$$\begin{aligned} &B(t_0) = 0, \\ &B(t_i) = B(t_{i-1}) + \sqrt{\Delta t}\, Z_i, \; i = 1, \ldots, N. \end{aligned}$$

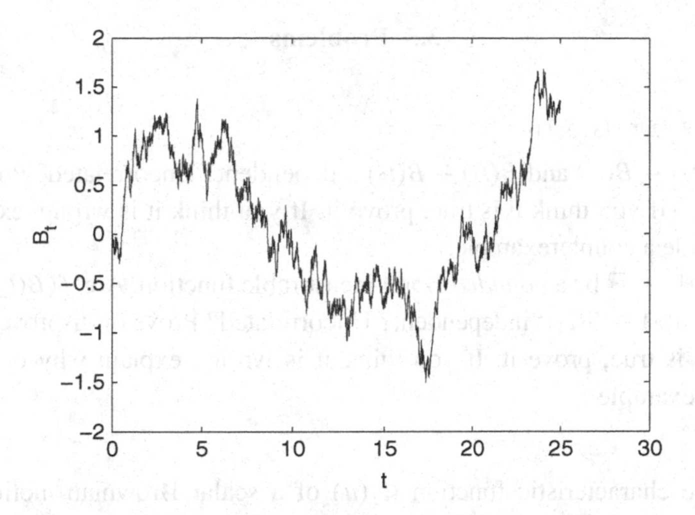

Figure 3.1 A sample path of Brownian motion $B(t)$.

Backward in time, for a partition of $[-T, 0]$, $-T = t_0 < t_1 < \cdots < t_i < t_{i+1} < \cdots < t_N = 0$ of equal subinterval length (or stepsize) $\Delta t > 0$ and

$$t_0 = -T, t_1 = -T + \Delta t, \ldots, t_i = t_{i-1} + \Delta t, \ldots, t_N = 0, \tag{3.25}$$

and

$$B(t_N) = 0$$

$$B(t_i) = B(t_{i+1}) + \sqrt{\Delta t}\, \xi_{i+1}, \; i = N-1, \ldots, 1, \tag{3.26}$$

where $\xi_1, \xi_2, \ldots, \xi_N$ are a sequence of independent standard normal random numbers (generated by the command `randn` in Matlab).

A sample path for a two-sided Brownian motion is shown in Figure 3.2.

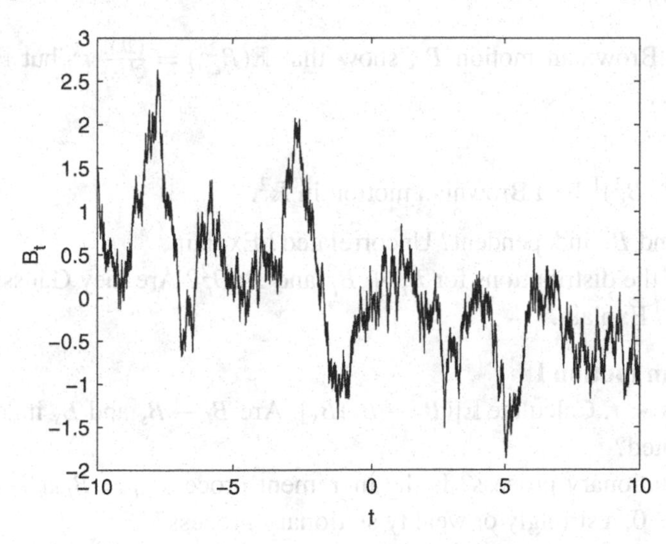

Figure 3.2 A sample path of a two-sided Brownian motion.

3.5 Problems

3.1

Let $0 < t_1 < t_2 < t_3 < t_4$.

(a) Are $B(t_2) - B(t_1)$ and $B(t_4) - B(t_3)$ independent? Uncorrelated? Prove or disprove it. (If you think it is true, prove it. If you think it is wrong, explain why or provide a counterexample.)

(b) Let $f : \mathbb{R} \to \mathbb{R}$ be a *bounded* Borel measurable function. Are $f(B(t_2) - B(t_1))$ and $f(B(t_4) - B(t_3))$ independent? Uncorrelated? Prove or disprove it. (If you think it is true, prove it. If you think it is wrong, explain why or provide a counterexample.)

3.2

(a) Find the characteristic function $\Phi_t(u)$ of a scalar Brownian motion B_t, for each t.

(b) Find the moment generating function $m_t(u)$ of a scalar Brownian motion B_t, for each t.

3.3

(a) Find the characteristic function $\Phi_t(u)$ of Brownian motion B_t in $\mathbb{R}^n$, for each t.

(b) Find the moment-generating function $M_t(u)$ of Brownian motion B_t in $\mathbb{R}^n$, for each t.

3.4

Calculate $\mathbb{E}e^{2B_t}$ and $\mathbb{E}e^{2B_t^2}$.

3.5

Calculate $\mathbb{E}|B_t|$.

3.6 For the Brownian motion B_t, show that $\mathbb{E}(B_t^{2k}) = \frac{(2k)!}{2^k \cdot k!} t^k$ but $\mathbb{E}(B_t^{2k-1}) = 0$, $k = 1, 2, \ldots$

3.7

Let $B_t = (B_t^1, B_t^2)^{\mathrm{T}}$ be a Brownian motion in $\mathbb{R}^2$.

(a) Are B_t^1 and B_t^2 independent? Uncorrelated? Explain.

(b) What are the distributions for $B_t^1 + B_t^2$ and $B_t^1 B_t^2$? Are they Gaussian random variables? Explain.

3.8 Brownian motion I

(a) Let $0 < s < t$. Calculate $\mathbb{E}[(B_t - B_s)B_s]$. Are $B_t - B_s$ and B_s independent or uncorrelated?

(b) Is B_t a stationary process? Is the increment process $\xi_t = B_{t+h} - B_t$, for any given $h > 0$, a strongly or weakly stationary process?

3.9 Brownian motion II

Show that $\mathrm{Cov}\left(\int_0^s B_u du, \int_0^t B_v dv\right) = \int_0^s \int_0^t (u \wedge v) du dv$.

What are the variance and covariance of $\xi_t = \int_0^t B_s ds$?

3.10 Brownian motion III

(a) Starting from definition, show that $\frac{B_t}{t}$ converges to 0 in probability, as $t \to \infty$.
(b) Starting from definition, show that $\frac{B_t}{t}$ converges to 0 in distribution, as $t \to \infty$.
(c) Find a positive number r such that $\frac{B_t}{t^r}$ converges to 0 in mean square, as $t \to \infty$.

3.11 Brownian motion IV

Consider the Brownian motion $B_t(\omega)$ in $\mathbb{R}^1, \mathbb{R}^2, \mathbb{R}^3$, respectively.

(i) Numerically compute the likelihood that B_t stays inside the interval $D_1 = (-1, 1)$ for $t > 0$, that is, compute $r(t) = \mathbb{P}(\omega: B_t(\omega) \in (-1, 1))$. What is the value of $r(0)$? Plot $r(t)$ for $0 \leq t \leq 10$.
(ii) Do the same for the two-dimensional Brownian motion for the unit disk $D_2 = \{(x, y): x^2 + y^2 < 1\}$.
(iii) Do the same for the three-dimensional Brownian motion for the unit ball $D_3 = \{(x, y, z): x^2 + y^2 + z^2 < 1\}$.

3.12 White noise

Explain why $\dot{B}_t$ is an appropriate model for white noise.

3.13 Two-sided Brownian motion

Two-sided Brownian motion $B_t, t \in \mathbb{R}$, is also defined in terms of two independent Brownian motions B_t^1 and B_t^2, as follows:

$$B_t = \begin{cases} B_t^1, & t \geq 0, \\ B_{-t}^2, & t < 0. \end{cases}$$

Is this an appropriate definition for a two-sided Brownian motion?

3.14 Sample paths of two-sided Brownian motion

Discuss how to generate sample paths for two-sided Brownian motion B_t, $t \in [-T, T]$, for some given $T > 0$. Write a Matlab code to plot a few sample paths.

3.15 Brownian motion as a Fourier series

Brownian motion can also be expressed as a Fourier sine series (which is sometimes called the Karhunen-Loève theorem), as in [7]:

$$B_t = \sum_{k=1}^{\infty} z_k \sqrt{2} \sin \frac{\left(k - \frac{1}{2}\right)\pi t}{\left(k - \frac{1}{2}\right)\pi}, \quad t \in [0, 1],$$

where $\{z_k\}$ is a sequence of independent Gaussian random variables with mean zero and variance 1. The convergence of this series is in mean square sense

$$\mathbb{E}\left(B_t - \sum_{k=1}^{n} z_k \sqrt{2} \sin \frac{(k-\frac{1}{2})\pi t}{(k-\frac{1}{2})\pi}\right)^2 \to 0, \quad \text{as } n \to \infty$$

and is uniform for $t \in [0, 1]$. Is this series also convergent in mean square sense and uniform for $t \in [-1, 0]$? Discuss how to generate sample paths for two-sided Brownian motion $B_t, t \in [-T, T]$ for some given $T > 0$, using this series representation. Write a Matlab code to plot a few sample paths.

4
A Crash Course in Stochastic Differential Equations

In this chapter, we first review stochastic integration, then discuss basic issues for stochastic differential equations, and finally discuss possible impacts of noise on the solutions of stochastic differential equations. Stochastic integration, Itô's formula, and stochastic differential equations follow [213, Chapters 3–5], while Section 4.7 is adopted from [152, Section 4.9].

4.1 Differential Equations with Noise

In the previous chapter, we see that $\frac{dB_t}{dt}$ is a mathematical model for white noise. A double-well system, in Example 1.1, subject to an external white noise force that is proportional to the system state, may be modeled by

$$\dot{x} = x - x^3 + kx\frac{dB_t}{dt},$$

where kx (with k constant) is the noise intensity. In stochastic calculus, it is customary to write differential equations in differential form and unknowns in capital letters. Thus the preceding equation becomes

$$dX_t = \left(X_t - X_t^3\right)dt + kX_t dB_t,$$

together with initial condition X_0. This is an example of stochastic differential equations (SDEs). Inspired by what we have learned in solving elementary ordinary differential equations, we rewrite the earlier SDE in integral form:

$$X_t = x_0 + \int_0^t (X_s - X_s^3)ds + k\int_0^t X_s dB_s. \tag{4.1}$$

The term $\int_0^t (X_s - X_s^3)ds$ is a Riemann integral (or Lebesgue integral), but $\int_0^t X_s dB_s$ will be a stochastic integral. Thus, to interpret an SDE, we need to define this stochastic integral appropriately (whatever it may be). At a first glance, it appears to be a Riemann-Stieltjes integral. To examine this issue, we recall the definition of Riemann-Stieltjes integrals and then define stochastic integrals in the next two sections, respectively.

4.2 Riemann-Stieltjes Integration

Recall the definition of a Riemann-Stieltjes integral $\int_0^T h(t)dg(t)$, with integrand h and integrator g. Partition the time interval $[0, T]$ into small subintervals of maximal subinterval length δ, $0 = t_0 < t_1 < \cdots < t_i < t_{i+1} < \cdots < t_n = T$, and then consider the limit of the following sum as $\delta \to 0$,

$$\sum_{i=0}^{n-1} h(\tau_i)(g(t_{i+1}) - g(t_i)), \tag{4.2}$$

where $\tau_i \in [t_i, t_{i+1}]$. If this limit exists, then the Riemann-Stieltjes integral $\int_0^T h(t)dg(t)$ exists, and the limit is defined to be the value of this integral.

We recall the concept of "finite variation" for a function f defined on $[0, T]$, introduced in Section 3.1.1. Consider all possible partitions of the time interval $[0, T]$ into small subintervals: $0 = t_0 < t_1 < \cdots < t_i < t_{i+1} < \cdots < t_n = T$. If there exists a positive number M such that

$$\sum_{i=0}^{n-1} |g(t_{i+1}) - g(t_i)| \le M,$$

for all partitions of $[0, T]$, then g is said to be of finite variation (or bounded variation) on the time interval $[0, T]$.

If f is of finite variation on the time interval $[0, T]$, it is necessarily bounded. Moreover, f is finite variation on the time interval $[0, T]$ if and only if f can be expressed as the difference of two increasing functions. An increasing function has left and right limits at every point, therefore any discontinuity is a jump discontinuity. In fact, it can have at most countably many jump discontinuities. Thus, a function of finite variation can have at most countably many discontinuities, and all discontinuities are jump discontinuities [151]. Moreover, a function of finite variation on $[0, T]$ is differentiable almost everywhere. A Lipschitz continuous function is of finite variation and is thus differentiable almost everywhere.

A sufficient condition for the limit (4.2) to exist, that is, the Riemann-Stieltjes integral $\int_0^T h(t)dg(t)$ to exist, is that the integrator $g(t)$ be of "finite variation" [6, Chapter 6].

Theorem 4.1 (Sufficient condition) *If the integrand h is continuous on $[0, T]$ and the integrator g is of finite variation on $[0, T]$, then the Riemann-Stieltjes integral $\int_0^T h(t)dg(t)$ exists.*

In fact, this "finite variation" condition is *close* to being necessary because of the following result [232, Chapter 1].

Theorem 4.2 (Necessary condition) *If the sum* (4.2) *with* $\tau_i \equiv t_i$ *converges to a limit for every continuous integrand h, then g is of finite variation.*

4.3 Stochastic Integration and Stochastic Differential Equations

Recall that a deterministic ordinary differential equation may be interpreted as an integral equation, while the integral is in Riemann-Stieltjes sense. To consider stochastic differential equations, we also need a concept of integration for stochastic functions, that is, integration with respect to Brownian motion. Indeed, the following stochastic differential equation

$$dX_t = b(X_t)dt + \sigma(X_t)dB_t, \; X_0 = x \tag{4.3}$$

may be interpreted as

$$X_t = x + \int_0^t b(X_u)du + \int_0^t \sigma(X_u)dB_u.$$

This requires a meaning for $\int_0^t \sigma(X_u(\omega))dB_u(\omega)$.

4.3.1 Motivation

We try to define the integration of a stochastic process (called integrand), $f(t, \omega)$, with respect to Brownian motion B_t (called integrator), in a probability space $(\Omega, \mathcal{F}, \mathbb{P})$,

$$\int_0^{\mathrm{T}} f(t, \omega)dB_t(\omega). \tag{4.4}$$

This appears to be a Riemann-Stieltjes integral for each fixed sample ω (sample-wise). But let us take a careful look.

Note that almost all sample paths of B_t are not of finite variation on $[0, T]$. Thus $\int_0^{\mathrm{T}} f(t, \omega)dB_t(\omega)$ may not be defined pathwise in the Riemann-Stieltjes sense.

Therefore, we have to define the integral (4.4) differently. At least we require the integrand f to be measurable in both t and ω, that is measurable with respect to the σ-field $\mathcal{B}(\mathbb{R}) \times \mathcal{F}$. It is also natural to require that f not depend on the future; that is, it is nonanticipating or adapted to $\mathcal{F}^B$, the σ-field generated by Brownian motion. We still partition the time interval $[0, T]$ into small subintervals of maximal subinterval length δ: $0 = t_0 < t_1 < \cdots < t_i < t_{i+1} < \cdots < t_n = T$. We use the following notations:

$$\delta \triangleq \max_i \{t_{i+1} - t_i\},$$
$$B_i \triangleq B(t_i),$$
$$\Delta B_i \triangleq B(t_{i+1}) - B(t_i).$$

Then, we consider the sum

$$\sum_{i=0}^{n-1} f(\tau_i, \omega)(B_{i+1}(\omega) - B_i(\omega)), \tag{4.5}$$

where $\tau_i \in [t_i, t_{i+1}]$, and its convergence is $\delta \to 0$, in an appropriate sense. Because these sums are random quantities, we take advantage of various convergence concepts in probability space as defined in Section 2.9.

Let us look at a simple example.

Example 4.3 (Riemann-Stieltjes integration is inappropriate) We consider the integral $\int_0^T B_t dB_t$. The integrand is B_t itself. Let us take τ_i to be either the left end point or the right end point on each subinterval $[t_i, t_{i+1}]$. The sum becomes, respectively,

$$s_n \triangleq \sum_{i=0}^{n-1} B_i(\omega)(B_{i+1}(\omega) - B_i(\omega)),$$
$$S_n \triangleq \sum_{i=0}^{n-1} B_{i+1}(\omega)(B_{i+1}(\omega) - B_i(\omega)).$$

Before examining the convergence of these sums as $\delta \to 0$, in an appropriate sense, we compute the mean of these sums. For s_n,

$$\mathbb{E}s_n = \sum_{i=0}^{n-1} \mathbb{E}[B_i(\omega)(B_{i+1}(\omega) - B_i(\omega))]$$
$$= \sum_{i=0}^{n-1} \mathbb{E}[(B_i(\omega) - B_0(\omega))(B_{i+1}(\omega) - B_i(\omega))] = 0,$$

because B_t has independent increments. Similarly, for S_n,

$$\begin{aligned}\mathbb{E}S_n &= \sum_{i=0}^{n-1} \mathbb{E}[B_{i+1}(\omega)(B_{i+1}(\omega) - B_i(\omega))] \\ &= \sum_{i=0}^{n-1} \mathbb{E}[(B_{i+1}(\omega) - B_i(\omega) + B_i(\omega))(B_{i+1}(\omega) - B_i(\omega))] \\ &= \sum_{i=0}^{n-1} \mathbb{E}[(B_{i+1}(\omega) - B_i(\omega))^2 + B_i(\omega)(B_{i+1}(\omega) - B_i(\omega))] \\ &= \sum_{i=0}^{n-1} \mathbb{E}(B_{i+1}(\omega) - B_i(\omega))^2 + 0 = \sum_{i=0}^{n-1}(t_{i+1} - t_i) = T.\end{aligned}$$

Hence the limit (say, in mean square sense) of these sums as $\delta \to 0$, even when it exists, depends on the choice of τ_i. Thus, if we define $\int_0^T B_t dB_t$ as a limit of a sum in (4.5), then τ_is cannot be arbitrarily chosen. This is very much unlike the the Riemann-Stieltjes integrals. This implies that stochastic integrals need to be defined differently.

In fact, it turns out that τ_is in the sum (4.5) have to be chosen in a fixed *pattern* in *each* subinterval $[t_i, t_{i+1}]$. The integral value depends on this specific *pattern* of choices of τ_i. This is a consequence of the unbounded variation of the Brownian motion paths. For example, for $\tau_i = t_i$, we obtain the Itô integral $\int_0^T f(t, \omega)dB_t(\omega)$, whereas for $\tau = \frac{1}{2}(t_i + t_{i+1})$, we have Stratonovich integral $\int_0^T f(t, \omega) \circ dB_t(\omega)$.

More generally, for $\tau_i = (1 - \lambda)t_i + \lambda t_{i+1}$ with $\lambda \in [0, 1]$, we have λ-integral: $(\lambda) \int_0^T f(t, \omega)dB_t(\omega)$. Note that the Itô integral and Stratonovich integral correspond to $\lambda = 0$ and $\lambda = \frac{1}{2}$, respectively.

4.3.2 Definition of Itô Integral

The Itô integral $\int_0^T f(t, \omega)dB_t(\omega)$ is defined for a class of integrands as follows [213, Chapter 3]. We do this for scalar integrand f and scalar Brownian motion B_t, as in vector case, and we define the Itô integral component by component. Let $(\Omega, \mathcal{F}, \mathbb{P})$ be a probability space and let $\mathcal{F}_t \triangleq \sigma(B_s, s \leq t)$ be the filtration generated by Brownian motion up to time t. In other words, $\mathcal{F}_t$ is the smallest σ-field containing events of the form

$$\{\omega\colon B_{t_1}(\omega) \in A_1, \ldots, B_{t_k}(\omega) \in A_k\},$$

for all $t_1, \ldots, t_k \leq t$ and all Borel sets $A_1, \ldots, A_k$ in $\mathbb{R}^1$. Note that $\mathcal{F}_t \subset \mathcal{F}$, and $\mathcal{F}_s \subset \mathcal{F}_t$ when $s < t$ (i.e., $\mathcal{F}_t$ is increasing).

First, introduce a class of stochastic integrands. Define $\mathbb{S}(T_0, T_f)$ to be a class of measurable functions

$$f : [0, \infty) \times \Omega \to \mathbb{R}^1,$$
$$(t, \omega) \to f(t, \omega),$$

such that

(i) f is $\mathcal{F}_t$-adapted, that is, $f(t, \cdot)$ is measurable with respect to the σ-field $\mathcal{F}_t$ (or $f(t, \cdot)$ is $\mathcal{F}_t$-measurable);
(ii) f is mean square (Lebesgue) integrable in the sense that $\mathbb{E} \int_{T_0}^{T_f} f^2(t, \omega)dt < \infty$.

Then, consider elementary functions in $\mathbb{S}(T_0, T_f)$ in the form

$$h(t, \omega) = \sum_i e_i(\omega)\, I_{[t_i, t_{i+1})}(t),$$

where e_i is a random variable and $I_{[t_i, t_{i+1})}$ is the (deterministic) indicator function for the subinterval $[t_i, t_{i+1})$, for each i. Such an elementary function is "randomly" constant on each subinterval $[t_i, t_{i+1})$ and the random constant $e_i(\omega)$ "starts" at the left end point (not including the right end point t_{i+1}). It is indeed adapted to $\mathcal{F}_t$. Naturally, its Itô integral is defined by

$$\int_{T_0}^{T_f} h(t, \omega)dB_t = \sum_i e_i(\omega)(B(t_{i+1}) - B(t_i)).$$

Third, for each $f \in \mathbb{S}(T_0, T_f)$, it can be shown that there exists a sequence of elementary functions f_n in $\mathbb{S}(T_0, T_f)$ such that f_n converges to f in the following "integrated mean square" sense:

$$\mathbb{E} \int_{T_0}^{T_f} (f(t, \omega) - f_n(t, \omega))^2 dt \to 0, \quad \text{as } n \to \infty. \tag{4.6}$$

Finally, define

$$\int_{T_0}^{T_f} f(t, \omega)dB_t = \lim_{n\to\infty} \int_{T_0}^{T_f} f_n(t, \omega)dB_t,$$

where the limit is taken in $L^2(\Omega)$. We summarize this in the following theorem [213, p. 29].

Theorem 4.4 *For $f \in \mathbb{S}(T_0, T_f)$, the Itô integral $\int_{T_0}^{T_f} f(t, \omega)dB_t$ exists, and its value can be evaluated by*

$$\int_{T_0}^{T_f} f(t, \omega)dB_t = \lim_{n\to\infty} \int_{T_0}^{T_f} f_n(t, \omega)dB_t$$

for a sequence of elementary functions $f_n(t,\omega)$ that approximates f in the integrated mean square sense (4.6). The value of the Itô integral does not depend on the specific choice of the elementary sequence f_n.

We will not develop a rigorous stochastic integration theory here, and interested readers may refer to, for example, [84, 147, 213], or [127].

Stochastic integrals for vector functions are defined component by component, although enlarging the family of integrands is necessary (see [213, Section 3.3]).

4.3.3 Practical Calculations

By Theorem 4.4, when an Itô stochastic integral $\int_{T_0}^{T_f} f(t,\omega)dB_t$ is known to exist, we could evaluate its value by $\lim_{n\to\infty}$ in ms. $\int_{T_0}^{T_f} f_n(t,\omega)dB_t$, for one specific sequence of elementary functions f_n that approximates f in the integrated mean square sense.

For example, if the integrand $f(t,\omega)$ is continuous in t (almost surely), we take a sequence of partitions $\mathfrak{P}^n$ of the time interval $[T_0, T_f]$, of equal subinterval length $\delta^n = \frac{T_f - T_0}{n}$:

$$T_0 = t_0^n < t_1^n < \cdots < t_i^n < t_{i+1}^n < \cdots < t_n^n = T_f,$$

for $n = 1, 2, \ldots$. Note that δ^n converges to 0 as $n \to \infty$. Then we choose a sequence of elementary functions as follows:

$$f_n(t,\omega) = \sum_{i=0}^{n-1} f\left(t_i^n, \omega\right) I_{[t_i^n, t_{i+1}^n)}(t), \tag{4.7}$$

where f is evaluated at the left end point on each subinterval $[t_i^n, t_{i+1}^n]$. The value of the Itô integral is thus obtained by the limit

$$\begin{aligned} \int_{T_0}^{T_f} f(t,\omega)dB_t &= \lim_{n\to\infty} \text{ in m.s.} \int_{T_0}^{T_f} f\left(t_i^n,\omega\right)dB_t \\ &= \lim_{n\to\infty} \text{ in m.s.} \sum_{i=0}^{n-1} f\left(t_i^n,\omega\right)\left(B\left(t_{i+1}^n\right) - B\left(t_i^n\right)\right), \end{aligned} \tag{4.8}$$

where f is evaluated at the left end point on each subinterval $[t_i^n, t_{i+1}^n]$.

4.3.4 Stratonovich Integral

Inspired by the evaluation formula (4.8) for Itô integral, we define Stratonovich integral $\int_{T_0}^{T_f} f(t,\omega) \circ dB_t(\omega)$, when the integrand f is continuous in t, by the following limit whenever it exists:

$$\int_{T_0}^{T_f} f(t,\omega) \circ dB_t(\omega)$$

$$= \lim_{n\to\infty} \text{ in m.s.} \sum_{i=0}^{n-1} f\left(\frac{1}{2}(t_i^n + t_{i+1}^n), \omega\right)\left(B(t_{i+1}^n) - B(t_i^n)\right), \quad (4.9)$$

where f is evaluated at the middle point on each subinterval $[t_i^n, t_{i+1}^n]$.

An interesting observation is useful here. When the integrand $f(t,\omega)$ is continuously differentiable in time (almost surely), we apply Taylor expansions at t_i^n and t_{i+1}^n, respectively, to get

$$f\left(\frac{1}{2}(t_i^n + t_{i+1}^n), \omega\right) = f(t_i^n, \omega) + O(t_{i+1}^n - t_i^n),$$

$$f\left(\frac{1}{2}(t_i^n + t_{i+1}^n), \omega\right) = f(t_{i+1}^n, \omega) + O(t_{i+1}^n - t_i^n).$$

Adding half of each of both equations together, we conclude that

$$f\left(\frac{1}{2}(t_i^n + t_{i+1}^n), \omega\right) = \frac{1}{2} f(t_i^n, \omega) + \frac{1}{2} f(t_{i+1}^n, \omega) + o(t_{i+1}^n - t_i^n).$$

Thus, by (4.9), the Stratonovich integral is also defined by

$$\int_{T_0}^{T_f} f(t,\omega) \circ dB_t(\omega)$$

$$= \lim_{n\to\infty} \text{ in m.s.} \sum_{i=0}^{n-1} \left[\frac{1}{2} f(t_i^n, \omega) + \frac{1}{2} f(t_{i+1}^n, \omega)\right]\left(B(t_{i+1}^n) - B(t_i^n)\right), \quad (4.10)$$

whenever the limit exists. In fact, this is also often taken as the definition of Stratonovich integral even when the integrand is not differentiable in time, as long as the limit in (4.10) exists. This may offer an advantage, as we do not need to evaluate f at the middle point of each subinterval; instead, we evaluate the average of f values at the end points of each subinterval.

Remark 4.5 If the integrand $f(t,\omega)$ is sufficiently smooth in time (e.g., Hölder continuous in time in mean square norm, with exponent larger than 1, then both Itô and Stratonovich integrals are identical; see [213, p. 39]). But in general, Itô and Stratonovich integrals differ. Note that B_t is only Hölder continuous in time ([151, Chapter 2] and [165]) with exponent less than $\frac{1}{2}$.

More generally, the λ-integral

$$(\lambda)\int_{T_0}^{T_f} f(t,\omega)dB_t(\omega)$$

for $\lambda \in [0,1]$ is defined by

$$(\lambda)\int_0^T f(t,\omega)dB_t(\omega) = \lim_{n\to\infty} \text{ in m.s.} \sum_{i=0}^{n-1} \left[(1-\lambda)f\left(t_i^n,\omega\right) + \lambda f\left(t_{i+1}^n,\omega\right)\right] \cdot \left(B\left(t_{i+1}^n\right) - B\left(t_i^n\right)\right) \tag{4.11}$$

whenever the limit exists. For $\lambda = 0$, this is the Itô integral, whereas for $\lambda = \frac{1}{2}$, it corresponds to the Stratonovich integral. When $\lambda = 1$ (i.e., evaluate f at the right end point on each subinterval $[t_i, t_{i+1}]$), it is called the backward Itô integral.

4.3.5 Examples

We now present a couple of examples of practical evaluation of stochastic integrals.

Remark 4.6 From Problem 2.12, we know that if $X_n \to X$ and $Y_n \to Y$ in mean square as $n \to \infty$, then $X_n + Y_n \to X + Y$ in mean square. Therefore, when considering convergence in mean square for a sum of finite terms, we can evaluate the convergence in mean square term by term (when each term converges).

Example 4.7 Let us calculate the following Itô integral [213, p. 29]:

$$\int_S^T B_t dB_t = \frac{1}{2}(B_T^2 - B_S^2) - \frac{1}{2}(T-S). \tag{4.12}$$

Introduce a sequence of partitions $\mathfrak{P}^n$ of the time interval $[S,T]$ of equal subinterval length $\delta^n = \frac{T-S}{n}$:

$$S = t_0^n < t_1^n < \cdots < t_i^n < t_{i+1}^n < \cdots < t_n^n = T,$$

for $n = 1, 2, \ldots$. Consider the sum

$$\begin{aligned}
&\sum_i B\left(t_i^n\right)\left(B\left(t_{i+1}^n\right) - B\left(t_i^n\right)\right) \\
&= \frac{1}{2}\sum_i \left(B^2\left(t_{i+1}^n\right) - B^2\left(t_i^n\right)\right) - \frac{1}{2}\sum_i \left(B\left(t_{i+1}^n\right) - B\left(t_i^n\right)\right)^2 \\
&= \frac{1}{2}(B_T^2 - B_S^2) - \frac{1}{2}\sum_i \left(B\left(t_{i+1}^n\right) - B\left(t_i^n\right)\right)^2.
\end{aligned}$$

Taking the mean square limit as $n \to \infty$ and by Theorem 3.6 (iii), we conclude that $\sum_i (B(t_{i+1}^n) - B(t_i^n))^2$ has mean square limit $T - S$. This verifies (4.12).

Example 4.8 We can also calculate the following Stratonovich integral:

$$\int_S^T B_t \circ dB_t = \frac{1}{2}(B_T^2 - B_S^2). \tag{4.13}$$

Introduce a sequence of partitions $\mathfrak{P}^n$ of the time interval $[S, T]$ of equal subinterval length $\delta^n = \frac{T-S}{n}$:

$$S = t_0^n < t_1^n < \cdots < t_i^n < t_{i+1}^n < \cdots < t_n^n = T,$$

for $n = 1, 2, \ldots$. Consider the sum

$$\begin{aligned}
&\sum_i B\left(\frac{1}{2}(t_i^n + t_{i+1}^n)\right)(B(t_{i+1}^n) - B(t_i^n)) \\
&= \sum_i \left[B\left(\frac{1}{2}(t_i^n + t_{i+1}^n)\right) - B(t_i^n) + B(t_i^n)\right] \\
&\quad \cdot \left[B(t_{i+1}^n) - B\left(\frac{1}{2}(t_i^n + t_{i+1}^n)\right) + B\left(\frac{1}{2}(t_i^n + t_{i+1}^n)\right) - B(t_i^n)\right] \\
&= \sum_i \left[B\left(\frac{1}{2}(t_i^n + t_{i+1}^n)\right) - B(t_i^n)\right]\left[B(t_{i+1}^n) - B\left(\frac{1}{2}(t_i^n + t_{i+1}^n)\right)\right] \\
&\quad + \sum_i \left[B\left(\frac{1}{2}(t_i^n + t_{i+1}^n)\right) - B(t_i^n)\right]^2 \\
&\quad + \sum_i B(t_i^n)[B(t_{i+1}^n) - B(t_i^n)] \\
&=: A_n + B_n + C_n.
\end{aligned}$$

Let us examine the convergence of random sequences A_n, B_n, and C_n, one by one. The random sequence C_n corresponds to the Itô integral and so its mean square limit as $n \to \infty$ is $\frac{1}{2}(B_T^2 - B_S^2) - \frac{1}{2}(T - S)$.

For the random sequence B_n, note that its mean $\mathbb{E}B_n = \sum_i \frac{1}{2}(t_{i+1}^n - t_i^n) = \frac{1}{2}(T - S)$. The mean square limit of B_n is also $\frac{1}{2}(T - S)$. Indeed, noting the property of independent increments for the Brownian motion, we have

$$\begin{aligned}
\mathbb{E}\left\{B_n - \frac{1}{2}(T - S)\right\}^2 &= \mathbb{E}\left\{\sum_i \left(\left[B\left(\frac{1}{2}(t_i^n + t_{i+1}^n)\right) - B(t_i^n)\right]^2 - \frac{1}{2}(t_{i+1}^n - t_i^n)\right)\right\}^2 \\
&= \mathbb{E}\sum_i \left\{\left[B\left(\frac{1}{2}(t_i^n + t_{i+1}^n)\right) - B(t_i^n)\right]^2 - \frac{1}{2}(t_{i+1}^n - t_i^n)\right\}^2.
\end{aligned}$$

Following an argument in [13, Theorem 9.3.4], denote the standard normal distribution $\mathcal{N}(0, 1)$ by Z and then $[B(\frac{1}{2}(t_i^n + t_{i+1}^n)) - B(t_i^n)]^2 \sim Z^2 \frac{1}{2}(t_{i+1}^n - t_i^n)$.

Therefore,

$$\mathbb{E}\left\{B_n - \frac{1}{2}(T-S)\right\}^2 = \mathbb{E}(Z^2-1)^2\frac{1}{4}\sum_i \left(t_{i+1}^n - t_i^n\right)^2$$

$$\leq \frac{1}{4}\mathbb{E}(Z^2-1)^2 \max_i \left(t_{i+1}^n - t_i^n\right)\sum_i \left(t_{i+1}^n - t_i^n\right)$$

$$= \frac{1}{4}\mathbb{E}(Z^2-1)^2 \max_i \left(t_{i+1}^n - t_i^n\right)\cdot(T-S) \to 0,$$

as $n \to \infty$.

As for A_n, we calculate

$$\mathbb{E}(A_n - 0)^2$$

$$= \mathbb{E}\sum_{i,j}\left\{\left[B\left(\frac{1}{2}(t_i^n + t_{i+1}^n)\right) - B(t_i^n)\right]\left[B(t_{i+1}^n) - B\left(\frac{1}{2}(t_i^n + t_{i+1}^n)\right)\right]\right.$$

$$\left.\cdot\left[B\left(\frac{1}{2}(t_j^n + t_{j+1}^n)\right) - B(t_j^n)\right]\left[B(t_{j+1}^n) - B\left(\frac{1}{2}(t_j^n + t_{j+1}^n)\right)\right]\right\}$$

$$= \mathbb{E}\sum_i\left[B\left(\frac{1}{2}(t_i^n + t_{i+1}^n)\right) - B(t_i^n)\right]^2\cdot\left[B(t_{i+1}^n) - B\left(\frac{1}{2}(t_i^n + t_{i+1}^n)\right)\right]^2$$

$$= \sum_i\left[\frac{1}{2}(t_{i+1}^n - t_i^n)\right]^2$$

$$= n\left(\frac{1}{2}\frac{T-S}{n}\right)^2 = \frac{(T-S)^2}{4n} \to 0,$$

as $n \to \infty$. We have used the following observation: the terms for $i \neq j$ have mean zero due to the independence of increments for Brownian motion. Also note that $[B(\frac{1}{2}(t_i^n + t_{i+1}^n)) - B(t_i^n)]^2$ and $[B(t_{i+1}^n) - B(\frac{1}{2}(t_i^n + t_{i+1}^n))]^2$ are independent. Therefore, $A_n + B_n + C_n$ has mean square limit $\frac{1}{2}(B_T^2 - B_S^2)$ as $n \to \infty$, and this verifies (4.13).

By taking the mathematical expectation for these two integrals (4.12) and (4.13), we see that

$$\mathbb{E}\int_S^{\mathrm{T}} B_t dB_t = \frac{1}{2}(\mathbb{E}B_T^2 - \mathbb{E}B_S^2) - \frac{1}{2}(T-S) = 0, \tag{4.14}$$

$$\mathbb{E}\int_S^{\mathrm{T}} B_t \circ dB_t = \frac{1}{2}(\mathbb{E}B_T^2 - \mathbb{E}B_S^2) = \frac{1}{2}(T-S) \neq 0. \tag{4.15}$$

In these two simple examples, the Itô integral has mean zero, while the Stratonovich integral does not. In fact, this property of the mean of stochastic integrals is true in general.

Remark 4.9 Similarly, we can calculate the λ-integral [152, Section 3.5]:

$$
\begin{aligned}
(\lambda)\int_S^T B_t dB_t &= \lim_{n\to\infty} \text{ in m.s.} \sum_i B\big((1-\lambda)t_i^n + \lambda t_{i+1}^n)\big)\big(B\big(t_{i+1}^n\big) - B\big(t_i^n\big)\big) \\
&= \lim_{n\to\infty} \text{ in m.s.} \sum_i B\big(t_i^n + \lambda\big(t_{i+1}^n - t_i^n\big)\big)\big(B\big(t_{i+1}^n\big) - B\big(t_i^n\big)\big) \\
&= \frac{1}{2}\big(B_T^2 - B_S^2\big) + \left(\lambda - \frac{1}{2}\right)(T - S).
\end{aligned}
$$

4.3.6 Properties of Itô Integrals

These properties of Itô integrals are useful for analyzing SDEs, and they hold when the involved Itô integrals exist, that is, when the integrands are in $\mathbb{S}(S, T)$. These properties are proved first for elementary functions in $\mathbb{S}(S, T)$, then approximating other functions in $\mathbb{S}(S, T)$ by elementary functions, and finally passing the limits [213, Chapter 3].

Linearity

$$
\int_S^T (f(t,\omega) + g(t,\omega))dB_t = \int_S^T f(t,\omega)dB_t + \int_S^T g(t,\omega)dB_t, \text{ a.s.} \tag{4.16}
$$

$$
\int_S^T c\, f(t,\omega)dB_t = c\int_S^T f(t,\omega)dB_t, \text{ a.s.} \tag{4.17}
$$

$$
\int_S^T f(t,\omega)dB_t = \int_S^\tau f(t,\omega)dB_t + \int_\tau^T f(t,\omega)dB_t, \text{ a.s.,} \tag{4.18}
$$

where c is a constant and $S < \tau < T$.

Zero Mean Property

$$
\mathbb{E}\int_S^T f(t,\omega)dB_t = 0. \tag{4.19}
$$

Itô Isometry in Scalar Case

$$
\mathbb{E}\left(\int_S^T f(t,\omega)dB_t\right)^2 = \mathbb{E}\int_S^T f^2(t,\omega)dt. \tag{4.20}
$$

More generally,

$$\mathbb{E}\left(\int_S^a f(t,\omega)dB_t \int_S^b g(t,\omega)dB_t\right) = \mathbb{E}\int_S^{a\wedge b} f(t,\omega)g(t,\omega)dt, \quad (4.21)$$

where $a \wedge b \triangleq \min\{a, b\}$.

Proof For the case $a = b$: Denote $I_1 = \int_S^a f(t,\omega)dB_t$ and $I_2 = \int_S^a g(t,\omega)dB_t$. Note that $I_1 I_2 = \frac{1}{2}[(I_1 + I_2)^2 - I_1^2 - I_2^2]$, and use the isometry property.

For the case $a \neq b$: say $a < b$, that is, $\min\{a, b\} = a$. Extend f to the time interval $[a, b]$ by setting it zero there. Then apply the preceding proof. □

Itô Isometry in Vector Case

Let $F(t,\omega)$ and $G(t,\omega)$ be $n \times n$ matrixes and B_t be n-dimensional Brownian motion. Then

$$\mathbb{E}\left(\int_S^a F(t,\omega)dB_t \cdot \int_S^b G(t,\omega)dB_t\right) = \mathbb{E}\int_S^{a\wedge b} \mathrm{Tr}(GF^{\mathrm{T}})(t,\omega)dt, \quad (4.22)$$

where $\cdot$ denotes the usual scalar product in $\mathbb{R}^n$ and Tr denotes the trace of a matrix (i.e., the sum of diagonal entries of a matrix).

In particular,

$$\mathbb{E}\left\|\int_S^a F(t,\omega)dB_t\right\|^2 = \mathbb{E}\int_S^a \mathrm{Tr}(FF^{\mathrm{T}})(t,\omega)dt, \quad (4.23)$$

and

$$\mathbb{E}\left(\int_S^a F(t,\omega)dB_t \cdot \int_S^b F(t,\omega)dB_t\right) = \mathbb{E}\int_S^{a\wedge b} \mathrm{Tr}(FF^{\mathrm{T}})(t,\omega)dt. \quad (4.24)$$

Inequalities Involving Itô Integrals

By the Itô isometry and the Doob martingale inequality [213, p. 33], for any constant $\lambda > 0$,

$$\mathbb{P}\left(\sup_{t_0\le t\le T}\left|\int_{t_0}^t f(s,\omega)dB_s\right| \ge \lambda\right) \le \frac{1}{\lambda^2}\mathbb{E}\int_{t_0}^{\mathrm{T}} |f(s,\omega)|^2 ds. \quad (4.25)$$

Moreover [8, p. 81],

$$\mathbb{E}\left(\sup_{t_0\le t\le T}\left|\int_0^t f(s,\omega)dB_s\right|^2\right) \le 4\mathbb{E}\int_{t_0}^{\mathrm{T}} |f(s,\omega)|^2 ds, \quad (4.26)$$

and for $k = 1, 2, \ldots$,

$$\mathbb{E}\left|\int_{t_0}^{t} f(s,\omega)dB_s\right|^{2k} \le (k(2k-1))^{k-1}(t-t_0)^{k-1}\mathbb{E}\int_{t_0}^{\mathrm{T}} |f(s,\omega)|^{2k}ds. \quad (4.27)$$

4.3.7 Stochastic Differential Equations

An Itô SDE (4.3) has a corresponding Stratonovich SDE

$$dX_t = b(X_t)dt + \sigma(X_t) \circ dB_t, \;\; X_0 = x_0. \quad (4.28)$$

In Section 4.7, we see how to convert an Itô SDE to a Stratonovich SDE, and vice versa. In fact, we could also have λ SDEs interpreted via λ-integrals, but they can be converted to Itô SDEs. Although mathematically we could focus on Itô SDEs, Stratonovich SDEs have been shown to have some merits in physical modeling.

Note that the Itô stochastic differential $\sigma(X)dB(t)$ in SDE (4.3) and the Stratonovich stochastic differential $\sigma(X) \circ dB(t)$ in SDE (4.28) are interpreted through their corresponding definitions of stochastic integrals $\int_0^{\mathrm{T}} \sigma(X)dB_t$ and $\int_0^{\mathrm{T}} \sigma(X) \circ dB_t$, respectively.

Thus, we have the two important kinds of SDEs, that is, Itô SDE

$$dX_t = b(X_t)dt + \sigma(X_t)dB_t, \;\; X_0 = x_0 \quad (4.29)$$

and Stratonovich SDE

$$dX_t = b(X_t)dt + \sigma(X_t) \circ dB_t, \;\; X_0 = x_0. \quad (4.30)$$

However, systems of Stratonovich SDEs can be converted to Itô SDEs, and vice versa; see Section 4.7 later in this chapter or [152, 213]. In the following we only consider Itô-type SDEs.

In (4.29) or (4.30), when σ does not depend on the system state X, the SDE is said to have an additive noise; otherwise, it is said to have a multiplicative noise. In these SDEs, $b(X)$ is called the vector field or the drift term, and $\sigma(X)$ is the diffusion coefficient or noise intensity.

4.3.8 SDEs in Engineering and Science Literature

In engineering and science literature, an SDE is sometimes written in the following form:

$$\dot{x} = x - x^3 + \dot{B}_t. \quad (4.31)$$

In this book, we occasionally use this form as well. We should realize that the SDE (4.31) is the same as

$$dX_t = \left(X_t - X_t^3\right)dt + dB_t. \quad (4.32)$$

4.3.9 SDEs with Two-Sided Brownian Motions

In Chapter 6, we discuss stochastic invariant manifolds and need to have a meaning of both future time and past time for a stochastic system. Thus we have two-sided Brownian motion B_t in SDEs and need to define the Itô integral for such a two-sided Brownian motion. In other words, we need a "backward Itô integral" as well as the "forward Itô integral" that we have discussed so far. This can be done by first introducing a two-parameter filtration $\mathcal{F}_s^t$ (in contrast to the one-parameter filtration $\mathcal{F}_t$, which we have used so far) and then define the Itô integral for both forward and backward time. This leads to the Itô integral for two-sided time. For more details, see [9, Chapter 2].

To analyze SDEs, we need a stochastic chain rule, that is Itô's formula. This is introduced in the next section.

4.4 Itô's Formula

In this section, we first motivate the need for a chain rule, and then present Itô's formula in scalar and vector cases, respectively.

4.4.1 Motivation for Stochastic Chain Rules

To analyze SDEs, we need to be able to manipulate stochastic differentials, which are interpreted via stochastic integrals. However, it is tedious and in general difficult to evaluate stochastic integrals by definition, as shown in the previous section. As in deterministic calculus, we need theoretical tools to manipulate integrals. One of the theoretical tools is Itô's formula, or the stochastic chain rule, which implies the stochastic product rule and integration by parts.

Remark 4.10 Recall that in the deterministic calculus, we can evaluate a simple integral $\int_0^1 t^2 dt$ by definition, as follows. For a sequence of partitions of [0, 1] with maximal length of subinterval $\delta^n = \frac{1}{n}$: $t_i^n = \frac{i}{n}$ for $i = 0, 1, 2, \ldots, n$, we consider the sum and its limit

$$\sum_i \frac{i^2}{n^2}\frac{1}{n} = \frac{1}{n^3}\frac{n(n+1)(2n+1)}{6} \to \frac{1}{3}$$

as $\delta^n = \frac{1}{n} \to 0$ or as $n \to \infty$. Thus, $\int_0^1 t^2 dt = \frac{1}{3}$. In general, it is difficult to evaluate integrals by definition. The chain rule, or more specifically the product rule and its integral form, the integration by parts, helps us to manipulate integrals:

$$\int u dv = uv - \int v du,$$

which could represent an integral $\int u dv$ in terms of a possibly simpler integral $\int v du$.

Before we review Itô's formula [213], let us recall the concept "differentiation" in deterministic calculus.

Let f be a scalar deterministic function in time and t_0 be a given time instant. To approximate the difference $\Delta f(t_0) = f(t_0 + \Delta t) - f(t_0)$ when Δt sufficiently small, we calculate the differential

$$df(t_0) = f'(t_0)dt. \tag{4.33}$$

The error for this approximation $df \approx \Delta f$ is $o(|\Delta t|^2)$, if f has a bounded second-order derivative. By Taylor expansion at t_0,

$$f(t_0 + \Delta t) - f(t_0) = f'(t_0)\Delta t + \frac{1}{2}f''(t_0)(\Delta t)^2 + \cdots,$$

or

$$df(t_0) = f'(t_0)dt + \frac{1}{2}f''(t_0)(dt)^2 + \cdots.$$

In other words, the differentiation for f at t_0 means we retain only the *first-order* (in Δt) terms in its Taylor expansion at t_0. Analytically, this means approximating a nonlinear function by a linear function. Geometrically, it means approximating a curve by its tangent line.

Itô's formula comes out in the same way, except we have to be sure what "first order" means in stochastic setting.

4.4.2 Itô's Formula in Scalar Case

Consider a scalar SDE

$$dX_t = b(X_t)dt + \sigma(X_t)dB_t, \tag{4.34}$$

where $b(\cdot)$, $\sigma(\cdot)$ are scalar functions and B_t is a scalar Brownian motion.

Let $g(t, x)$ be a given (deterministic) scalar smooth function. Let us try to apply Talyor expansion of g or the deterministic chain rule to obtain

$$\begin{aligned} dg(t, X_t) &= \frac{\partial g}{\partial t}dt + \frac{\partial g}{\partial x}dX_t \\ &\quad + \frac{1}{2}\left[\frac{\partial^2 g}{\partial t^2}(dt)^2 + 2\frac{\partial^2 g}{\partial t \partial x}dt dX_t + \frac{\partial^2 g}{\partial x^2}(dX_t)^2\right] + \text{h. o. t.} \\ &= \frac{\partial g}{\partial t}dt + \frac{\partial g}{\partial x}[b(X_t)dt + \sigma(X_t)dB_t] \\ &\quad + \frac{1}{2}\left[\frac{\partial^2 g}{\partial t^2}(dt)^2 + 2\frac{\partial^2 g}{\partial t \partial x}dt[b(X_t)dt + \sigma(X_t)dB_t\right] \\ &\quad + \frac{\partial^2 g}{\partial x^2}[b(X_t)dt + \sigma(X_t)dB_t]^2] + \text{h. o. t.}, \end{aligned} \tag{4.35}$$

where each partial derivative is evaluated at (t, X_t) and h. o. t. denotes higher-order terms. Note that $(dt)^2$ is a second-order term, and $dtdB_t$ is higher than first order, and so we discard them in the Itô stochastic differential. But how about the term with $(dB_t)^2$? Is it first order (retain) or higher than first order (discard)?

Denote $\Delta B_t = B(t + \Delta t) - B(t)$. Recall that $\mathbb{E}(\Delta B_t)^2 = \Delta t$ by the definition of Brownian motion. Thus, intuitively, at the level of mathematical expectation, $(dB_t)^2 = dt$. However, a stochastic differential needs to be interpreted in terms of a stochastic integral, and so let us examine $(dB_t)^2$ in the context of stochastic integral on $[0, T]$.

Introduce a sequence of partitions $\mathfrak{P}^n$ of the time interval $[0, T]$ of maximal subinterval length δ^n:

$$0 = t_0^n < t_1^n < \cdots < t_i^n < t_{i+1}^n < \cdots < t_n^n = T,$$

for $n = 1, 2, \ldots$. Applying Theorem 3.6 (iii), we have

$$\sum_i \left[B(t_{i+1}^n) - B(t_i^n)\right]^2 \to T = \int_0^T 1\, dt,$$

as $\delta^n \to 0$. The left-hand side may be thought as an integral $\int_0^T 1\,(dB_t)^2$, while the right-hand side is $\int_0^T 1\, dt$. It may be imagined that this would be true if the integrand 1 is replaced by a constant or by an elementary function and thus by a function that can be approximated (in mean square sense) by a sequence of elementary functions. In fact, for an appropriate integrand $a(t, \omega)$, it is indeed true that [213, Section 4.1]

$$\sum_i a(t_i^n, \omega)\left[B(t_{i+1}^n) - B(t_i^n)\right]^2 \to \int_0^T a(t, \omega)\, dt.$$

This could be written symbolically as $\int_0^T a(t,\omega)\,(dB_t)^2 = \int_0^T a(t, \omega)\, dt$, as if $(dB_t)^2 = dt$. Therefore, in the formula (4.35), the term with $(dB_t)^2$ is actually a first-order term in dt, and we thus need to retain it in the stochastic chain rule or Itô's formula.

Fortunately, the preceding formal derivation can be made rigorous (see [213, Section 4.1] or [147, Section 3.3]), but we will omit it here.

We thus have Itô's formula in differential form:

$$dg(t, X_t) = \left[\frac{\partial g}{\partial t}(t, X_t) + b(X_t)\frac{\partial g}{\partial x}(t, X_t) + \frac{1}{2}\sigma^2(X_t)\frac{\partial^2 g}{\partial x^2}(t, X_t)\right] dt + \frac{\partial g}{\partial x}(t, X_t)\sigma(X_t) dB_t. \tag{4.36}$$

The term $\frac{1}{2}\frac{\partial^2 g}{\partial x^2}(t, X_t)\sigma^2(X_t)$ is called the Itô correction term.

Remark 4.11 When either $\sigma(\cdot) = 0$ or $\frac{\partial^2 g}{\partial x^2} = 0$ (i.e., g is either linear in x or does not depend on x at all), the stochastic differential formally identifies with the deterministic differential.

Equivalently, Itô's formula is

$$dg(t, X_t) = \frac{\partial g}{\partial t}(t, X_t)dt + \frac{\partial g}{\partial x}(t, X_t)dX_t + \frac{1}{2}\frac{\partial^2 g}{\partial x^2}(t, X_t)(dX_t)^2, \quad (4.37)$$

where $(dX_t)^2$ is evaluated using the symbolic rules

$$dtdt = dtdB_t = 0, \quad dB_tdB_t = dt. \quad (4.38)$$

Itô's formula in integral form is

$$g(t, X_t) = g(0, X_0) + \int_0^t \left[\frac{\partial g}{\partial t}(s, X_s) + b(X_s)g_x(s, X_s) + \frac{1}{2}\sigma^2(X_s)\frac{\partial^2 g}{\partial x^2}(s, X_s)\right]ds$$
$$+ \int_0^t \frac{\partial g}{\partial x}(s, X_s)\sigma(X_s)dB_s. \quad (4.39)$$

Under quite general conditions on the coefficients in the SDE (4.34) as listed in Section 4.6, the solution process X_t is a Markov process [213, Section 7.1]. With an "observable" (i.e., a measurable function $f: \mathbb{R}^1 \to \mathbb{R}^1$), we could observe or measure the process X_t to obtain $f(X_t)$. Then we take the mean of our observations to get a time-dependent deterministic function $\mathbb{E}f(X_t)$ and it is called the semigroup for the process. The time derivative (at $t = 0$) of this semigroup is a linear operator A,

$$Af(x) \triangleq \left.\frac{d}{dt}\right|_{t=0} \mathbb{E}f(X_t) = \lim_{t \downarrow 0} \frac{\mathbb{E}f(X_t) - f(x)}{t}, \; x \in \mathbb{R}^1, \quad (4.40)$$

whenever the limit exists. The domain for A is the set of fs such that this limit exists. This linear operator A is called the (infinitesimal) generator for the SDE (4.34), or for its solution process X_t. It is the time derivative of the "mean observation of the solution process," and its representation is known as ([213, Section 7.3], [18, Chapter 3], [35], or [272, Chapter 9])

$$Af \triangleq bf_x + \frac{1}{2}\sigma^2 f_{xx}, \quad (4.41)$$

for $f \in H_0^2(\mathbb{R}^1)$. The Sobolev space $H_0^2(\mathbb{R}^1)$ is defined in Section 2.6.

For example, the scalar SDE, $dX_t = 0dt + dB_t$, with initial condition $X_0 = x$, has solution $X_t = x + B_t$ (a "Brownian motion starting at x"). Thus, by (4.41), the generator for this Brownian motion is $\frac{1}{2}\frac{d^2}{dx^2}$ (Laplacian operator). This fact can also be proved directly by the definition (4.40); see Problem 4.6.

With the generator A, Itô's formula (4.36) can be rewritten as

$$dg(t, X_t) = \left[\frac{\partial g}{\partial t}(t, X_t) + Ag(t, X_t)\right] dt + \frac{\partial g}{\partial x}(t, X_t)\sigma(X_t)dB_t. \quad (4.42)$$

Being the derivative of observation on the solution process and also a significant part of the Itô formula, the generator A carries dynamical information for the SDE system (4.34). In fact, it helps quantify certain aspects, such as mean exit time and escape probability in Chapter 5, of the dynamical evolution of stochastic systems.

We consider some examples.

Example 4.12 Let $g(x)$ be a scalar, deterministic smooth function and B_t be a scalar Brownian motion. B_t may be regarded as a solution for the SDE $dX_t = 0dt + dB_t$, with initial condition $X_0 = 0$. Thus, by Itô's formula,

$$dg(B_t) = \frac{1}{2}g''(B_t)dt + g'(B_t)dB_t. \quad (4.43)$$

In particular,

$$dB_t^3 = 3B_t dt + 3B_t^2 dB_t,$$

and thus,

$$\int_0^T B_t^2 dB_t = \frac{1}{3}B_T^3 - \int_0^T B_t dt.$$

Example 4.13 Let X_t be a solution of the scalar SDE $dX_t = X^2 dt + \sin(X_t)dB_t$. Let $g = t + x^2$. By Itô's formula,

$$dg(t, X_t) = \left[1 + 2X_t^3 + \sin^2(X_t)\right]dt + 2X_t \sin(X_t)dB_t. \quad (4.44)$$

4.4.3 Itô's Formula in Vector Case

Consider an SDE system in $\mathbb{R}^n$,

$$dX_t = b(X_t)dt + \sigma(X_t)dB_t, \quad (4.45)$$

where b is an n-dimensional vector function, σ is an $n \times m$ matrix function, and $B_t(\omega)$ is an m-dimensional Brownian motion.

Let $g(t, x)$ be a given (deterministic) scalar smooth function for $x \in \mathbb{R}^n$ and $t \in \mathbb{R}$. Then

$$\begin{aligned} dg(t, X_t) &= \frac{\partial g}{\partial t}dt + \nabla g \cdot dX_t + \frac{1}{2}(dX_t)^{\mathrm{T}} H(g) dX_t \\ &= \frac{\partial g}{\partial t}dt + \sum_{i=1}^{n} \frac{\partial g}{\partial x_i} dX_t^i + \frac{1}{2}\sum_{i,j=1}^{n} \frac{\partial^2 g}{\partial x_i \partial x_j} dX_t^i dX_t^j, \end{aligned} \tag{4.46}$$

where $\frac{\partial g}{\partial t}$, $\frac{\partial g}{\partial x_i}$, and $\frac{\partial^2 g}{\partial x_i \partial x_j}$ are evaluated at (t, X_t).

Symbolically, we may also use the following rules in manipulating the Itô differential:

$$dtdt = 0, \quad dtdB_t = 0, \quad dB_t \cdot dB_t = \mathrm{Tr}(Q)dt = ndt. \tag{4.47}$$

Note that the covariance matrix $Q = I$ for n-dimensional Brownian motion B_t, as mentioned in Remark 3.10.

With these symbolic operations, Itô's formula in differential form becomes

$$\begin{aligned} dg(t, X_t) = &\left\{\frac{\partial g}{\partial t}(t, X_t) + b^{\mathrm{T}} \nabla g(t, X_t) + \frac{1}{2}\mathrm{Tr}[\sigma\sigma^{\mathrm{T}} H(g)](t, X_t)\right\} dt \\ &+ (\nabla g(t, X_t))^{\mathrm{T}} \sigma(X_t)\, dB_t, \end{aligned} \tag{4.48}$$

where superscript T denotes transpose matrix, $H(g) = (g_{x_i x_j})$ is the $n \times n$ symmetric Hessian matrix which is also denoted as $D^2(g)$, and Tr denotes the trace of a matrix.

Remark 4.14 Note that $H(g)(x, y) = x^{\mathrm{T}} H(g) y$, $H(g)$ is bilinear, and $B_t = \sum_{i=1}^n B_i(t) e_i$.

We have the following interpretation:

$$\begin{aligned} H(g)(\sigma dB_t, \sigma dB_t) &= H(g)\left(\sigma \sum_{i=1}^{n} dB_i(t) e_i, \sigma \sum_{j=1}^{n} dB_j(t) e_j\right) \\ &= \sum_{i,j=1}^{n} dB_i(t) dB_j(t) H(g)(\sigma e_i, \sigma e_j) \\ &= \left[\sum_{i=1}^{n} e_i^{\mathrm{T}} \sigma^{\mathrm{T}} H(g) \sigma e_i\right] dt \\ &= \mathrm{Tr}[\sigma^{\mathrm{T}} H(g) \sigma] dt \\ &= \mathrm{Tr}[H(g) \sigma\sigma^{\mathrm{T}}] dt \\ &= \mathrm{Tr}[\sigma\sigma^{\mathrm{T}} H(g)] dt. \end{aligned}$$

The generator A for this SDE system (4.45), or for its solution process X_t, is

$$Ag = b \cdot \nabla g + \frac{1}{2}\mathrm{Tr}[\sigma\sigma^{\mathrm{T}} H(g)], \tag{4.49}$$

for $g \in H_0^2(\mathbb{R}^n)$. The Sobolev space $H_0^2(\mathbb{R}^n)$ is defined in Section 2.6.

For example, Brownian motion B_t in $\mathbb{R}^n$ is the solution for $dX_t = 0dt + dB_t$, $X_0 = x$. Thus, by (4.49), the generator for Brownian motion starting at x, $X_t = x + B_t$, in $\mathbb{R}^n$ is $\frac{1}{2}\Delta$ (Laplacian operator).

With the generator A, Itô's formula (4.48) can be rewritten as

$$dg(t, X_t) = \left\{\frac{\partial g}{\partial t}(t, X_t) + Ag(t, X_t)\right\} dt + (\nabla g(t, X_t))^{\mathrm{T}}\sigma(X_t)dB_t. \tag{4.50}$$

Itô's formula in integral form is

$$g(t, X_t) = g(0, X_0) + \int_0^t \left\{\frac{\partial g}{\partial t}(s, X_s) + b^{\mathrm{T}}\nabla g(s, X_s) + \frac{1}{2}\mathrm{Tr}[\sigma\sigma^{\mathrm{T}} H(g)](s, X_s)\right\}ds$$

$$+ \int_0^t (\nabla g(s, X_s))^{\mathrm{T}}\sigma(X_s)dB_s. \tag{4.51}$$

Example 4.15 Let X_t, Y_t satisfy the following SDE system in $\mathbb{R}^2$:

$$dX_t = Y_t dt + aX_t dB_t^1, \tag{4.52}$$

$$dY_t = \left(-X_t + X_t^3\right) dt + bY_t dB_t^2. \tag{4.53}$$

Let $g = \frac{1}{2}(x^2 + y^2)$. By Itô's formula,

$$dg(X_t, Y_t) = \left[X_t^3 Y_t + \frac{1}{2}\left(a^2 X_t^2 + b^2 Y_t^2\right)\right]dt + aX_t^2 dB_t^1 + bY_t^2 dB_t^2. \tag{4.54}$$

4.4.4 Stochastic Product Rule and Integration by Parts

Let X_t, Y_t be solutions of two scalar SDEs, respectively. Then, by applying the two-dimensional Itô's formula to $g(x, y) = xy$, we get the stochastic product rule

$$\begin{aligned} d(X_t Y_t) &= \frac{\partial g}{\partial x}dX_t + \frac{\partial g}{\partial y}dY_t + \frac{1}{2}(dX_t dY_t + dY_t dX_t) \\ &= X_t dY_t + Y_t dX_t + dX_t dY_t. \end{aligned} \tag{4.55}$$

The corresponding integral form is the stochastic integration by parts

$$\int_0^{\mathrm{T}} X_t dY_t = X_T Y_T - X_0 Y_0 - \int_0^{\mathrm{T}} Y_t dX_t - \int_0^{\mathrm{T}} dX_t dY_t. \tag{4.56}$$

Let us consider a few examples.

Example 4.16 Taking $X_t = \sin(t)$ and $Y_t = B_t$ (a scalar Brownian motion), and using the stochastic product rule, we have

$$\begin{aligned} d(\sin(t)B_t) &= d(\sin(t))B_t + \sin(t)dB_t + d(\sin(t))dB_t \\ &= \cos(t)B_t dt + \sin(t)dB_t + \cos(t)dt dB_t \\ &= \cos(t)B_t dt + \sin(t)dB_t. \end{aligned} \tag{4.57}$$

Example 4.17 Taking $X_t = t^2$ and $Y_t = e^{B_t}$, with B_t a scalar Brownian motion, and using the stochastic product rule, we have

$$\begin{aligned} d(t^2 e^{B_t}) &= d(t^2)e^{B_t} + t^2 de^{B_t} + d(t^2)de^{B_t} \\ &= 2te^{B_t}dt + t^2\left[\frac{1}{2}e^{B_t}dt + e^{B_t}dB_t\right] + 0 \\ &= \left(2t + \frac{1}{2}t^2\right)e^{B_t}dt + t^2 e^{B_t}dB_t. \end{aligned} \tag{4.58}$$

Example 4.18 Taking $X_t = f(t, \omega)$ to be continuous and of finite variation in t (and thus it is differentiable in t almost everywhere) and $Y_t = B_t$, we obtain the following integration by parts formula from the stochastic product rule:

$$\int_0^T f(t, \omega)dB_t = f(T, \omega)B_T(\omega) - \int_0^T B_t df(t, \omega). \tag{4.59}$$

4.5 Linear Stochastic Differential Equations

We consider a few examples of linear stochastic differential equations.

Example 4.19 (Langevin equation) Consider the following scalar SDE with additive noise:

$$dX_t = -bX_t dt + a dB_t, \quad X_0 = x_0, \tag{4.60}$$

where a, b are positive deterministic parameters.

Recall the solution of a deterministic nonhomogeneous linear differential equation $\dot{x} = kx + f(t)$ with initial condition $x(0) = x_0$:

$$x(t) = x_0 e^{kt} + \int_0^t e^{k(t-s)} f(s)ds.$$

Thus, the solution for (4.60) is

$$X_t = e^{-bt}x_0 + ae^{-bt}\int_0^t e^{bs}dB_s, \tag{4.61}$$

and it is called an Ornstein-Uhlenbeck process. Indeed, this can be verified as follows:

$$dX_t = d\left[e^{-bt}x_0 + ae^{-bt}\int_0^t e^{bs}dB_s\right]$$
$$= de^{-bt}x_0 + ad\left[e^{-bt}\int_0^t e^{bs}dB_s\right]$$
$$= -be^{-bt}x_0dt + a\left[-be^{-bt}dt\int_0^t e^{bs}dB_s + e^{-bt}e^{bt}dB_t\right]$$
$$= -bX_tdt + adB_t.$$

When the initial value X_0 is a deterministic value in $\mathbb{R}^1$, the solution process X_t is a Gaussian process, with the mean

$$\mathbb{E}X_t = e^{-bt}x_0$$

and variance

$$\mathrm{Var}(X_t) = \mathbb{E}[X_t - \mathbb{E}X_t]^2$$
$$= \mathbb{E}\left[ae^{-bt}\int_0^t e^{bs}dB_s\right]^2$$
$$= a^2e^{-2bt}\mathbb{E}\int_0^t e^{2bs}ds$$
$$= \frac{a^2}{2b}[1 - e^{-2bt}],$$

where we have used the Itô isometry.

When the initial value is a normal distribution, $X_0 \sim \mathbb{N}(0, \sigma^2)$, the solution is still a Gaussian process, with the mean $\mathbb{E}X_t = 0$. The covariance and variance of the solution are

$$\mathrm{Cov}(X_s, X_t) = \sigma^2e^{-b(s+t)} + \frac{a^2}{2b}[e^{-b|s-t|} - e^{-b(s+t)}], \tag{4.62}$$

$$\mathrm{Cov}(X_0, X_t) = \sigma^2e^{-bt} + \frac{a^2}{2b}[e^{-b|t|} - e^{-bt}], \tag{4.63}$$

$$\mathrm{Var}(X_t) = \sigma^2e^{-2bt} + \frac{a^2}{2b}[1 - e^{-2bt}]. \tag{4.64}$$

The correlation of the solution at two different time instants is

$$\mathrm{Corr}(X_s, X_t) = \frac{\mathrm{Cov}(X_s, X_t)}{\sqrt{\mathrm{Var}(X_s)}\sqrt{\mathrm{Var}(X_t)}}. \tag{4.65}$$

When $\sigma^2 = \frac{a^2}{2b}$, we have

$$\mathrm{Cov}(X_s, X_t) = \sigma^2 e^{-b|s-t|}, \tag{4.66}$$

$$\mathrm{Var}(X_t) = \sigma^2. \tag{4.67}$$

Namely, in this case, X_t is a stationary process.

Example 4.20 (Geometric Brownian motion) This is a linear scalar SDE with multiplicative noise

$$dX_t = rX_t dt + \alpha X_t dB_t, \tag{4.68}$$

where r and α are real constants, and $X_t > 0$, a.s. As in [213, Chapter 5], we rewrite the SDE as

$$\frac{dX_t}{X_t} = r dt + \alpha dB_t. \tag{4.69}$$

If this were a deterministic differential equation, the left-hand side would be $d \ln X_t$. This inspires us to consider the stochastic differential of $\ln X_t$. Applying Itô's formula to $\ln X_t$, we obtain

$$d(\ln X_t) = \frac{1}{X_t} dX_t + \frac{1}{2}\left(-\frac{1}{X_t^2}\right)(dX_t)^2 = \frac{dX_t}{X_t} - \frac{1}{2}\alpha^2 dt. \tag{4.70}$$

That is, $\frac{dX_t}{X_t} = d(\ln X_t) + \frac{1}{2}\alpha^2 dt$. Thus (4.69) becomes

$$d(\ln X_t) = \left(r - \frac{1}{2}\alpha^2\right) dt + \alpha dB_t. \tag{4.71}$$

Integrating from 0 to t,

$$\ln \frac{X_t}{X_0} = \left(r - \frac{1}{2}\alpha^2\right) t + \alpha B_t.$$

We hence get the final solution

$$X_t = X_0 \exp\left(\left(r - \frac{1}{2}\alpha^2\right) t + \alpha B_t\right). \tag{4.72}$$

This is called a geometric Brownian motion starting at X_0.

Example 4.21 (Stochastic exponential) Consider a linear scalar SDE with no drift term

$$dX_t = \alpha X_t dB_t. \tag{4.73}$$

As known from the previous example, the solution is

$$X_t = X_0 \exp\left(\left(-\frac{1}{2}\alpha^2\right)t + \alpha B_t\right). \tag{4.74}$$

More generally, for the linear SDE with no drift term

$$dX_t = \alpha(t,\omega)X_t dB_t, \tag{4.75}$$

the solution is

$$X_t = X_0 \exp\left(-\frac{1}{2}\int_0^t \alpha^2(s,\omega)ds + \int_0^t \alpha(s,\omega)dB_s\right), \tag{4.76}$$

where the random process $\alpha(t,\omega)$ is such that the integrals in (4.76) exist. This solution process is called a stochastic exponential process. This can be verified by Itô's formula to obtain dX_t and compare it with $\alpha(t,\omega)X_t dB_t$.

Example 4.22 (A general linear scalar SDE) Now we consider a more general linear scalar SDE

$$dX_t = [a_1(t)X_t + a_2(t)]dt + [b_1(t)X_t + b_2(t)]dB_t, \quad X_{t_0} \text{ given}, \tag{4.77}$$

where a_1, a_2, b_1, and b_2 are known deterministic functions. When the initial condition is Gaussian or deterministic, and $b_1 = 0$, the solution is still Gaussian [8, Chapter 8].

The fundamental solution is

$$\Phi_{t,t_0} = \exp\left[\int_{t_0}^t \left(a_1(s) - \frac{1}{2}b_1^2(s)\right)ds + \int_{t_0}^t b_1(s)dB_s\right]. \tag{4.78}$$

Thus the general solution is [152, Chapter 3]

$$X_t = \Phi_{t,t_0}\left\{X_{t_0} + \int_{t_0}^t [a_2(s) - b_1(s)b_2(s)]\,\Phi_{s,t_0}^{-1} ds + \int_{t_0}^t b_2(s)\,\Phi_{s,t_0}^{-1} dB_s\right\}. \tag{4.79}$$

Example 4.23 (A special linear system of SDEs) Consider a linear SDE system in $\mathbb{R}^n$:

$$dX_t = [AX_t + f(t)]dt + \sum_{k=1}^m g_k(t)dB_t^k, \quad X_{t_0} \text{ given}, \tag{4.80}$$

where A is a constant $n \times n$ matrix; $X(t)$, $f(t)$, and $g_k(t)$ are n-dimensional vector functions; and B_t^k are independent scalar Brownian motions. This is a system with constant coefficient matrix and additive noise. In this case, we can find out the solution completely with the help of the matrix exponential [213, Chapter 5].

The fundamental solution matrix for the corresponding linear system $dX_t = AX_t dt$ is

$$\Phi_{t,t_0} = e^{A(t-t_0)}. \tag{4.81}$$

The solution for the nonhomogeneous linear system with constant coefficient matrix (4.80) is

$$X_t = e^{A(t-t_0)} X_{t_0} + \int_{t_0}^{t} e^{A(t-s)} f(s) ds + \sum_{k=1}^{m} \int_{t_0}^{t} e^{A(t-s)} g_k(s) dB_s^k). \tag{4.82}$$

In particular, for the special linear SDE system in $\mathbb{R}^n$

$$dX_t = [AX_t + f(t)]dt + K dB_t, \quad X_{t_0} \text{ given}, \tag{4.83}$$

where K is a constant $n \times n$ matrix and B_t is a Brownian motion in $\mathbb{R}^n$, the solution is

$$X_t = e^{A(t-t_0)} X_{t_0} + \int_{t_0}^{t} e^{A(t-s)} f(s) ds + \int_{t_0}^{t} e^{A(t-s)} K dB_s. \tag{4.84}$$

Example 4.24 (Random oscillators) A random oscillator is a spring-mass system modeled by Newton's second law ([213, p. 77] or [192, p. 276]) as a second-order stochastic differential equation:

$$\ddot{x} + a\dot{x} + bx = \sigma \dot{B}_t, \tag{4.85}$$

where a, b are nonnegative constants, σ is a real constant and B_t is a scalar Brownian motion. The term $a\dot{x}$ describes a damping force. This second-order SDE may be rewritten as a first-order SDE system:

$$\dot{x} = y, \tag{4.86}$$

$$\dot{y} = -bx - ay + \sigma \dot{B}_t. \tag{4.87}$$

In matrix form this becomes

$$\dot{X}_t = AX_t + K\dot{B}_t, \tag{4.88}$$

where

$$A = \begin{pmatrix} 0 & 1 \\ -b & -a \end{pmatrix}$$

and

$$K = \begin{pmatrix} 0 \\ \sigma \end{pmatrix}.$$

The solution is

$$X_t = e^{At} X_0 + \int_0^t e^{A(t-s)} K \, dB_s. \tag{4.89}$$

Example 4.25 (Random harmonic oscillators) A special case of random oscillators is the random harmonic oscillator when the damping is absent:

$$\ddot{x} + kx = h\dot{B}_t, \tag{4.90}$$

where k, h are positive constants. The corresponding first-order SDE system becomes

$$\dot{x} = y, \tag{4.91}$$

$$\dot{y} = -kx + h\dot{B}_t. \tag{4.92}$$

In matrix form this becomes

$$\dot{X}_t = AX_t + K\dot{B}_t, \tag{4.93}$$

where

$$A = \begin{pmatrix} 0 & 1 \\ -k & 0 \end{pmatrix}$$

and

$$K = \begin{pmatrix} 0 \\ h \end{pmatrix}.$$

Noticing that $A^2 = -kI$ with I, the 2×2 identity matrix, we have

$$e^{At} = \sum_{n=0}^{\infty} \frac{A^n}{n!} t^n = \begin{pmatrix} \cos(\sqrt{k}t) & \frac{1}{\sqrt{k}} \sin(\sqrt{k}t) \\ -\sqrt{k}\sin(\sqrt{k}t) & \cos(\sqrt{k}t) \end{pmatrix}. \tag{4.94}$$

The final solution for the stochastic harmonic oscillator is

$$x(t) = x_0 \cos(\sqrt{k}t) + \frac{y_0}{\sqrt{k}} \sin(\sqrt{k}t) + \frac{h}{\sqrt{k}} \int_0^t \sin(\sqrt{k}(t-s)) dB_s, \tag{4.95}$$

$$y(t) = -x_0\sqrt{k}\sin(\sqrt{k}t) + y_0 \cos(\sqrt{k}t) + h \int_0^t \cos(\sqrt{k}(t-s)) dB_s. \tag{4.96}$$

4.6 Nonlinear Stochastic Differential Equations

In this section, we recall some basic results about existence and uniqueness for solutions of nonlinear stochastic differential equations.

4.6.1 Existence, Uniqueness, and Smoothness

We consider the following n-dimensional stochastic system:

$$dX_t = b(X_t)dt + \sigma(X_t)dB_t, \ X_0 = x \in \mathbb{R}^n, \ t \in \mathbb{R}. \tag{4.97}$$

The solution (or strong solution) is defined as a process X, adapted to the filtration $\mathcal{F}_t^B$ generated by B_t, satisfying the corresponding stochastic integral equation almost surely (a.s.):

$$X_t = x + \int_0^t b(X_s)ds + \int_0^t \sigma(X_s)dB_s, \ t \in \mathbb{R}. \tag{4.98}$$

The uniqueness means pathwise uniqueness, that is, any two solutions X_1 and X_2 are indistinguishable:

$$X_1(t, \omega) = X_2(t, \omega), \quad \text{for all } t, \quad \text{a.s.} \tag{4.99}$$

We refer to [147, 103, 128, 232, 213, 151] for sufficient conditions that guarantee existence and uniqueness of solutions. This usually requires that b and σ satisfy a local Lipschitz condition:

$$\|b(x) - b(y)\| + \|\sigma(x) - \sigma(y)\| \leq K_N \|x - y\|, \tag{4.100}$$

for $\|x\| \leq N$, $\|y\| \leq N$, and $N > 0$. Here the Lipschitz constant K_N depends on the positive number N. The local Lipschitz condition leads to existence and uniqueness of a local, continuous solution (i.e., defined on a time interval containing time 0) in $L^2(\Omega)$. An additional a priori estimate on the finiteness of the solution in $L^2(\Omega)$ permits the local solution to be extended to all time, and thus we have existence and uniqueness of the global solution for each initial condition. Note that a smooth function (with at least first-order continuous derivative) satisfies the local Lipschitz condition owing to the mean value theorem.

Moreover, a boundedness condition [267, Section 5.1], on b and σ, also allows the unique solution to exist for all time.

We now present a specific set of sufficient conditions to ensure global existence and uniqueness of solutions. For a natural number k and a positive real number $\delta \in [0, 1]$, as in [9, Appendix B], we define the function space $C^{k,\delta}$ to be the set of functions $f\colon \mathbb{R}^d \to \mathbb{R}^d$, which are k-times continuously differentiable and (for $\delta > 0$) whose kth derivative is locally δ-Hölder continuous (for $\delta = 1$, locally Lipschitz continuous).

Let $C_b^{k,\delta}$ be the Banach space of functions $f: \mathbb{R}^d \to \mathbb{R}^d$, which are in $C^{k,\delta}$ and for which the norm

$$\| f \|_{k,0} \triangleq \sup_{x\in\mathbb{R}^d} \frac{\|f(x)\|}{1+\|x\|} + \sum_{1\le|\alpha|\le k} \sup_{x\in\mathbb{R}^d} |D^\alpha f(x)|,$$

$$\| f \|_{k,\delta} \triangleq \| f \|_{k,0} + \sum_{|\alpha|=k} \sup_{x\neq y} \frac{\|D^\alpha f(x) - D^\alpha f(y)\|}{\|x-y\|^\delta}, \quad 0<\delta\le 1,$$

is finite. We state the following existence and uniqueness theorem [9, Theorem 2.3.39] or [163]).

Theorem 4.26 *Assume that* $b \in C_b^{k,\delta}$, $\sigma \in C_b^{k+1,\delta}$, *and* $\sum_{i=1}^n b_i \frac{\partial b_i}{\partial x_i} \in C_b^{k,\delta}$ *for some natural number* $k \ge 1$ *and* $\delta > 0$. *Then* $dX_t = b(X_t)dt + \sigma(X_t)dB_t$, $X_0 = x \in \mathbb{R}^n$, *has a pathwise unique global solution.*

Other sufficient conditions for global existence and uniqueness have also been obtained recently [90, 307].

4.6.2 Probability Measure $\mathbb{P}^x$ and Expectation $\mathbb{E}^x$ Associated with an SDE

We introduce some notations for convenience as in [18, p. 12]. Define Ω' to be the set of all continuous functions from $[0,\infty)$ to $\mathbb{R}^n$. We define a canonical process Z_t by identifying $Z_t(\omega) \triangleq \omega(t)$, for $\omega \in \Omega'$. For the solution $X_t(x,\omega)$ to an SDE starting at each x, define a probability measure $\mathbb{P}^x$ by

$$\mathbb{P}^x(Z_{t_1} \in A_1, \ldots, Z_{t_1} \in A_k) \triangleq \mathbb{P}(X_t(x,\omega) \in A_1, \ldots, X_t(x,\omega) \in A_k),$$

whenever $t_1, \ldots, t_k \in [0,\infty)$ and $A_1, \ldots, A_k$ are Borel sets in $\mathbb{R}^n$. The expectation with respect to this probability measure $\mathbb{P}^x$ is denoted by $\mathbb{E}^x$.

4.7 Conversion between Itô and Stratonovich Stochastic Differential Equations

We have two main stochastic integrals, Itô and Stratonovich integrals. Thus we have Itô SDEs and Stratonovich SDEs. These two types of SDEs can be converted from one to the other.

4.7.1 Scalar SDEs

First consider a scalar Stratonovich SDE

$$dX_t = b(t, X_t)dt + \sigma(t, X_t) \circ dB_t, \tag{4.101}$$

where b is the drift term and $\sigma(t, X_t)$ is the diffusion term. Using the Taylor expansion theorem and the mean value theorem in the sum for the definition of stochastic integrals, it is shown that [152, Chapter 4]

$$\int_0^T \sigma(t, X_t) \circ dB_t = \int_0^T \sigma(t, X_t) dB_t + \frac{1}{2} \int_0^T \sigma(t, X_t) \frac{\partial \sigma}{\partial x}(t, X_t) dt, \tag{4.102}$$

or, in differential form,

$$\sigma(t, X_t) \circ dB_t = \sigma(t, X_t) dB_t + \frac{1}{2} \sigma(t, X_t) \frac{\partial \sigma}{\partial x}(t, X_t) dt. \tag{4.103}$$

This also says that the Stratonovich integral may not have zero mean (unlike the Itô integral):

$$\mathbb{E} \int_0^T \sigma(t, X_t) \circ dB_t = \frac{1}{2} \mathbb{E} \int_0^T \sigma(t, X_t) \frac{\partial \sigma}{\partial x}(t, X_t) dt. \tag{4.104}$$

Thus we have the following conclusion.

Theorem 4.27 *The Stratonovich SDE*

$$dX_t = b(t, X_t) dt + \sigma(t, X_t) \circ dB_t \tag{4.105}$$

is converted to the following Itô SDE:

$$dX_t = \left[b(t, X_t) + \frac{1}{2} \sigma(t, X_t) \frac{\partial \sigma}{\partial x}(t, X_t) \right] dt + \sigma(t, X_t) dB_t, \tag{4.106}$$

with a new drift term $b(t, X_t) + \frac{1}{2}\sigma(t, X_t)\frac{\partial \sigma}{\partial x}(t, X_t)$.
Conversely, an Itô SDE

$$dX_t = b(t, X_t) dt + \sigma(t, X_t) dB_t \tag{4.107}$$

is equivalent to the following Stratonovich SDE:

$$dX_t = \left[b(t, X_t) - \frac{1}{2} \sigma(t, X_t) \frac{\partial \sigma}{\partial x}(t, X_t) \right] dt + \sigma(t, X_t) \circ dB_t, \tag{4.108}$$

with a modified drift term $b(t, X_t) - \frac{1}{2}\sigma(t, X_t)\frac{\partial \sigma}{\partial x}(t, X_t)$.

Example 4.28 The Stratonovich SDE

$$dX_t = (X_t - X_t^2) dt + X_t^3 \circ dB_t \tag{4.109}$$

is equivalent to the following Itô SDE:

$$dX_t = \left(X_t - X_t^2 + \frac{3}{2} X_t^5 \right) dt + X_t^3 dB_t. \tag{4.110}$$

4.7.2 SDE Systems

Similarly, it is also possible to convert SDE systems from Stratonovich to Itô forms, and vice versa. Consider a Stratonovich SDE system

$$dX_t = b(t, X_t)dt + \sigma(t, X_t) \circ dB_t, \tag{4.111}$$

where b is the drift term in $\mathbb{R}^n$, $\sigma(t, X_t)$ is an $n \times m$ matrix, X_t is in $\mathbb{R}^n$, and B_t is in $\mathbb{R}^m$. Again, it is known that [152, Chapter 4]

$$\int_0^T \sigma(t, X_t) \circ dB_t = \int_0^T \sigma(t, X_t)dB_t + \int_0^T c(t, X_t)dt, \tag{4.112}$$

where the vector c has the components

$$c_i = \frac{1}{2}\sum_{j=1}^{n}\sum_{k=1}^{m} \sigma_{j,k}(t, X_t)\frac{\partial \sigma_{i,k}}{\partial x_j}(t, X_t), \tag{4.113}$$

for $i = 1, \ldots, n$. Or, in differential form,

$$\sigma(t, X_t) \circ dB_t = \sigma(t, X_t)dB_t + c(t, X_t)dt. \tag{4.114}$$

Thus the Stratonovich SDE (4.111) is converted to the following Itô SDE:

$$dX_t = [b(t, X_t) + c(t, X_t)]dt + \sigma(t, X_t)dB_t, \tag{4.115}$$

with a new drift term $b(t, X_t) + c(t, X_t)$.

Conversely, an Itô SDE system

$$dX_t = b(t, X_t)dt + \sigma(t, X_t)dB_t \tag{4.116}$$

is equivalent to the following Stratonovich SDE:

$$dX_t = [b(t, X_t) - c(t, X_t)]dt + \sigma(t, X_t) \circ dB_t, \tag{4.117}$$

with a modified drift term $b(t, X_t) - c(t, X_t)$.

Let us look at an example.

Example 4.29 Consider a system of stochastic differential equations

$$dX_t = -X_t dt + \left(X_t Y_t^2 - X_t^3 - \frac{1}{2}X_t\right) dt + X_t \circ dB_t^1,$$

$$dY_t = 0Y_t dt + \left(-2 + X_t^2 Y_t - Y_t^3 - \frac{1}{2}Y_t\right) dt + Y_t \circ dB_t^2,$$

where B_t^1 and B_t^2 are independent scalar Brownian motions.

By the preceding discussions, the equivalent Itô's stochastic differential equations are

$$dX_t = -X_t dt + \left(X_t Y_t^2 - X_t^3\right)dt + X_t dB_t^1,$$
$$dY_t = 0Y_t dt + \left(-2 + X_t^2 Y_t - Y_t^3\right)dt + Y_t dB_t^2.$$

4.8 Impact of Noise on Dynamics

In some systems, noise is not negligible [28, 279], whereas in some other situations, noise could even be beneficial [119]. Thus, it is desirable to better understand the impact of noise on the dynamical evolution of complex systems.

Let us look at a few examples to acquire some impressions of the impact of noise on dynamics.

Example 4.30 (Impact of noise on mean and variance) Consider the Langevin equation whose solution is also called an Ornstein-Uhlenbeck process:

$$dX_t = -bX_t dt + a dB_t, \tag{4.118}$$

where a, b are real parameters and the initial value X_0 is deterministic. The solution is

$$X_t = e^{-bt} X_0 + a e^{-bt} \int_0^t e^{bs} dB_s. \tag{4.119}$$

Because the Itô integral has zero mean, we see that $\mathbb{E}X_t = e^{-bt} X_0$, which is the solution when noise is absent ($a = 0$). So the noise does not affect the solution at the level of mean (first moment). But it indeed has an impact on the solution at the level of variance, as shown subsequently.

We calculate the covariance and variance as follows:

$$\mathrm{Cov}(X_s, X_t) = \frac{a^2}{2b}\left[e^{-b|s-t|} - e^{-b(s+t)}\right], \tag{4.120}$$

$$\mathrm{Cov}(X_0, X_t) = \frac{a^2}{2b}\left[e^{-b|t|} - e^{-bt}\right], \tag{4.121}$$

$$\mathrm{Var}(X_t) = \frac{a^2}{2b}\left[1 - e^{-2bt}\right]. \tag{4.122}$$

Hence the noise ($a \neq 0$) affects the variance (and thus the second moment) of the solution. At the level of sample paths for the solution, we immediately observe the impact of noise as in Figure 4.1.

Interested readers may examine how the noise affects higher moments.

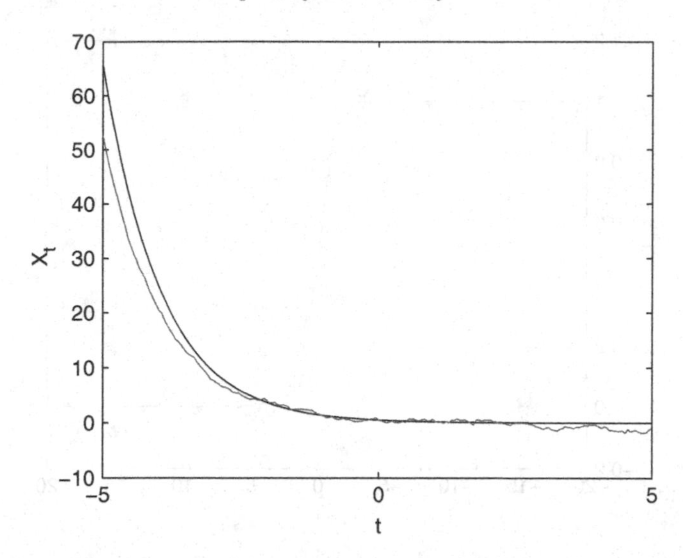

Figure 4.1 Example 4.30 with $b = 1$, $a = 0.985$, and $X_0 = 0.5$: a solution path with "wigglings," on top of the deterministic solution (i.e., when $a = 0$).

Example 4.31 (Impact of noise on growth rate) Consider the following linear scalar SDE with multiplicative noise:

$$dX_t = rX_t dt + \alpha X_t dB_t, \tag{4.123}$$

where r and α are real constants and $X_t > 0$, a.s. The solution of this SDE is called a geometric Brownian motion.

By Example 4.20, the solution for SDE is

$$X_t = X_0 \exp\left(\left(r - \frac{1}{2}\alpha^2\right)t + \alpha B_t\right). \tag{4.124}$$

When the noise is absent ($\alpha = 0$), the solution is $X_0 \exp(rt)$ with the usual, deterministic exponential growth rate. With noise, the deterministic growth rate is reduced by $\frac{1}{2}\alpha^2$ (although the overall growth rate is compensated by the amount αB_t, which has mean zero). In fact, in Chapter 6, we see that the corresponding deterministic system has the Lyapunov exponent r, whereas the SDE has Lyapunov exponent $r - \frac{1}{2}\alpha^2$.

Example 4.32 (Impact of noise on solution paths) Consider

$$dX_t = (-X_t + X_t^3)dt + \varepsilon dB_t, \quad X_0 = 0.5. \tag{4.125}$$

A solution path showing the impact of noise on the solution is in Figure 4.2.

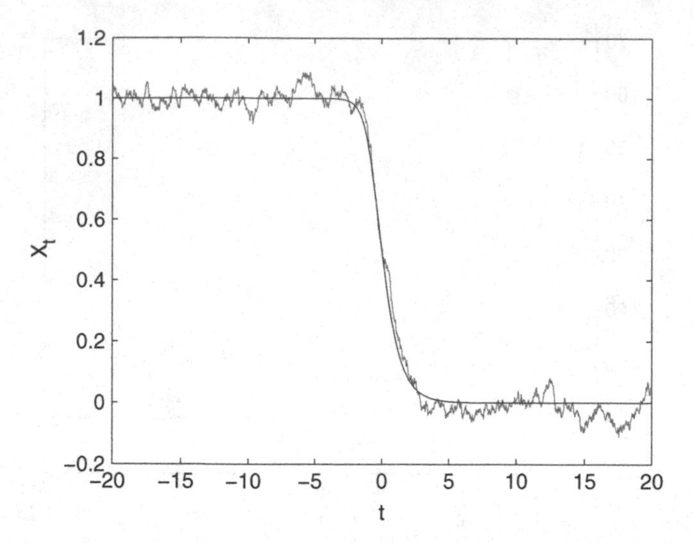

Figure 4.2 Example 4.32 and Example 4.34: a solution path with "wigglings" when $\varepsilon = 0.05$, on top of the deterministic solution (i.e., when $\varepsilon = 0$).

Example 4.33 (Noise may "enhance" pathwise uniqueness of solutions) For a deterministic ordinary differential equation $\dot{x} = f(x)$, a sufficient condition for uniqueness of solutions is a local Lipschitz continuity condition for vector field f. Without this condition, the uniqueness is often violated.

However, for $\dot{x} = f(x) + \dot{B}_t$ with f being Borel measurable and bounded (but not Lipschitz continuous), the uniqueness holds [282]. For more recent developments on this issue, see [91, 161, 19, 307].

4.9 Simulation

The Euler method and Milstein method are usually used to simulate SDEs [152, 275]. See [122] for Matlab codes for simulating SDEs with these two methods.

Let us discuss how to simulate a solution path via the Euler method. An Itô SDE with initial condition is

$$dX_t = b(X_t)dt + \sigma(X_t)dB_t, \, X_0 = x, \quad t \in \mathbb{R}, \tag{4.126}$$

both forward and backward in time, $t \in [-T, T]$.

Forward in Time

For a partition of $[0, T]$: $0 = t_0 < t_1 < \cdots < t_i < t_{i+1} < \cdots < t_n = T$ of equal subinterval length (or stepsize) Δt (for convenience), and denote $X_i \triangleq X_{t_i}$,

$$X_{i+1} = X_i + b(X_i)\Delta t + \sigma(X_i)(B_{t_{i+1}} - B_{t_i}), \; i = 0, 1, \ldots, n-1. \tag{4.127}$$

Backward in Time

For a partition of $[-T, 0]$: $-T = t_0 < t_1 < \cdots < t_i < t_{i+1} < \cdots < t_n = 0$ of equal subinterval length (or stepsize) Δt (for convenience), and denote $X_i \triangleq X_{t_i}$,

$$X_i = X_{i+1} - b(X_{i+1})\Delta t - \sigma(X_{i+1})(B_{t_{i+1}} - B_{t_i}),$$
$$i = n-1, n-2, \ldots, 1. \tag{4.128}$$

Example 4.34 Numerically compute the solution paths for the following scalar SDE:

$$dX_t = (-X_t + X_t^3)dt + \varepsilon dB_t, \quad X_0 = 0.5.$$

A solution path is shown in Figure 4.2.

4.10 Problems

4.1 Explain why the usual definition of the Riemann-Stieltjes integral is not appropriate for defining the stochastic integral $\int_0^T f(t,\omega)dB_t$.

4.2 What does the symbolic notation $(dB_t)^2 = dt$ really mean in stochastic calculus?
Hint: See Section 4.1 in [213].

4.3 If $\int_0^T f(t,\omega)dB_t = 0$, then is it true that $f = 0$? Discuss when this may be true.

4.4 Evaluate the following stochastic integral ("taking the right end point" on each subinterval):

$$\int_0^T B_t \heartsuit dB_t(\omega) \triangleq \lim_{\delta^n \to 0} \text{ in m.s.} \sum_{i=0}^{n-1} B(t_{i+1}^n)\big(B(t_{i+1}^n) - B(t_i^n)\big).$$

4.5 Does the following stochastic integral ("taking the left and right end points alternately" on each subinterval) make sense, that is, does the following limit exist?

$$\int_0^T B_t \Box dB_t(\omega) \triangleq \lim_{\delta^n \to 0} \text{ in m.s.}$$
$$\times \left[\sum_{\substack{\text{Odd} \\ \text{subintervals}}} B(t_i^n)\big(B(t_{i+1}^n) - B(t_i^n)\big) + \sum_{\substack{\text{Even} \\ \text{subintervals}}} B(t_{i+1}^n)\big(B(t_{i+1}^n) - B(t_i^n)\big) \right],$$

where the subinterval $[t_i^n, t_{i+1}^n]$ is called 'odd' ('even'), when i is odd (even).

4.6 Starting from the definition $Af(x) = \frac{d}{dt}|_{t=0}\mathbb{E}f(X_t)$, show that the generator for a scalar Brownian motion starting at x, $X_t = x + B_t$, is $A = \frac{1}{2}\frac{d^2}{dx^2}$.

4.7 Verify that $\int_0^T B_t dB_t = \frac{1}{2}B_T^2 - \frac{1}{2}T$ by applying Itô's formula to B_t^2.

4.8 For the Langevin equation $dX_t = -X_t dt + 2dB_t$, with the initial condition $X_0 \sim \mathcal{N}(0, 1)$ (and X_0 being independent of B_t for $t \geq 0$), what is the mean $\mathbb{E}X_t$, and what is the covariance $\mathrm{Cov}(X_t, X_s)$ (for $s < t$)? Is X_t a stationary process (i.e., the covariance $\mathrm{Cov}(X_t, X_s)$ depends only on $t - s$)?

4.9 For a deterministic mean square integrable function $f(t)$, show that $\mathbb{E}e^{\int_0^t f(s)dB_s} = e^{\frac{1}{2}\int_0^t f^2(s)ds}$. In particular, what is $\mathbb{E}e^{B_t}$? Is it possible to get this result by using the Taylor expansion $e^x = 1 + \frac{1}{1!}x + \frac{1}{2!}x^2 + \cdots$? Is it true that $\mathbb{E}e^{B_t} = e^{\mathbb{E}B_t} = e^0 = 1$?

4.10 Find the solution for the following scalar stochastic differential equation:

$$dX_t = (t + 3X_t)dt + (4X_t + \sin(t))dB_t, \quad X_0 = x.$$

4.11 Itô's formula

(i) Consider an SDE $dX_t = 3X_t dt + \sin(X_t)dB_t$. What is $d(X_t^2)$? What is $d\sin(X_t)$?

(ii) Consider a system of SDEs

$$dX_t = -8X_t dt + X_t Y_t dB_t^1 + 5dB_t^2, \; dY_t = X_t^2 Y_t dt - dB_t^1 + \left(X_t^2 + Y_t^2\right)dB_t^2.$$

What is $d(X_t^2 + Y_t^2)$? What is $d(X_t Y_t)$? Is it true or false that $d(X_t Y_t) = X_t dY_t + Y_t dX_t$?

4.12 What is the Itô form for the Stratonovich SDE $dX_t = (\sin(X_t) - X_t^4)dt + 5X_t^2 \circ dB_t$?

4.13 What is the Stratonovich form for the following Lorenz system of Itô SDEs:

$$dx = (-sx + sy)dt + \sqrt{\varepsilon}\, x dB_t^1,$$

$$dy = (rx - y - xz)dt + \sqrt{\varepsilon}\, y dB_t^2,$$

$$dz = (-bz + xy)dt + \sqrt{\varepsilon}\, z dB_t^3,$$

where b, r, s, ε are real parameters? Simulate a few solution paths by the Euler method.

4.14 Examine the impact of noise on the motion of a simple pendulum. First rewrite the noise-driven pendulum equation, $\ddot{X} = -\sin(X) + \varepsilon \dot{B}_t$, $t > 0$, into a system of SDEs

$$dX_t = Y_t dt,$$
$$dY_t = -\sin(X_t)dt + \varepsilon dB_t,$$

where ε is a small positive noise intensity. Then simulate a few solution paths by the Euler method and the Milstein method. What can you observe about the impact of noise in this system?

4.15 How does one numerically compute sample paths for a solution of the scalar SDE $dX_t = a(X_t)dt + b(X_t)dB_t$ with $X_0 = x$, $-T \le t \le T$? What is the Euler scheme in this case?

4.16 Matrix exponential
For the matrix A in Example 4.25,

$$A = \begin{pmatrix} 0 & 1 \\ -k & 0 \end{pmatrix},$$

verify that

$$e^{At} = \sum_{n=0}^{\infty} \frac{A^n}{n!} t^n = \begin{pmatrix} \cos(\sqrt{k}t) & \frac{1}{\sqrt{k}} \sin(\sqrt{k}t) \\ -\sqrt{k}\sin(\sqrt{k}t) & \cos(\sqrt{k}t) \end{pmatrix}. \tag{4.129}$$

4.17 Random harmonic oscillator
Consider a harmonic oscillator under random forcing. First rewrite the dynamical system, $\ddot{X} = -X + \dot{B}_t$, with B_t a scalar Brownian motion, into a system of first-order SDEs:

$$dX_t = Y_t dt,$$
$$dY_t = -X_t dt + dB_t,$$

with the initial condition $(X(0), Y(0)) = (x_0, y_0)^{\mathrm{T}} \in \mathbb{R}^2$. Find the solution $(X(t), Y(t))$. Are the position $X(t)$ and velocity $Y(t)$ asymptotically independent? You may answer this question by calculating the covariance $\mathrm{Cov}(X(t), Y(t))$ and examine it for $t \gg 1$. Simulate a few solution paths by the Euler method.

4.18 Simulate a few solution paths by the Euler method for the following random harmonic oscillator:

$$dX_t = Y_t dt,$$
$$dY_t = -X_t dt + dB_t,$$

with various initial conditions.

4.19 Random harmonic oscillator again

Consider a system of SDEs

$$dX_t = Y_t dt + a dB_t^1,$$
$$dY_t = -X_t dt + b dB_t^2,$$

with a, b real parameters. Find the solution $(X(t), Y(t))$ with the initial condition $(X(0), Y(0)) = (x_0, y_0)^{\mathrm{T}} \in \mathbb{R}^2$. Are the position $X(t)$ and velocity $Y(t)$ independent? You may answer this question by calculating the covariance $\mathrm{Cov}(X(t), Y(t))$. Simulate a few solution paths by the Milstein method.

4.20 Simulate a few solution paths by the Milstein method for the following random harmonic oscillator:

$$dX_t = Y_t dt + a dB_t^1,$$
$$dY_t = -X_t dt + b dB_t^2,$$

with various initial conditions and different noise intensities a and b.

5

Deterministic Quantities for Stochastic Dynamics

Eigenvalues, Poincaré index, Conley index, Hausdorff dimension of attractors, and other quantities have been used to study deterministic dynamical systems. Deterministic quantities, such as mean, variance, and characteristic functions, have also been used to quantify random variables and stochastic processes.

In this chapter, we consider deterministic quantities that carry dynamical information or that can be used to quantify dynamical behavior of SDEs. These deterministic quantities include moments for solution paths, probability density functions for solution paths, mean exit time, and escape probability. The mean exit time for a stochastic system quantifies how long the system stays in a region in the state space, and the escape probability describes the likelihood of a transition of the system from one regime to another. Fortunately, probability density functions, mean exit time, and escape probability all are solutions of deterministic partial differential equations. In fact, probability density functions satisfy the well-known Fokker-Planck equations, which are a class of parabolic partial differential equations, while mean exit time and escape probability satisfy elliptic partial differential equations. The main theme in this chapter is stochastic dynamics via deterministic partial differential equations.

Some basic knowledge in deterministic partial differential equations will be helpful. Elementary methods for solving linear partial differential equations are discussed in, for example, [111], [228], [265], [205]. More advanced topics, such as well-posedness and solution estimates, for partial differential equations may be found in popular textbooks such as [89], [236], [141], [198].

We consider an SDE system in $\mathbb{R}^n$,

$$dX_t = b(X_t)dt + \sigma(X_t)dB_t, \tag{5.1}$$

where b is an n-dimensional vector function, σ is an $n \times m$ matrix function, and $B_t(\omega)$ is an m-dimensional Brownian motion. Assume that b and σ satisfy an appropriate local Lipschitz condition as well as smoothness and growth conditions as specified in Section 4.6, so that the unique solution exists. The generator for this SDE system is a linear second-order differential operator

$$Ag = b \cdot (\nabla g) + \frac{1}{2}\mathrm{Tr}[\sigma\sigma^{\mathrm{T}} H(g)], \quad g \in H_0^2(\mathbb{R}^n), \tag{5.2}$$

where H denotes the Hessian matrix of a multivariate function and Tr denotes the trace of a matrix. We view A as a linear differential operator in Hilbert space $L^2(\mathbb{R})$ with the domain of definition $D(A) = H_0^2(\mathbb{R}^n)$. The adjoint operator of the generator A is

$$A^* h = -\nabla \cdot (bh) + \frac{1}{2}\mathrm{Tr}[H(\sigma\sigma^{\mathrm{T}} h)], \quad h \in H_0^2(\mathbb{R}^n), \tag{5.3}$$

where $H(\sigma\sigma^{\mathrm{T}} h)$ is interpreted as matrix multiplications of H, σ and σ^{T}, and then acts on a scalar function h. See Section 5.2.

We assume that the generator for the system (5.1), $Ag = b^{\mathrm{T}}\nabla g + \frac{1}{2}\mathrm{Tr}[\sigma\sigma^{\mathrm{T}} H(g)]$, is uniformly elliptic. That is, there exists a positive constant λ such that

$$\sum_{i,j=1}^{n} (\sigma(x)\sigma^{\mathrm{T}}(x))_{ij}\xi_i\xi_j \geq \lambda|\xi|^2, \tag{5.4}$$

for $x \in D$ and all $\xi \in \mathbb{R}^n$. A simple example of a uniformly elliptic operator is the Laplace operator $A = \frac{1}{2}\Delta$, where $b = 0$ and σ is the identity matrix.

We discuss moments, probability density, mean exit time, and escape probability as deterministic quantities that carry dynamical information for solution orbits of SDE systems.

5.1 Moments

For a random variable, we usually try to compute its mean, variance, or higher moments to have some quantitative understanding of its behavior. In this section, we discuss the impact of noise on system evolution by estimating moments of solutions using Itô's formula. Moments are classical deterministic quantities that offer some information for a solution process. Let us first look at the Langevin equation in Example 4.19.

Example 5.1 (Langevin equation) Consider the following scalar SDE with additive noise:

$$dX_t = -bX_t dt + a dB_t, \quad X_0 = x_0, \tag{5.5}$$

where a, b are positive deterministic parameters and x_0 is a deterministic initial value.

The solution for (5.5) is

$$X_t = e^{-bt}x_0 + ae^{-bt}\int_0^t e^{bs}dB_s, \tag{5.6}$$

and it is called an Ornstein-Uhlenbeck process. The mean for the solution is

$$\mathbb{E}X_t = e^{-bt}x_0$$

and the variance is

$$\begin{aligned}\mathrm{Var}(X_t) &= \mathbb{E}[X_t - \mathbb{E}X_t]^2 \\ &= \mathbb{E}\left[ae^{-bt}\int_0^t e^{bs}dB_s\right]^2 \\ &= a^2e^{-2bt}\mathbb{E}\int_0^t e^{2bs}ds \\ &= \frac{a^2}{2b}\left[1 - e^{-2bt}\right],\end{aligned}$$

where we have used Itô isometry. We observe that the mean of the solution is not affected by the noise term adB_t (i.e., the noise intensity, a, does not appear in the mean). In other words, at the level of mean, the Langevin system does not feel the impact of noise. We could also say that we are unable to detect the effect of noise using the deterministic quantity "mean" for this simple system. However, the noise is felt by the variance (i.e., the noise intensity, a, appears in the variance). The noise thus affects the evolution of the variance for the solution.

Consider an SDE system in $\mathbb{R}^n$,

$$dX_t = b(X_t)dt + \sigma(X_t)dB_t. \tag{5.7}$$

A typical application of Itô's formula for such an SDE is to apply the formula to the system "energy" $\frac{1}{2}\|X_t\|^2$,

$$\frac{1}{2}d\|X_t\|^2 = \left[b \cdot X_t + \frac{1}{2}\mathrm{Tr}(\sigma\sigma^{\mathrm{T}})\right]dt + X_t\sigma(X_t)dB_t. \tag{5.8}$$

Taking mean on both sides, we get

$$\frac{1}{2}\frac{d}{dt}\mathbb{E}\|X_t\|^2 = \mathbb{E}(b \cdot X_t) + \frac{1}{2}\mathbb{E}\ \mathrm{Tr}(\sigma\sigma^{\mathrm{T}}). \tag{5.9}$$

This describes how the noise term affects the mean energy evolution.

The following example demonstrates how to estimate the mean energy evolution for a well-known dynamical system under uncertainty.

Example 5.2 (Lorenz system under uncertainty) Consider the Lorenz system with multiplicative noise

$$
\begin{aligned}
dX_t &= (-sX_t + sY_t)dt + \sqrt{\varepsilon}\, X_t\, dB_t^1, \\
dY_t &= (rX_t - Y_t - X_t Z_t)dt + \sqrt{\varepsilon}\, Y_t\, dB_t^2, \\
dZ_t &= (-bZ_t + X_t Y_t)dt + \sqrt{\varepsilon}\, Z_t\, dB_t^3,
\end{aligned}
$$

where B_t^1, B_t^2, and B_t^3 are independent scalar Brownian motions, and r, s, b, ε are positive parameters. The famous "chaos" case corresponds to $r = 28,\ s = 10,\ b = 8/3$. Let $U_t = (X_t, Y_t, Z_t)^{\mathrm{T}}$. Then, by Itô's formula, we obtain the following estimate for the mean energy evolution:

$$
\begin{aligned}
\frac{1}{2}\frac{d}{dt}\mathbb{E}\|U_t\|^2 &= \mathbb{E}\left[-sX_t^2 - Y_t^2 - bZ_t^2 + (r+s)X_tY_t + \frac{1}{2}\varepsilon\left(X_t^2 + Y_t^2 + Z_t^2\right)\right] \\
&\leq \left[-\min\{s, 1, b\} + \frac{1}{2}(r + s + \varepsilon)\right] \mathbb{E}\|U_t\|^2,
\end{aligned}
$$

where we have used the fact that $xy \leq \frac{1}{2}(x^2 + y^2) \leq \frac{1}{2}(x^2 + y^2 + z^2)$. By the Gronwall inequality in Section 2.5, we conclude that

$$
\mathbb{E}\|U_t\|^2 \leq \mathbb{E}\|U_0\|^2\, e^{[-2\min\{s,1,b\}+r+s+\varepsilon]t}. \tag{5.10}
$$

In the famous "chaos" case with $r = 28,\ s = 10,\ b = 8/3$, the mean energy estimate is

$$
\mathbb{E}\|U_t\|^2 \leq \mathbb{E}\|U_0\|^2\, e^{(36+\varepsilon)t}. \tag{5.11}
$$

5.2 Probability Density Functions

The probability density function is another quantity helpful for understanding a random variable. In fact, once a probability density function is known, we can compute its mean, variance, and moments. Because a solution for an SDE depends on time, its probability density function also depends on time. In this section, we consider the evolution of the probability density function for the solution paths. This evolution is governed by a Fokker-Planck equation. We further discuss the likelihood for transitions from one dynamical regime to another.

We start with a motivating example.

Example 5.3 Recall that the solution for $dX_t = dB_t$ with a deterministic initial condition $X_0 = \xi$ is $X_t = \xi + B_t$. This solution has normal distribution $\mathcal{N}(\xi, t)$, with the probability density function

$$p(x,t) = \frac{1}{\sqrt{2\pi t}}\, e^{-\frac{(x-\xi)^2}{2t}}. \tag{5.12}$$

Readers familiar with the Fourier transform method for solving partial differential equations ([265, Section 12.4] or [198, Section 5.2]) may notice that this probability density function p satisfies the following linear partial differential equation:

$$p_t = \frac{1}{2} p_{xx}. \tag{5.13}$$

This observation can also be verified directly, by inserting (5.12) into (5.13). In fact,

$$\partial_t p = \frac{1}{\sqrt{2\pi t}}\, e^{-\frac{(x-\xi)^2}{2t}} \left[-\frac{1}{2t} + \frac{(x-\xi)^2}{2t^2} \right],$$

and

$$\partial_{xx} p = \frac{1}{\sqrt{2\pi t}}\, e^{-\frac{(x-\xi)^2}{2t}} \left[-\frac{1}{t} + \frac{(x-\xi)^2}{t^2} \right].$$

Therefore, $p_t = \frac{1}{2} p_{xx}$. We will see soon that the partial differential equation (5.13) is actually the Fokker-Planck equation for the SDE $dX_t = dB_t$. Moreover, p in (5.12) also satisfies the initial condition $p(x,0) = \delta(x-\xi)$, which is the probability density function version of the initial condition $X_0 = \xi$. Indeed, one of Dirac's original intuitive definitions for $\delta(x-\xi)$ is [255, p. 217]:

$$\delta(x-\xi) \triangleq \lim_{t\to 0+} \frac{1}{\sqrt{2\pi t}} e^{-\frac{(x-\xi)^2}{2t}}.$$

This can be justified in the context of generalized functions (or "distributions"); see Section 3.3.

5.2.1 Scalar Fokker-Planck Equations

First we consider the one-dimensional Fokker-Planck equation for a scalar SDE:

$$dX_t = b(X_t)dt + \sigma(X_t)dB_t, \quad X_0 = x_0 \in \mathbb{R}^1. \tag{5.14}$$

Let f: $\mathbb{R}^1 \to \mathbb{R}^1$ be a given function. Assume that X_t has a conditional probability density $p(x,t) \triangleq p(x,t|x_0,0)$. Our goal here is to derive a partial differential equation satisfied by $p(x,t)$. On one hand,

$$\mathbb{E} f(X_t) = \int_{\mathbb{R}^1} f(x)p(x,t)dx, \tag{5.15}$$

and thus

$$\frac{d}{dt}\mathbb{E}f(X_t) = \int_{\mathbb{R}^1} f(x)\frac{\partial}{\partial t}p(x,t)dx. \tag{5.16}$$

On the other hand, by Itô's formula,

$$\begin{aligned} df(X_t) &= f_x(X_t)dX_t + \frac{1}{2}f_{xx}(X_t)(dX_t)^2 \\ &= \left[bf_x + \frac{1}{2}\sigma^2 f_{xx}\right]dt + \sigma f_x dB_t. \end{aligned} \tag{5.17}$$

Taking expectation on both sides, we get

$$d\mathbb{E}f(X_t) = \mathbb{E}\left(bf_x + \frac{1}{2}\sigma^2 f_{xx}\right)dt. \tag{5.18}$$

Noting that the generator for this SDE is $Af(x) = bf_x + \frac{1}{2}\sigma^2 f_{xx}$, for $f \in H_0^2(\mathbb{R}^1)$, the preceding equation is rewritten as

$$\begin{aligned} \frac{d}{dt}\mathbb{E}f(X_t) &= \mathbb{E}\left(bf_x + \frac{1}{2}\sigma^2 f_{xx}\right) \\ &= \int_{\mathbb{R}^1}\left(bf_x + \frac{1}{2}\sigma^2 f_{xx}\right)p(x,t)dx \\ &= \int_{\mathbb{R}^1} f(x)\left[-\partial_x(bp) + \frac{1}{2}\partial_{xx}(\sigma^2 p)\right]dx. \end{aligned}$$

Here we have used the fact that p is zero at infinity. Thus

$$\frac{d}{dt}\mathbb{E}f(X_t) = \int_{\mathbb{R}^1} f(x)A^*p\,dx, \tag{5.19}$$

where A^* is the adjoint operator of A,

$$A^*p = -\partial_x(bp) + \frac{1}{2}\partial_{xx}(\sigma^2 p), \quad p \in H_0^2(\mathbb{R}^1). \tag{5.20}$$

Both A^* and A are linear differential operators. Comparing equations (5.16) and (5.19), we get

$$\frac{\partial}{\partial t}p = A^*p. \tag{5.21}$$

This is called the Fokker-Planck equation, or Kolmorgrov forward equation, for the probability density function p of the solution process X_t. We also simply say it is the Fokker-Planck equation for the SDE (5.14). It is a deterministic linear partial differential equation. The initial value for p is the Dirac delta function at x_0,

$$p(x,0) = \delta(x - x_0). \tag{5.22}$$

Note that $p(x, 0)$ may also be a given initial probability density profile $p_0(x)$ at time $t = 0$. An implicit condition is

$$p(x, t) \geq 0 \quad \text{and} \quad \int_{\mathbb{R}^1} p(x, t)dx = 1, \tag{5.23}$$

for $t > 0$.

Example 5.4 The Fokker-Planck equation for the simple scalar SDE

$$dX_t = 0dt + dB_t$$

is

$$p_t = \frac{1}{2}p_{xx}.$$

This is the well-known diffusion equation.

Example 5.5 The Fokker-Planck equation for the Langevin SDE

$$dX_t = -bX_t dt + a dB_t$$

is

$$p_t = \frac{1}{2}a^2 p_{xx} + b(xp)_x,$$

where a, b are positive real parameters.

Example 5.6 Consider a scalar SDE

$$dX_t = \left(X_t - X_t^3\right)dt + \sigma(X_t)dB_t, \quad X_0 = x_0.$$

The Fokker-Planck equation is

$$p_t = -((x - x^3)p)_x + \frac{1}{2}(\sigma^2(x)p)_{xx},$$

with the initial condition

$$p(x, 0) = \delta(x - x_0).$$

5.2.2 Multidimensional Fokker-Planck Equations

Now we consider the multidimensional Fokker-Planck equation for the following SDE system:

$$dX_t = b(X_t)dt + \sigma(X_t)dB_t, \quad X_0 = x_0 \in \mathbb{R}^n, \tag{5.24}$$

where b is an n-dimensional vector function, σ is an $n \times m$ matrix function, and $B_t(\omega)$ is an m-dimensional Brownian motion.

Let f: $\mathbb{R}^n \to \mathbb{R}^1$ be a given function. Assume that X_t has a conditional probability density $p(x,t) \triangleq p(x,t|x_0,0)$. Then, on one hand,

$$\mathbb{E}f(X_t) = \int_{\mathbb{R}^n} f(x)p(x,t)dx,$$

and thus

$$\frac{d}{dt}\mathbb{E}f(X_t) = \int_{\mathbb{R}^n} f(x)\frac{\partial}{\partial t}p(x,t)dx. \tag{5.25}$$

On the other hand, according to Itô's formula,

$$\begin{aligned} df(X_t) &= \sum_i \frac{\partial}{\partial x_i} f(X_t)dX_i + \frac{1}{2}\sum_{i,j}\frac{\partial^2}{\partial x_i \partial x_j} f(X_t)dX_i dX_j \\ &= \sum_i b_i \frac{\partial f}{\partial x_i}dt + \frac{1}{2}\sum_{i,j}\frac{\partial^2 f}{\partial x_i \partial x_j}(\sigma dB)_i(\sigma dB)_j + \sum_i \frac{\partial f}{\partial x_i}(\sigma dB)_i. \end{aligned}$$

Note that

$$\begin{aligned} (\sigma dB)_i(\sigma dB)_j &= \left(\sum_k \sigma_{ik} dB_k\right)\left(\sum_n \sigma_{jn} dB_n\right) \\ &= \left(\sum_k \sigma_{ik}\sigma_{jk}\right)dt = (\sigma\sigma^{\mathrm{T}})_{ij}dt. \end{aligned}$$

Thus,

$$df(X_t) = \left(\sum_i b_i \frac{\partial f}{\partial x_i} + \frac{1}{2}\sum_{i,j}(\sigma\sigma^{\mathrm{T}})_{ij}\frac{\partial^2 f}{\partial x_i \partial x_j}\right)dt + \sum_{i,k}\sigma_{ik}\frac{\partial f}{\partial x_i}dB_k. \tag{5.26}$$

Taking expectation on both sides, we obtain

$$d\mathbb{E}f(X_t) = \mathbb{E}\left(\sum_i b_i \frac{\partial f}{\partial x_i} + \frac{1}{2}\sum_{i,j}(\sigma\sigma^{\mathrm{T}})_{ij}\frac{\partial^2 f}{\partial x_i \partial x_j}\right)dt,$$

and therefore,

$$\frac{d}{dt}\mathbb{E}f(X_t) = \mathbb{E}\left(\sum_i b_i \frac{\partial f}{\partial x_i} + \frac{1}{2}\sum_{i,j}(\sigma\sigma^{\mathrm{T}})_{ij}\frac{\partial^2 f}{\partial x_i \partial x_j}\right). \tag{5.27}$$

Then, (5.27) becomes

$$\frac{d}{dt}\mathbb{E}f(X_t) = \int_{\mathbb{R}^n}\left(\sum_i b_i\frac{\partial f}{\partial x_i} + \frac{1}{2}\sum_{i,j}(\sigma\sigma^{\mathrm{T}})_{ij}\frac{\partial^2 f}{\partial x_i \partial x_j}\right) p(x,t)dx. \quad (5.28)$$

After integration by parts, and noticing that p is zero at infinity, we obtain

$$\begin{aligned}&\frac{d}{dt}\mathbb{E}f(X_t)\\ &= \int_{\mathbb{R}^n} f(x)\left[-\sum_i \frac{\partial}{\partial x_i}(b_i p) + \frac{1}{2}\sum_{i,j}\frac{\partial^2}{\partial x_i \partial x_j}((\sigma\sigma^{\mathrm{T}})_{ij}\, p)\right] dx. \quad (5.29)\end{aligned}$$

Recall that the generator for this SDE system is

$$\begin{aligned}Af(X_t) &= \sum_i b_i\frac{\partial f}{\partial x_i} + \frac{1}{2}\sum_{i,j}(\sigma\sigma^{\mathrm{T}})_{ij}\frac{\partial^2 f}{\partial x_i \partial x_j}\\ &= b\cdot\nabla f + \frac{1}{2}\mathrm{Tr}(\sigma\sigma^{\mathrm{T}}H(f)),\quad f\in H_0^2(\mathbb{R}^n), \quad (5.30)\end{aligned}$$

where $H(f)$ is the Hessian matrix for f. The adjoint operator for the generator A in $L^2(\mathbb{R}^n)$ is

$$A^*p = -\sum_i \frac{\partial}{\partial x_i}(b_i p) + \frac{1}{2}\sum_{i,j}\frac{\partial^2}{\partial x_i \partial x_j}((\sigma\sigma^{\mathrm{T}})_{ij}\, p),\quad p\in H_0^2(\mathbb{R}^n). \quad (5.31)$$

Note that A^* is often called the Fokker-Planck operator. It may be rewritten as

$$\begin{aligned}A^*p &= \frac{1}{2}\sum_{i,j}\partial_{x_i x_j}(\sigma\sigma^{\mathrm{T}})_{ij}\,p - \sum_i \partial_{x_i}(b_i(x)p)\\ &= \frac{1}{2}\mathrm{Tr}(\nabla\nabla^{\mathrm{T}}(\sigma\sigma^{\mathrm{T}}p)) - \nabla\cdot(bp)\\ &= \frac{1}{2}\mathrm{Tr}(H(\sigma\sigma^{\mathrm{T}}p)) - \nabla\cdot(bp), \quad (5.32)\end{aligned}$$

where we have used the fact for the (symmetric) Hessian matrix that

$$H = \nabla\nabla^{\mathrm{T}} = (\partial_{x_i x_j}).$$

Note that p is a scalar function, and we interpret $H(\sigma\sigma^{\mathrm{T}}p)$ as matrix multiplications of H, σ and $\sigma^{\mathrm{T}}p$. The adjoint operator (5.32) may be further rewritten as

$$\begin{aligned} A^*p &= \frac{1}{2}\nabla\cdot(\nabla\cdot(\sigma\sigma^{\mathrm{T}}p)) - \nabla\cdot(bp) \\ &= \frac{1}{2}\mathrm{div}(\mathrm{div}(\sigma\sigma^{\mathrm{T}}p)) - \mathrm{div}(bp), \end{aligned} \tag{5.33}$$

where $\nabla\cdot(\nabla\cdot(\sigma\sigma^{\mathrm{T}}p))$ is explained as follows as a two-step operation: first, take divergence for each of the n row vectors of the matrix $n\times n$ symmetric matrix $\sigma\sigma^{\mathrm{T}}p$; this gives us an n-vector, namely, $V = \nabla\cdot(\sigma\sigma^{\mathrm{T}}p)$. Then, take the divergence of V.

In terms of the adjoint operator A^*, the equation (5.29) becomes

$$\frac{d}{dt}\mathbb{E}f(X_t) = \int_{\mathbb{R}^n} f(x)A^*p\,dx. \tag{5.34}$$

Comparing equation (5.25) with equation (5.34), we get

$$\frac{\partial}{\partial t}p(x,t) = A^*p(x,t). \tag{5.35}$$

This is the multidimensional Fokker-Planck equation for the probability density function p of the solution process X_t or for the SDE system (5.24). It is a deterministic linear partial differential equation. This equation is supplemented with an initial condition, that is, a Dirac delta function at x_0,

$$p(x,0) = \delta(x - x_0). \tag{5.36}$$

Note that $p(x,0)$ may also be a given initial probability density function at time $t = 0$. An implicit condition is

$$p(x,t) \geq 0 \quad \text{and} \quad \int_{\mathbb{R}^n} p(x,t)dx = 1, \tag{5.37}$$

for $t > 0$.

Example 5.7 Consider a SDE system

$$dX_t = f(X_t, Y_t)dt + \sigma_1(X_t)dB_t^1, \quad X_0 = x_0, \tag{5.38}$$

$$dY_t = g(X_t, Y_t)dt + \sigma_2(Y_t)dB_t^2, \quad Y_0 = y_0, \tag{5.39}$$

where σ_1, σ_2 are noise intensities and B_t^1, B_t^2 are independent scalar Brownian motions. The Fokker-Planck equation is

$$p_t = \frac{1}{2}\left((\sigma_1^2(x)p)_{xx} + (\sigma_2^2(y)p)_{yy}\right) - [(f(x,y)p)_x + (g(x,y)p)_y],$$

with the initial condition

$$p(x, y, 0) = \delta(x - x_0, y - y_0).$$

In particular, for the SDE system,

$$dX_t = -X_t dt + \varepsilon dB_t^1, \quad X_0 = x_0, \tag{5.40}$$

$$dY_t = Y_t dt + \varepsilon dB_t^2, \quad Y_0 = y_0, \tag{5.41}$$

where ε is a constant, the Fokker-Planck equation is

$$p_t = \frac{1}{2}\varepsilon^2(p_{xx} + p_{yy}) - [-(xp)_x + (yp)_y].$$

5.2.3 Existence and Uniqueness for Fokker-Planck Equations

In numerical simulations, we have to take x in a large but bounded doma: in $D \subset \mathbb{R}^n$, and we could impose an absorbing boundary condition on ∂D; that is, as long as a "particle" or a solution path reaches the boundary, it is removed from the system. So we are led to consider the following system:

$$\frac{\partial}{\partial t} p(x, t) = A^* p(x, t), \tag{5.42}$$

$$p|_{\partial D} = 0, \quad p(x, 0) = p_0(x), \tag{5.43}$$

for $x \in D$ and $t > 0$. Owing to the absorbing boundary condition (5.43), the quantity $\int_D p(x, t)dx$ is nonincreasing in time t.

As seen in (5.32), the second-order derivative term in the Fokker-Planck operator is

$$\frac{1}{2}\sum_{i,j}(\sigma\sigma^{\mathrm{T}})_{ij}\partial_{x_i x_j}.$$

Thus A^* is a uniformly elliptic operator if there exists a positive constant γ such that

$$\sum_{i,j}(\sigma(x)\sigma^{\mathrm{T}}(x))_{ij}\xi_i\xi_j \geq \gamma \sum_i \xi_i^2 \tag{5.44}$$

for all ξ in $\mathbb{R}^n$.

We assume that the differential operator A^* is uniformly elliptic on D, that is, the uniform ellipticity condition (5.44) holds for all $\xi \in D$. Now we state the following result.

Theorem 5.8 *Assume that the operator A^* is a uniformly elliptic operator on a bounded domain D.*

(i) If b, σ, together with the first-order derivatives of b and second-order derivatives of σ, are bounded on D and the initial datum p_0 is in $L^2(D)$, then the unique solution p to (5.42)–(5.43) exists and is in $L^2(0, T; H_0^1(D))$. Moreover, if, additionally, p_0 is in $H_0^1(D)$, then the solution p is actually in $L^2(0, T; H^2(D)) \bigcap L^\infty(0, T; H_0^1(D))$.

(ii) If b, σ, the first-order derivatives of b, second-order derivatives of σ, and p_0 are all uniformly Hölder continuous with exponent γ in D and are all bounded in $C^\gamma(D)$, then the unique solution p to (5.42)–(5.43) exists and is in $C^{2,\gamma}(D)$.

Proof

(i) These results are from Theorems 3, 4, and 5 in [89, Section 7.1].

(ii) This follows from Theorem 9 in [94, Chapter 3]. □

We also consider the so-called reflecting boundary condition [99, Section 5.1.1]. By the formula for the adjoint operator (5.33), the Fokker-Planck equation (5.35) may be rewritten as

$$\frac{\partial}{\partial t} p(x, t) = \nabla \cdot \left[\frac{1}{2} (\nabla \cdot (\sigma \sigma^{\mathrm{T}} p)) - bp(x, t) \right]. \tag{5.45}$$

The reflecting boundary condition means, particles, or solution paths cannot leave a bounded domain D, and hence there is zero net flow of p crossing the boundary ∂D. Thus we impose the following reflecting boundary condition:

$$\frac{1}{2} (\nabla \cdot (\sigma \sigma^{\mathrm{T}} p)) - bp = 0 \quad \text{on} \quad \partial D. \tag{5.46}$$

Integrating (5.45) over D and using the boundary condition (5.46) together with the divergence theorem, we indeed have conservation of probability

$$\frac{\partial}{\partial t} \int_D p(x, t) dx = 0, \tag{5.47}$$

that is, $\int_D p(x, t) dx = 1$, for all $t > 0$.

As discussed in [178, Theorem 5.18], [94, Chapter 5, Theorem 2], or [272, Theorem 8.4.1], we also have existence and uniqueness of the solution for the Fokker-Planck equation (5.35), under the reflecting boundary condition (5.46) and an appropriate initial condition $p(x, 0) = p_0(x)$. See also Problem 5.7 at the end of this chapter.

Grasman and van Herwaarden [107] and Risken [238] have discussed various analytical methods for examining the solutions of Fokker-Planck equations.

5.2.4 Likelihood for Transitions between Different Dynamical Regimes Under Uncertainty

A dynamical regime is a physical concept. For our purpose, we consider a dynamical regime as a stochastic state distributed around a deterministic equilibrium state. More specifically, we treat a dynamical regime as a small neighborhood U (i.e., with small Lebesgue measure) containing a deterministic equilibrium state, on which the system is distributed (i.e., with probability density), either with a uniform distribution on U or a distribution with compact support on U. How does this stochastic state evolve as time goes on?

Suppose the system starts at one regime U_0 with distribution density $p_0(x)$. The probability density $p(x,t)$ for the system state is governed by the Fokker-Planck equation (5.42). The likelihood of the system reaching U at a future time T is

$$\tilde{P}(U_0, U, T)) \triangleq \mathbb{P}(X_T \in U) = \int_U p(x, T)dx. \tag{5.48}$$

This is computable once we numerically solve the Fokker-Planck equation.

We can also compare the likelihood for transition, at a future time T, from U_0 to U_1, or U_0 to U_2, by computing $\tilde{P}(U_0, U_1, T))$ and $\tilde{P}(U_0, U_2, T))$.

5.3 Most Probable Phase Portraits

For deterministic dynamical systems, phase portraits provide geometric pictures of dynamical orbits, at least for lower-dimensional systems. Note that a better term for a phase portrait is "state portrait," although we still use the former as it is so much more common. For a stochastic dynamical system, can we have a sort of phase portrait? The situation, however, is quite different from the deterministic case. Two apparent options are the mean phase portraits and the almost sure phase portraits, but they are not very useful.

In this section, we propose a deterministic geometric tool (not quite a deterministic quantity), most probable phase portraits, to visualize stochastic dynamics. These deterministic phase portraits provide geometric pictures of most probable or maximal likely orbits of stochastic dynamical systems. They are based on Fokker-Planck equations discussed in the previous section.

5.3.1 Mean Phase Portraits

How about plotting the orbits for the mean of the system state?

Let us consider a simple linear SDE system $dX_t = 3X_t\,dt + dB_t$. The mean $\mathbb{E}X_t$ evolves according to the linear deterministic system $\frac{d}{dt}\mathbb{E}X_t = 3\mathbb{E}X_t$, which is the original system without noise. In other words, the mean phase portrait will not capture the impact of noise in this simple linear SDE system.

The situation is even worse for nonlinear SDE systems. For example, consider $dX_t = (X_t - X_t^3)dt + dB_t$. Take the mean on both sides of this SDE to get $\frac{d}{dt}\mathbb{E}X_t = \mathbb{E}X_t - \mathbb{E}(X_t^3)$. Thus, we do not have a "closed" differential equation for the evolution of mean $\mathbb{E}X_t$, because $\mathbb{E}(X_t^3) \neq (\mathbb{E}X_t)^3$. This is a theoretical difficulty for analyzing mean phase portraits for stochastic systems. The same difficulty arises for mean square phase portraits and higher-moment phase portraits.

5.3.2 Almost Sure Phase Portraits

Another possible option is to plot sample solution orbits for an SDE system, mimicking deterministic phase portraits. If we plot representative sample orbits in the state space, we will see complicated Wuhan noodles,[1] which could hardly offer useful information for understanding dynamics.

5.3.3 Most Probable Phase Portraits

Now we consider most probable phase portraits. Note that each sample orbit is a possible "outcome" of a realistic orbit X_t of the system. But which sample orbit is most probable or maximal likely? This is determined by the maximizers of the probability density function $p(x, t)$ of X_t, at every time t.

For an SDE system in $\mathbb{R}^n$,

$$dX_t = b(X_t)dt + \sigma(X_t)dB_t, \quad X_0 = \xi, \tag{5.49}$$

the Fokker-Planck equation for the probability density function $p(x, t)$ of X_t is

$$\frac{\partial}{\partial t}p(x, t) = A^* p(x, t), \tag{5.50}$$

with initial condition $p(x, 0) = \delta(x - \xi)$. Recall that the Fokker-Planck operator is

$$A^* p = \frac{1}{2}\mathrm{Tr}(H(\sigma\sigma^{\mathrm{T}} p)) - \nabla \cdot (bp), \tag{5.51}$$

where $H = (\partial_{x_i x_j})$ is the Hessian matrix, Tr evaluates the trace of a matrix, and $H(\sigma\sigma^{\mathrm{T}} p)$ denotes the matrix multiplication of $H, \sigma, \sigma^{\mathrm{T}} p$.

Consider the sample orbit starting at the initial point ξ in state space (more popularly, though not quite correctly, called phase space) $\mathbb{R}^n$. The probability density function p is a surface in the (x, t, p)-space. At a given time instant t, the maximizer $x_m(t)$ for $p(x, t)$ indicates the most probable (i.e., maximal likely)

[1] Wuhan noodles are a famous specialty food in Wuhan, China.

location of this orbit at time t. The orbit traced out by $x_m(t)$ is called a most probable orbit starting at ξ. Thus, $x_m(t)$ follows the top ridge or plateau of the surface in the (x, t, p)-space, as time goes on. Like the Monkey King[2] walking on the top of clouds, he only walks along the highest ridges or following the highest plateau.

It may be possible to have more than one maximizer at a given time t. In that case, we could have more than one "most probable orbit" starting at ξ. This is like the situation when the chance of raining is 50 percent, the chance of not raining is also 50 percent, and both are "most probable."

We now consider a few examples.

Example 5.9 Consider a scalar SDE with additive noise

$$dX_t = X_t dt + dB_t, \quad X_0 = x_0 \in \mathbb{R}^1. \tag{5.52}$$

The solution for this SDE is

$$X(t) = e^t x_0 + \int_0^t e^{t-s} dB_s.$$

This is a Gaussian process. Once we have its mean and variance, we then have its probability density function $p(x, t)$. Its mean is $\mathbb{E}X(t) = e^t x_0$ and the variance is

$$\begin{aligned}\mathrm{Var}(X(t)) &= \mathbb{E}\left[X(t) - e^t x_0\right]^2 = \mathbb{E}\left[\int_0^t e^{t-s} dB_s\right]^2 \\ &= \int_0^t e^{2(t-s)} ds = \frac{1}{2}\left(e^{2t} - 1\right).\end{aligned}$$

So the probability density function for the solution $X(t)$ is

$$\begin{aligned}p(x, t) &= \frac{1}{\sqrt{2\pi}\,\frac{1}{\sqrt{2}}\sqrt{e^{2t} - 1}} \exp\left(-\frac{(x - e^t x_0)^2}{e^{2t} - 1}\right) \\ &= \frac{1}{\sqrt{\pi}\,\sqrt{e^{2t} - 1}} \exp\left(-\frac{(x - e^t x_0)^2}{e^{2t} - 1}\right).\end{aligned} \tag{5.53}$$

By setting $\partial_x p = 0$ or just observing $p(x, t)$, we obtain the maximizer at time t: $x_m(t) = e^t x_0$ for every $x_0 \in \mathbb{R}^1$. Thus, the most probable dynamical system is

$$\dot{x}_m = x_m.$$

This is the same as the corresponding deterministic dynamical system $\dot{x} = x$ for this SDE system with additive noise.

The phase portrait for the corresponding deterministic system and the most probable phase portrait are in Figure 5.1.

[2] The Monkey King is a main character in the Chinese classic novel *The Journey to the West*.

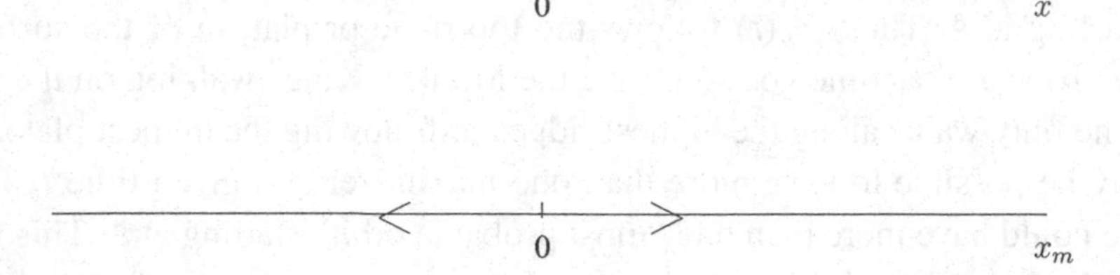

Figure 5.1 (top) Phase portrait for the corresponding deterministic system (noise absent) in Example 5.9. (bottom) Most probable phase portrait for Example 5.9.

Example 5.10 Consider a scalar SDE with multiplicative noise

$$dX_t = X_t dt + X_t dB_t, \quad X_0 = x_0 \in \mathbb{R}^1. \tag{5.54}$$

By Example 4.20, the solution for this SDE is

$$X(t) = e^{\frac{1}{2}t} e^{B_t} x_0.$$

Let us find out its probability distribution function $F(x,t)$ and then its probability density function $p(x,t)$.

When the initial state $x_0 = 0$, the solution is 0 (an equilibrium state). When the initial state x_0 is positive (negative), the solution is also positive (negative), and the distribution function is calculated as

$$\begin{aligned} F(x,t) &= \mathbb{P}(X_t \le x) = \mathbb{P}\big(e^{\frac{1}{2}t} e^{B_t} x_0 \le x\big) \\ &= \mathbb{P}\left(e^{B_t} \le \frac{x}{x_0}\, e^{-\frac{1}{2}t}\right) \\ &= \mathbb{P}\left(B_t \le \ln\left(\frac{x}{x_0}\right) - \frac{1}{2}t\right) \\ &= \int_{-\infty}^{\ln\left(\frac{x}{x_0}\right) - \frac{1}{2}t} \frac{1}{\sqrt{2\pi t}} e^{-\frac{\xi^2}{2t}}\, d\xi. \end{aligned} \tag{5.55}$$

For $x_0 \neq 0$, the distribution function F is nonzero only if x and x_0 have the same sign (i.e., $\frac{x}{x_0}$ is positive). The probability density function is then

$$\begin{aligned} p(x,t) &= \partial_x F(x,t) \\ &= \frac{1}{x} \frac{1}{\sqrt{2\pi t}} \exp\left(-\frac{\left(\ln\left(\frac{x}{x_0}\right) - \frac{1}{2}t\right)^2}{2t}\right). \end{aligned} \tag{5.56}$$

Setting $\partial_x p(x,t) = 0$, we obtain

$$x_m(t) = x_0 e^{-\frac{1}{2}t}. \tag{5.57}$$

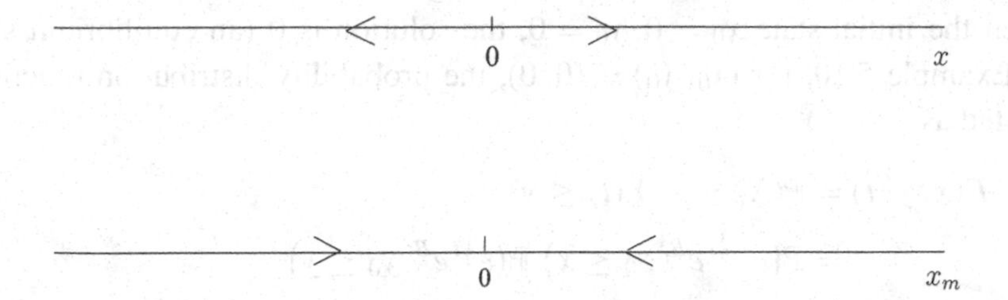

Figure 5.2 (top) Phase portrait for the corresponding deterministic system (noise absent) in Example 5.10. (bottom) Most probable phase portrait for Example 5.10.

It can be checked that this is indeed the maximizer of $p(x,t)$ at time t. Thus, the most probable dynamical system is

$$\dot{x}_m(t) = -\frac{1}{2}x_m. \tag{5.58}$$

This is different from the corresponding deterministic dynamical system $\dot{x} = x$ for this SDE system with multiplicative noise. In fact, the state 0 is unstable for the corresponding deterministic dynamical system but stable for the most probable dynamical system. The most probable phase portrait and the phase portrait of the corresponding deterministic system are in Figure 5.2.

For a more general scalar SDE $dX_t = rX_t dt + \alpha X_t dB_t$ in Example 4.20, the most probable phase portrait may or may not differ from that of the corresponding deterministic differential equation $\dot{x} = rx$, depending on the relative magnitudes of r and α.

Example 5.11 Consider a scalar SDE with multiplicative noise

$$dX_t = -X_t dt + X_t dB_t^1, \quad X_0 = x_0 \in \mathbb{R}^1, \tag{5.59}$$

$$dY_t = Y_t dt + Y_t dB_t^2, \quad Y_0 = y_0 \in \mathbb{R}^1, \tag{5.60}$$

where B_t^1, B_t^2 are scalar independent Brownian motions. By Example 4.20, the solution for this SDE system is

$$X(t) = e^{-\frac{3}{2}t}e^{B_t^1}x_0, \quad Y(t) = e^{\frac{1}{2}t}e^{B_t^2}y_0.$$

Note that $X(t), Y(t)$ are independent. Let us find out its probability distribution function $F(x,y,t)$ and then its probability density function $p(x,y,t)$.

When the initial state $x_0 = 0$, $y_0 = 0$, the solution is 0 (an equilibrium state). As in Example 5.10, for $(x_0, y_0) \neq (0, 0)$, the probability distribution function is calculated as

$$
\begin{aligned}
F(x, y, t) &= \mathbb{P}(X_t \leq x, Y(t) \leq y) \\
&= \mathbb{P}\big(e^{-\frac{3}{2}t} e^{B_t^1} x_0 \leq x\big)\, \mathbb{P}\big(e^{\frac{1}{2}t} e^{B_t^2} y_0 \leq y\big) \\
&= \mathbb{P}\left(B_t^1 \leq \ln\left(\frac{x}{x_0}\right) + \frac{3}{2}t\right) \mathbb{P}\left(B_t^2 \leq \ln\left(\frac{y}{y_0}\right) - \frac{1}{2}t\right) \\
&= \int_{-\infty}^{\ln(\frac{x}{x_0})+\frac{3}{2}t} \frac{1}{\sqrt{2\pi t}} e^{-\frac{\xi^2}{2t}} d\xi \cdot \int_{-\infty}^{\ln(\frac{y}{y_0})-\frac{1}{2}t} \frac{1}{\sqrt{2\pi t}} e^{-\frac{\eta^2}{2t}} d\eta.
\end{aligned} \tag{5.61}
$$

For $(x_0, y_0) \neq (0, 0)$, the distribution function F is nonzero only if x and x_0 and y and y_0 have the same sign. The probability density function is then

$$
\begin{aligned}
p(x, y, t) &= \partial_{xy}^2 F(x, y, t) \\
&= \frac{1}{x} \frac{1}{\sqrt{2\pi t}} \exp\left(-\frac{\left(\ln\left(\frac{x}{x_0}\right) + \frac{3}{2}t\right)^2}{2t}\right) \\
&\quad \cdot \frac{1}{y} \frac{1}{\sqrt{2\pi t}} \exp\left(-\frac{\left(\ln\left(\frac{y}{y_0}\right) - \frac{1}{2}t\right)^2}{2t}\right).
\end{aligned} \tag{5.62}
$$

Setting $\partial_x p(x, y, t) = \partial_y p(x, y, t) = 0$, we obtain

$$
x_m(t) = x_0 e^{-\frac{5}{2}t}, \quad y_m(t) = y_0 e^{-\frac{1}{2}t}. \tag{5.63}
$$

It can be checked that (x_m, y_m) is indeed the maximizer of $p(x, y, t)$ at time t. Thus, the most probable dynamical system is

$$
\dot{x}_m(t) = -\frac{5}{2} x_m, \quad \dot{y}_m(t) = -\frac{1}{2} y_m. \tag{5.64}
$$

This is different from the corresponding deterministic dynamical system $\dot{x} = -x$, $\dot{y} = y$. In fact, the state $(0, 0)$ is a saddle and thus unstable for the corresponding deterministic dynamical system but is a sink and thus stable for the most probable dynamical system. See Figures 5.3 and 5.4 for the phase portrait of the corresponding deterministic system and the most probable phase portrait, respectively.

For a more general system of SDEs

$$
\begin{aligned}
dX_t &= r_1 X_t dt + \alpha_1 X_t dB_t^1, \\
dY_t &= r_2 Y_t dt + \alpha_2 Y_t dB_t^2,
\end{aligned}
$$

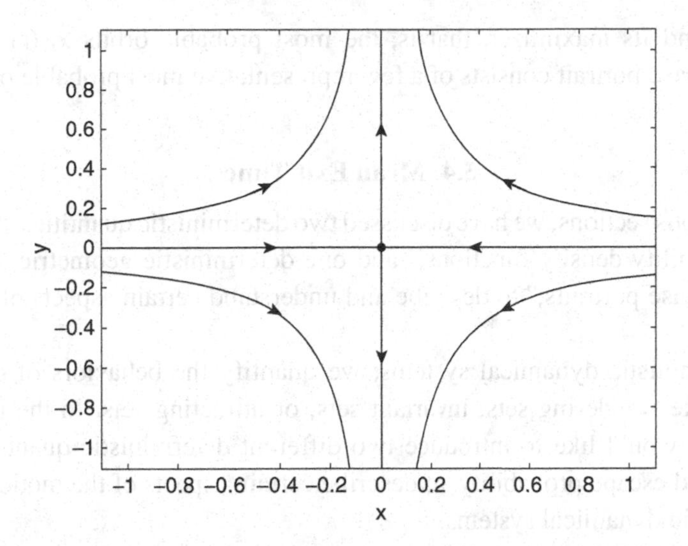

Figure 5.3 Phase portrait for the corresponding deterministic system (noise absent) in Example 5.11: $\dot{x} = -x$, $\dot{y} = y$.

the most probable phase portrait may or may not differ from that of the corresponding deterministic system $\dot{x} = r_1 x$, $\dot{y} = r_2 y$, depending on the relative magnitudes of r_1, r_2 and α_1, α_2.

For a more complicated SDE system, we will have to numerically solve the Fokker-Planck equation (5.50) to obtain the probability density function $p(x, t)$

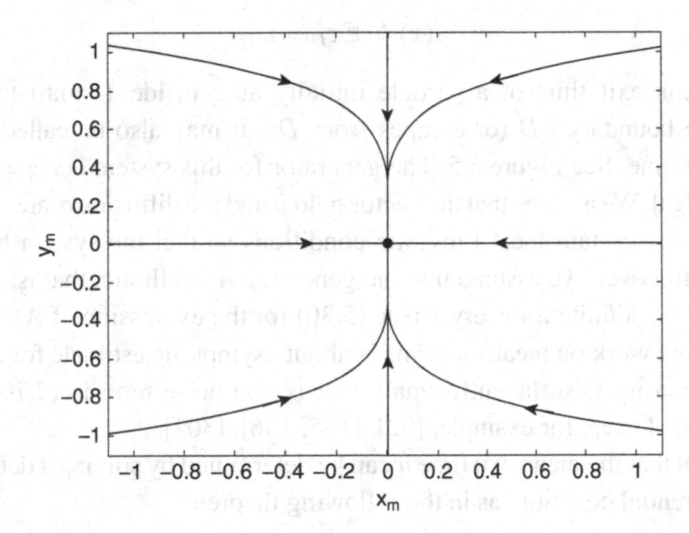

Figure 5.4 Most probable phase portrait for Example 5.11: $\dot{x}_m(t) = -\frac{5}{2}x_m$, $\dot{y}_m(t) = -\frac{1}{2}y_m$.

and then find its maximizer, that is, the most probable orbits $x_m(t)$. The most probable phase portrait consists of a few representative most probable orbits.

5.4 Mean Exit Time

In the previous sections, we have discussed two deterministic quantities, "moments" and "probability density functions," and one deterministic geometric tool, "most probable phase portraits," to describe and understand certain aspects of stochastic dynamics.

In deterministic dynamical systems, we quantify the behaviors of orbits with concepts like wandering sets, invariant sets, or attracting sets. In the rest of this chapter, we would like to introduce two different deterministic quantities, mean exit time and escape probability, to describe certain aspects of the motion of orbits for stochastic dynamical systems.

Consider an SDE system in $\mathbb{R}^n$,

$$dX_t = b(X_t)dt + \sigma(X_t)dB_t, \; X_0 = x. \tag{5.65}$$

For a bounded domain $D \subset \mathbb{R}^n$ (with boundary ∂D), the first exit time for a solution orbit starting at $x \in D$ is defined as

$$\tau_D(\omega) \triangleq \inf\{t > 0 : \quad X_0 = x, \; X_t \in \partial D\}. \tag{5.66}$$

By [213, p. 117], it is known that τ_D is a stopping time (see Definition 2.15). The mean exit time or mean residence time is defined as

$$u(x) \triangleq \mathbb{E}\tau_D(\omega). \tag{5.67}$$

It is the mean exit time of a particle initially at x inside D until the particle first hits the boundary ∂D (or escapes from D). It may also be called the mean first passage time. See Figure 5.5. The generator for this system is $Ag = b \cdot \nabla g + \frac{1}{2}\mathrm{Tr}[\sigma\sigma^{\mathrm{T}} H(g)]$. We assume that the vector field b and the diffusion σ are continuous and also satisfy certain local Lipschitz conditions so that the system has unique solutions. Moreover, we assume that the generator A is elliptic, that is, the matrix $\sigma\sigma^{\mathrm{T}}$ is positive definite for every x (see (5.30) for the expression of A).

The existing work on mean exit time is about asymptotic estimate for $u(x)$ when the noise intensity is sufficiently small, that is, the noise term in (7.70) is $\varepsilon \, dL_t$ with $0 < \varepsilon \ll 1$. See, for example, [92], [135, 136], [302].

It turns out that the mean exit time u can be determined by solving a deterministic partial differential equation, as in the following theorem.

Theorem 5.12 (Mean exit time) *Under the uniform ellipticity condition* (5.4), *the mean exit time* $u(x) = \mathbb{E}\tau_D(\omega)$ *of the stochastic system* (5.65), *for an orbit (i.e.,*

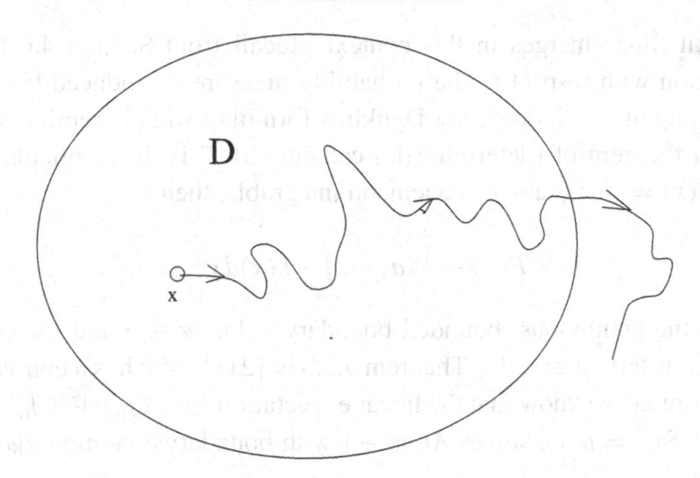

Figure 5.5 Mean exit time.

a trajectory) starting at $x \in D$, satisfies the following elliptic partial differential equation:

$$Au = -1, \tag{5.68}$$

$$u|_{\partial D} = 0, \tag{5.69}$$

where A is the generator

$$Au = b \cdot \nabla u + \frac{1}{2} Tr[\sigma\sigma^{\mathrm{T}} H(u)]. \tag{5.70}$$

Moreover,

(i) If the domain D has $C^{2,\gamma}$ boundary and the drift b and diffusion σ are in $C^{\gamma}(\bar{D})$ for some $\gamma \in (0, 1)$, then the mean exit time u uniquely exists and is in $C^{2,\gamma}(\bar{D})$.

(ii) If the drift b and diffusion σ are locally integrable and bounded on D, then the solution u uniquely exists and is in $H^1(D)$. If, additionally, σ is uniformly Lipschitz continuous in D, then for every subdomain $D' \subset\subset D$, u is in $H^2(D')$ (local regularity). If, furthermore, the boundary ∂D is C^2, then u is in $H^2(D)$ (global regularity).

Remark 5.13 To prove this theorem, we recall the Dynkin's formula [213, p. 124]: assume that τ is a stopping time (see Definition 2.15) with $\mathbb{E}^x \tau < \infty$. Then

$$\mathbb{E}^x f(X_\tau) - f(x) = \mathbb{E}^x \left[\int_0^\tau Af(X_s)ds \right], \tag{5.71}$$

for f in the domain of definition of the generator A. This may be regarded as a stochastic version of the fundamental theorem of calculus. The preceding theorem on mean exit time emerges in this context. Recall from Section 4.6 that $\mathbb{E}^x$ is the expectation with respect to the probability measure $\mathbb{P}^x$ induced by a solution process starting at x. Indeed, the Dynkin's formula (5.71) is reminiscent of the fundamental theorem of (deterministic) calculus: if F is differentiable on $[a, b]$ such that $F'(x) = f(x)$, and f is Riemann integrable, then

$$F(b) - F(a) = \int_a^b f(x)dx.$$

Proof Take the continuous, bounded boundary value $\psi \equiv 0$ and the continuous inhomogeneous term $g \equiv 1$. By Theorem 9.3.3 in [213], which is a consequence of Dynkin's formula, we know that the linear expectation $\mathbb{E}\psi(X_{\tau_D}) + \mathbb{E}\int_0^{\tau_D} g(X_s)ds$, which is just $\mathbb{E}\tau_D = u(x)$, solves $Au = -1$ with boundary condition $u|_{\partial D} = 0$.

(i) The existence, uniqueness, and regularity of the mean exit time u follow from the result for classical solutions of linear elliptic partial differential equations in Theorem 6.8 in [105, p. 100].
(ii) The existence and uniqueness of the solution u in $H^1(D)$ follow from the result for weak solutions of linear elliptic partial differential equations in Theorem 8.3 in [105, p. 181]. The local regularity follows from Theorem 8.8 in [105, p. 183], and the global regularity follows from Theorem 8.12 in [105, p. 186].

This completes the proof of this theorem. □

Remark 5.14 The formulation of mean exit times as solutions of elliptic partial differential equations are considered in, for example, [196, 197, 206, 253, 254] in the context of asymptotic analysis and in [40, 41] in the context of numerical analysis.

Remark 5.15 See also [280, p. 305] for an intuitive derivation of this result or [196, 206, 253] for a similar proof.

The average mean exit time on D is defined as

$$\bar{u}(D) \triangleq \frac{1}{|D|}\int_D u(x)dx, \tag{5.72}$$

where $|D|$ is the Lebesgue measure of the bounded domain D.

We can numerically compute, in Matlab, mean exit time u by solving elliptic partial differential equations or, in the one-dimensional case, solving ordinary differential equations, with zero Dirichlet boundary conditions. We now present a few examples.

Example 5.16 For the scalar stochastic dynamical system

$$dX_t = b(X_t)dt + \sigma(X_t)dB_t,\ X_0 = x, \tag{5.73}$$

the mean residence time $u(x)$ in an interval (α, β) satisfies

$$bu'(x) + \frac{1}{2}\sigma^2(x)u''(x) = -1, \quad u(\alpha) = u(\beta) = 0. \tag{5.74}$$

(i) For a simple scalar SDE $dX_t = dB_t$, $X_0 = x$, when does the solution orbit $X_t(\omega) = \varphi(t, \omega, x)$ exit interval $D = (0, 1)$? In this case, the generator is $A = \frac{1}{2}\frac{d^2}{dx^2}$. Thus the mean exit time $u(x)$ satisfies

$$\frac{1}{2}u''(x) = -1, \quad u(0) = u(1) = 0.$$

Solving this boundary value problem, we obtain $u(x) = x - x^2$. The maximal mean exit time $u_{\text{max}} = \frac{1}{4}$ is attained at the middle of the interval D, that is, at $x = 0.5$.

More generally, the mean exit time from an interval $D = (-r, r)$ with $r > 0$ is

$$u(x) = r^2 - x^2.$$

(ii) For $dX_t = dt + dB_t$, $X_0 = x$, when does the solution orbit $X_t(\omega) = \varphi(t, \omega, x)$ exit interval $D = (0, 1)$? In this case, the generator is $A = \frac{1}{2}\frac{d^2}{dx^2} + \frac{d}{dx}$. Thus the mean exit time $u(x)$ satisfies

$$\frac{1}{2}u''(x) + u'(x) = -1, \quad u(0) = u(1) = 0.$$

Solving this boundary value problem, we obtain $u(x) = -x - \frac{1}{e^{-2}-1} + \frac{1}{e^{-2}-1}e^{-2x}$.

The maximal mean exit time u_{max} is attained within the left half of the interval D, that is, at $x = -\frac{1}{2}\ln\frac{1-e^{-2}}{2} < 0.5$. As we see by comparing with case (i), the vector field (or drift) 1 has an impact on the mean exit time.

(iii) For the Ornstein-Uhlenbeck equation $dX_t = -\beta X_t dt + \alpha dB_t$ with α, β constants, when does the solution orbit $X_t(\omega) = \varphi(t, \omega, x)$ exit interval $D = (0, 1)$? The generator is $A = -\beta x\frac{d}{dx} + \frac{1}{2}\alpha^2\frac{d^2}{dx^2}$. The mean exit time $u(x)$ satisfies

$$-\beta x u'(x) + \frac{1}{2}\alpha^2 u''(x) = -1, \; u(0) = u(1) = 0.$$

Example 5.17 Consider a scalar SDE

$$dX_t = \big(-X_t + X_t^3\big)dt + \varepsilon dB_t,$$

with noise intensity $\varepsilon > 0$.

The mean exit time $u(x)$ from the interval $(-0.5, 0.5)$ for an orbit starting at x satisfies

$$\frac{1}{2}\varepsilon^2 u''(x) + (-x + x^3)u'(x) = -1, \tag{5.75}$$

$$u(-0.5) = u(0.5) = 0. \tag{5.76}$$

For $\varepsilon = 0.5$, the mean exit time is shown in Figure 5.6.

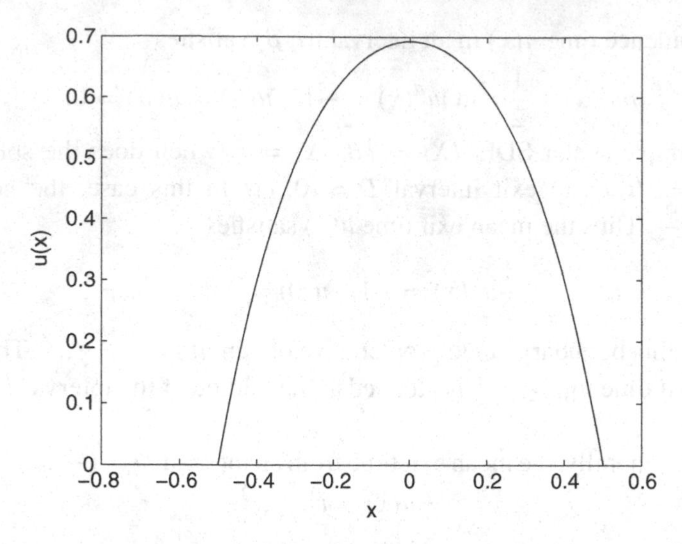

Figure 5.6 Example 5.17: mean exit time from the interval $(-0.5, 0.5)$ for $\varepsilon = 0.5$.

Example 5.18 Consider the following two-dimensional stochastic dynamical system

$$\dot{X}_t = a_1(X_t, Y_t) + b_1(X_t, Y_t)\dot{B}_t^1, \ X_0 = x, \tag{5.77}$$

$$\dot{Y}_t = a_2(X_t, Y_t) + b_2(X_t, Y_t)\dot{B}_t^2, \ Y_0 = y, \tag{5.78}$$

where B_t^1, B_t^2 are two real independent Brownian motions and a_1, a_2, b_1, b_2 are given deterministic functions.

The mean exit time $u(x, y)$ from a domain D satisfies

$$\frac{1}{2}b_1^2(x, y)u_{xx} + \frac{1}{2}b_2^2(x, y)u_{yy} + a_1(x, y)u_x + a_2(x, y)u_y = -1, \tag{5.79}$$

$$u|_{\partial D} = 0. \tag{5.80}$$

Especially, for the SDE system

$$dX_t = Y_t dt,$$
$$dY_t = -X_t dt + \sqrt{\epsilon}dB_t,$$

the mean exit time from the domain $D = \{(x, y) \colon x^2 + y^2 \leq 1\}$ satisfies

$$yu_x - xu_y + \frac{1}{2}\epsilon u_{yy} = -1$$
$$u \mid_{\partial D} = 0.$$

5.5 Escape Probability

Almost all sample paths of a Brownian motion are continuous in time. For a dynamical system driven by Brownian motion, almost all orbits (or paths or trajectories) are thus continuous in time. The *escape probability* is the likelihood that an orbit, starting inside a domain D, exits this domain first through a specific part Γ of the boundary ∂D. This concept helps understand various phenomena in complex systems. One example is in molecular genetics [260]. The frequency of collisions of two single strands of long helical DNA molecules that leads to a double-stranded molecule is of interest and can be computed by solving an escape probability problem. It turns out that the escape probability satisfies an elliptic partial differential equation with properly chosen boundary conditions [260, 180, 253, 40].

This section is adopted from our earlier works [40, 234].

More precisely, let X_t be the solution of the SDE system in $\mathbb{R}^n$

$$dX_t = b(X_t)dt + \sigma(X_t)dB_t, X_0 = x, \tag{5.81}$$

defined on a probability space $(\Omega, \mathcal{F}, \mathbb{P})$. Let D be an open domain in $\mathbb{R}^d$. Define the *first exit time*

$$\tau_{D^c} := \inf\{t > 0 : X_t \in D^c\}, \tag{5.82}$$

where D^c is the complement of D in $\mathbb{R}^d$. Namely, τ_{D^c} is the first time when X_t hits D^c.

When X_t has almost surely continuous paths, that is, X_t is a solution process for a dynamical system (5.81), a solution orbit starting at $x \in D$ will hit D^c by hitting ∂D first (assume for the moment that ∂D is smooth). Thus $\tau_{D^c} = \tau_{\partial D}$. Let Γ be a subset of the boundary ∂D. The likelihood that X_t, starting at x exits from D first through Γ is called the escape probability from D through Γ, denoted as $p(x)$. That is,

$$p(x) = \mathbb{P}\{X_{\tau_{\partial D}} \in \Gamma\}. \tag{5.83}$$

We will show that the escape probability $p(x)$ solves a linear elliptic partial differential equation, with a specifically chosen Dirichlet boundary condition.

The following theorem about the characterization of mean exit time holds for an annular domain D in Figure 5.7, with Γ either the inner boundary or the outer boundary.

Theorem 5.19 (Escape probability) *Assume that the generator for the stochastic system* (5.81) *is uniformly elliptic, that is, the condition* (5.4) *holds. Let D be an annular domain in Figure 5.7, with Γ either the inner boundary or the outer boundary. Then the escape probability $p(x)$ from the domain D through Γ, for the*

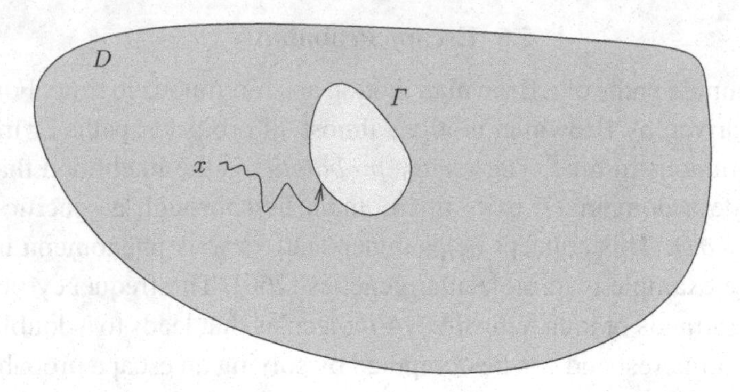

Figure 5.7 Escape probability from an annular domain D, together with a subset Γ of its boundary ∂D.

system (5.81)*, is the solution to the following Dirichlet boundary value problem:*

$$\begin{cases} Ap = 0, \\ p|_\Gamma = 1, \\ p|_{\partial D \setminus \Gamma} = 0, \end{cases}$$

where A is the generator

$$Ag = b \cdot \nabla g + \frac{1}{2} Tr[\sigma \sigma^{\mathrm{T}} H(g)]. \tag{5.84}$$

Moreover, if the domain D has $C^{2,\gamma}$ boundary for some $\gamma \in (0, 1)$, in the sense of Definition 2.4, and the drift b and diffusion σ are in $C^\gamma(\bar{D})$, then the escape probability p uniquely exists and is in $C^{2,\gamma}(\bar{D})$.

Proof Taking

$$\psi(x) = \begin{cases} 1, & x \in \Gamma, \\ 0, & x \in \partial D \setminus \Gamma, \end{cases}$$

we have

$$\begin{aligned} \mathbb{E}[\psi(X_{\tau_{\partial D}}(x))] &= \int_{\{\omega: X_{\tau_{\partial D}}(x) \in \Gamma\}} \psi(X_{\tau_{\partial D}}(x)) \mathrm{d}\mathbb{P}(\omega) \\ &\quad + \int_{\{\omega: X_{\tau_{\partial D}}(x) \in \partial D \setminus \Gamma\}} \psi(X_{\tau_{\partial D}}(x)) \mathrm{d}\mathbb{P}(\omega) \\ &= \mathbb{P}\{\omega : X_{\tau_{\partial D}}(x) \in \Gamma\} \\ &= p(x). \end{aligned}$$

This means that $\mathbb{E}[\psi(X_{\tau_{\partial D}}(x))]$ is the escape probability $p(x)$ that we are looking for.

By [213, Theorem 9.2.14] or [58], we know that the escape probability $p(x)$ is a harmonic function with respect to X_t. But this requires the boundary data ψ to be bounded and continuous on ∂D. Therefore the domain D is taken as an annular domain in Figure 5.7 with Γ either the inner boundary or outer boundary. Thus, $Ap = 0$ together with $p|_\Gamma = 1$ and $p|_{\partial D \setminus \Gamma} = 0$, by [213, Theorem 9.2.14] or [234].

For this annular domain, the boundary condition ψ is in $C^\gamma(\bar{D})$. Thus the existence, uniqueness, and regularity of the escape probability p follow from the result for classical solutions of linear elliptic partial differential equations in Theorem 6.8 in [105, p. 100]. □

Suppose that particles are initially distributed in D with probability density function $f(x)$. The average escape probability $\bar{P}$ that a particle will leave D through a part of the boundary Γ, before leaving through the rest of the boundary, is given by (e.g., [180, 253])

$$\bar{P} = \int_D p(x) f(x) dx. \tag{5.85}$$

In particular, if particles are initially uniformly distributed in D, that is,

$$f(x) = \begin{cases} \frac{1}{|D|}, & x \in D, \\ 0, & \text{otherwise}, \end{cases} \tag{5.86}$$

then the average escape probability $\bar{P}$ that a particle will leave D through Γ is

$$\bar{P} = \frac{1}{|D|} \int_D p(x) dx. \tag{5.87}$$

We now consider a few examples.

Example 5.20 In the one-dimensional case, take $D = (-r, r)$ and $\Gamma = \{r\}$. For each $x \in D$, the escape probability $p(x)$ for the solution paths of $dX_t = dB_t$ from D through Γ satisfies the following differential equation:

$$\begin{cases} \frac{1}{2} p''(x) = 0, & x \in (-r, r), \\ p(r) = 1, \\ p(-r) = 0. \end{cases}$$

We obtain that $p(x) = \frac{x+r}{2r}$ for $x \in [-r, r]$. It is a straight line (see Figure 5.8).

Example 5.21 In the two-dimensional case, take $D = \{x \in \mathbb{R}^2; r < |x| < R\}$ and $\Gamma = \{x \in \mathbb{R}^2; |x| = r\}$. For every $x \in D$, the escape probability $p(x)$ for the solution paths of $dX_t = dB_t$ from D through Γ satisfies the following elliptic partial

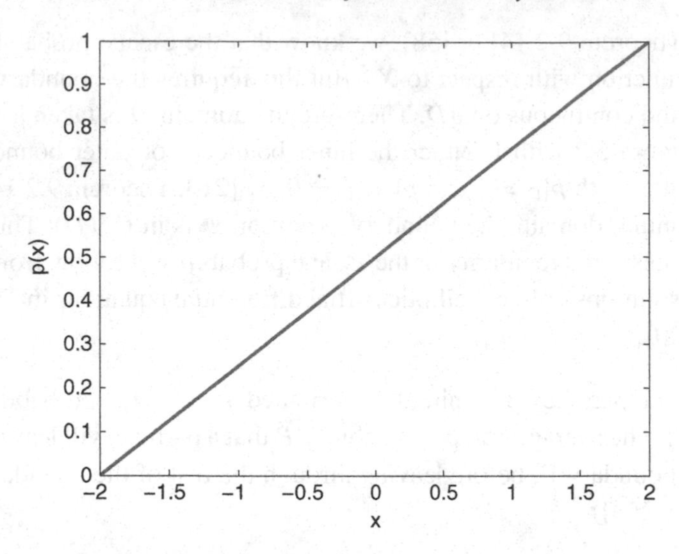

Figure 5.8 Escape probability for Example 5.20, $r = 2$.

differential equation:

$$\begin{cases} \frac{1}{2}\Delta p(x) = 0, & x \in D, \\ p(x)|_{|x|=r} = 1, \\ p(x)|_{|x|=R} = 0. \end{cases}$$

By solving this elliptic partial differential equation, we obtain that $p(x) = \frac{\log R - \log |x|}{\log R - \log r}$. It is plotted in Figure 5.9.

Example 5.22 Consider the following scalar SDE:

$$dX_t = b(X_t)dt + \sigma(X_t)dW_t,$$

where b and (nonzero) σ are real functions. We take $D = (-r, r)$ and $\Gamma = \{r\}$. For each $x \in D$, the escape probability $p(x)$ satisfies

$$\begin{cases} \frac{1}{2}\sigma^2(x)p''(x) + b(x)p'(x) = 0, & x \in (-r, r), \\ p(r) = 1, \\ p(-r) = 0. \end{cases}$$

The solution is

$$p(x) = \frac{\int_{-r}^{x} e^{-2\int_{-r}^{y} \frac{b(z)}{\sigma^2(z)} dz} dy}{\int_{-r}^{r} e^{-2\int_{-r}^{y} \frac{b(z)}{\sigma^2(z)} dz} dy},$$

for $x \in (-r, r)$. See Figure 5.10.

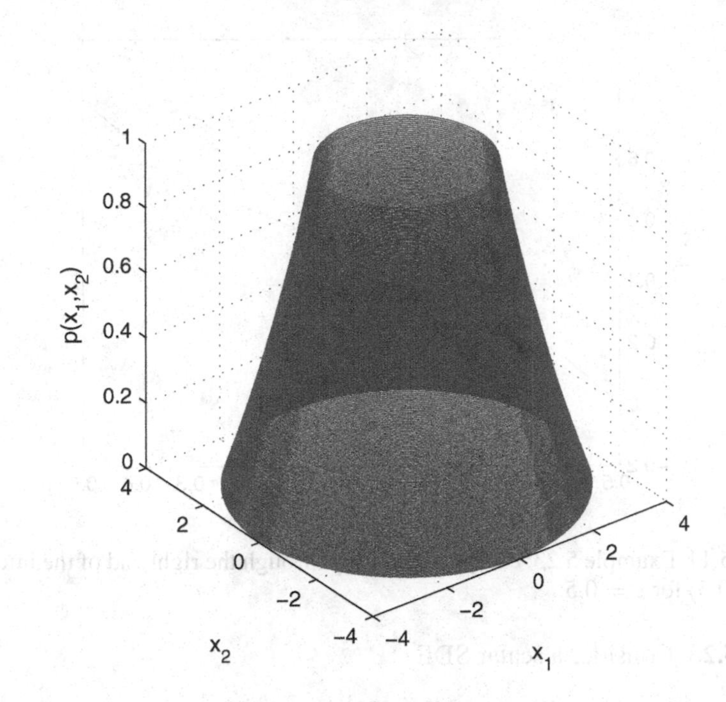

Figure 5.9 Escape probability for Example 5.21: $r = 2$, $R = 4$.

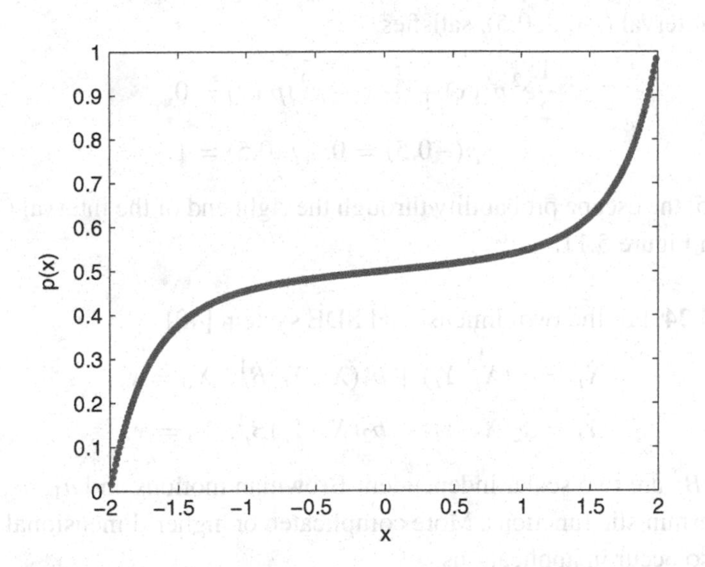

Figure 5.10 Escape probability in Example 5.22: $b(x) = -x$, $\sigma(x) = 1$, $r = 2$.

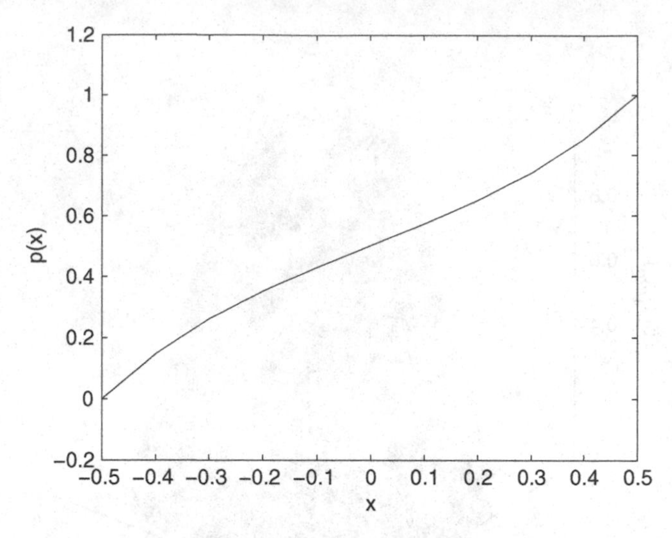

Figure 5.11 Example 5.23: escape probability through the right end of the interval $(-0.5, 0.5)$ for $\varepsilon = 0.5$.

Example 5.23 Consider a scalar SDE

$$dX_t = \big(-X_t + X_t^3\big)dt + \varepsilon dB_t,$$

with noise intensity $\varepsilon > 0$.

The escape probability $p(x)$, for an orbit starting at x to exit through the right end of the interval $(-0.5, 0.5)$, satisfies

$$\frac{1}{2}\varepsilon^2 p''(x) + (-x + x^3)p'(x) = 0, \tag{5.88}$$

$$p(-0.5) = 0, \quad p(0.5) = 1. \tag{5.89}$$

For $\varepsilon = 0.5$, the escape probability through the right end of the interval $(-0.5, 0.5)$ is shown in Figure 5.11.

Example 5.24 For the two-dimensional SDE system [40]

$$\dot{X}_t = a_1(X_t, Y_t) + b_1(X_t, Y_t)\dot{B}_t^1, \; X_0 = x, \tag{5.90}$$

$$\dot{Y}_t = a_2(X_t, Y_t) + b_2(X_t, Y_t)\dot{B}_t^2, \; Y_0 = y, \tag{5.91}$$

where B_t^1, B_t^2 are two scalar independent Brownian motions and a_1, a_2, b_1, b_2 are known deterministic functions. More complicated or higher-dimensional stochastic systems also occur in applications.

Let D be a planar annular bounded domain D with ∂D boundary composed of an inner boundary Γ_1 and an outer boundary Γ_2. The escape probability $p(x, y)$ for a particle or solution orbit starting at (x, y) in D to escape from D by first hitting the inner boundary Γ_1, before hitting the outer boundary Γ_2, satisfies

$$\frac{1}{2}b_1^2(x,y)p_{xx} + \frac{1}{2}b_2^2(x,y)p_{yy} + a_1(x,y)p_x + a_2(x,y)p_y = 0, \quad (5.92)$$

$$p|_{\Gamma_1} = 1, \quad (5.93)$$

$$p|_{\Gamma_2} = 0. \quad (5.94)$$

We need to numerically compute (e.g., in Matlab) escape probability p by solving this elliptic partial differential equation with zero Dirichlet boundary conditions.

If the initial particles are distributed over D with probability distribution function $f(x,y)$, the average escape probability $\bar{P}$ that a particle will leave D through the inner boundary Γ_1, before leaving through the outer boundary Γ_2, is given by (e.g., [180, 253])

$$\bar{P} = \iint_D p(x,y)f(x,y)dxdy. \quad (5.95)$$

5.6 Problems

5.1 Moment estimation for a Langevin equation
Consider the Langevin equation

$$dX_t = -bX_t dt + a dW_t,$$

where a, b are nonnegative parameters and the initial condition $X_0 \sim \mathbb{N}(0, \sigma^2)$. Find or estimate $\mathbb{E}X_t$, $\mathbb{E}X_t^2$ and $\mathbb{E}\sin(X_t)$. For this linear system, is the mean $\mathbb{E}X_t$ affected by noise? How about $\mathbb{E}X_t^2$ and $\mathbb{E}\sin(X_t)$?

Note: The solution is

$$X_t = e^{-bt}X_0 + ae^{-bt}\int_0^t e^{bs}dW_s.$$

But this problem does not require the use of this exact solution.

5.2 Stochastic Duffing–van der Pol system
Consider the stochastic system

$$dX_t = Y_t dt,$$
$$dY_t = (aX_t + bY_t - X_t^3 - X_t^2 Y_t)dt + c\, X_t\, dB_t,$$

where B_t is a standard scalar Brownian motion and a, b, c are real parameters.

(a) Apply Itô's formula in differential form to the energy $E(t) = \frac{1}{2}(X_t^2 + Y_t^2)$.
(b) Estimate the mean of the energy $E(t)$ in terms of parameters a, b, c.

Hint: You may need to use Young's inequality and Gronwall inequality.

5.3 Moment estimation for a double-well system
Consider a linear SDE $dX_t = (X_t - X_t^3)dt + \epsilon dB_t$ with initial data X_0. Here $\epsilon > 0$ is a parameter. Estimate the first and second moments $\mathbb{E}X_t$ and EX_t^2 for $t \in [0, T]$, with T a fixed positive constant. For this nonlinear system, is the mean $\mathbb{E}X_t$ affected by noise? How about $\mathbb{E}X_t^2$?

5.4 Fokker-Planck equation
Consider a nonlinear SDE

$$dX_t = \left(X_t - X_t^3\right)dt + \epsilon dB_t.$$

What is the associated Fokker-Planck equation? How does one compute the mean exit time of solution orbits from the interval $D = (-1, 1)$?

5.5 Consider a simple pendulum under small external random forcing $\ddot{X}_t + \sin(X_t) = \varepsilon dB_t$. It may be written as a system of two nonlinear SDEs

$$dX_t = Y_t dt, \quad dY_t = -\sin(X_t)dt + \varepsilon dB_t, \quad (X_0, Y_0) = (x, y),$$

with small parameter $0 < \varepsilon \ll 1$. What is the associated Fokker-Planck equation? How does one compute the mean exit time of solution orbits from a bounded domain D?

5.6 A Langevin equation and Ornstein-Uhlenbeck process
Consider the scalar Langevin equation $dX_t = -bX_t dt + a dB_t$, where a, b are nonnegative real parameters and the initial condition $X(t_0) = x_0 \in \mathbb{R}^1$.

(a) Write down the Fokker-Plack equation for $p(x, t)$ and impose initial condition $p(x, 0) = f(x) \in C_0^2(\mathbb{R})$. What does $p(t, x)$ mean?
(b) A special case: $b = 0, a = 1$ and $p(x, 0) = \delta(x - x^*)$ (with x^* a fixed real number). Here the Dirac δ function satisfies $\int \mathbb{R}^1 \delta(y - x^*)h(y)dy = h(x^*)$. What is the probability density $p(x, t)$ in this case (can you get it explicitly)?
By the way, what is the solution (and its probability density function) of the SDE $dX_t = dB_t, X_0 = x^*$? Do the results look familiar to you?

5.7 Reflecting boundary condition
Consider a scalar SDE $dX_t = b(X_t)dt + dB_t, \ X_0 = x \in (0, 1)$ with a constant b. The associated Fokker-Planck equation is

$$p_t = \frac{1}{2}p_{xx} - (b(x)p)_x.$$

We impose the reflecting boundary condition $b(x)p - \frac{1}{2}p_x = 0$ at $x = 0, 1$ and the initial condition $p(x, 0) = p_0(x)$. Discuss the existence and uniqueness of appropriate solutions.

5.8 Mean exit time for a Langevin equation
Consider the following Langevin equation:

$$dX_t = -bX_t dt + a dW_t,$$

where a, b are nonnegative parameters and the initial condition is X_0.

(i) How does one calculate the mean exit time of solution orbits from the interval $D = (-1, 1)$?
(ii) Is it more likely that the solution orbits to exit the interval $D = (-1, 1)$ from the left or from right?

5.9 Mean exit time and probability density evolution
Consider the following scalar stochastic differential equation:

$$dX_t = (\alpha X_t - X_t^2)dt + X_t dB_t,$$

where α is real parameter.

(a) Compute and plot the mean exit time $u(x)$ for x in the domain $D = (-1, 2)$ for various α values. What is the impact of α on the mean exit time?
(b) Compute and plot the probability density function $p(x, t)$ for the system starting at time zero with $X_0 \sim \mathcal{N}(0, 1)$, for $\alpha = 1$ at $t = 0.1, 0.2, \ldots$.
(c) Is it more likely that the solution orbits to exit the interval $D = (-1, 2)$ from the left or from the right? Answer this question by computing appropriate escape probabilities.

5.10 Escape probability and mean exit time
Consider the following system of stochastic differential equations:

$$dX_t = (9X_t - X_t^3)dt + \varepsilon dB_t^1,$$
$$dY_t = -Y_t dt + \varepsilon dB_t^2,$$

where B_t^1, B_t^2 are independent scalar Brownian motions and $\varepsilon \in (0, 1)$ is the positive noise intensity. For the corresponding deterministic dynamical system, $\mathcal{A} = \{(x, y): x \in [-3, 3],\ y \in [-3, 3]\}$ is an attracting set, as shown in Figure 5.12.

(a) Compute and plot the probability density function $p(x, y, t)$ for the system starting at time zero with (X_0, Y_0) in the uniform distribution on the domain $\bar{D} = (-0.1, 0.1) \times (-0.1, 0.1)$, for various ε at $t = 0.1, 0.8, \ldots$.
(b) Compute and plot the mean exit time $u(x, y)$ for (x, y) in the domain $D = (-0.75, 0.75) \times (-1, 1)$. What is the impact of ε on the mean exit time?

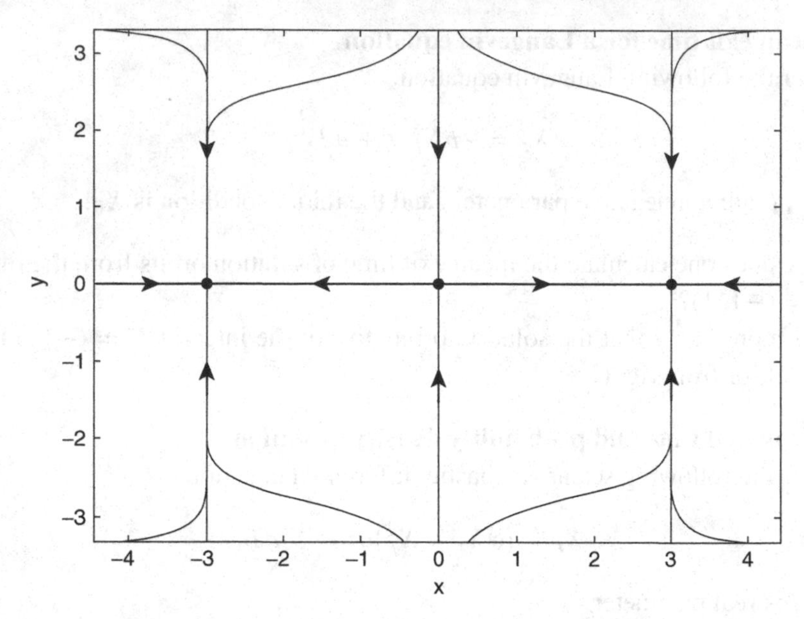

Figure 5.12 Phase portrait for the deterministic dynamical system $\dot{x} = 9x - x^3$, $\dot{y} = -y$.

(c) If the "particles" (X_t, Y_t) are initially distributed uniformly in a small domain $\bar{D} = (-0.1, 0.1) \times (-0.1, 0.1)$, what is the average probability that these particles will leave this domain through the left side boundary before any other side boundary?

6

Invariant Structures for Stochastic Dynamics

Invariant geometric structures, such as stable or unstable eigenspaces and stable or unstable manifolds, have played a significant role in our understanding of dynamics of linear and nonlinear deterministic dynamical systems, respectively. A foundation for the investigation of these geometric invariant structures is actually an analytical structure, the so-called flow property of ordinary differential equations ([226], or [60]). It is a property for the solution mappings of ordinary differential equations.

The stochastic flow property, or cocycle property, for SDEs considered in this book was discovered in the 1980s. The discovery of this stochastic analytical structure opened the door for dynamical systems methods for studying dynamical behavior of stochastic differential equations. Since the 1990s, the investigation of dynamical systems approaches for SDEs has emerged in various contexts, including slow and inertial manifolds, stable and unstable manifolds, stochastic bifurcation, random attractors, monotone random dynamical systems, and topological aspects of random dynamics.

In this chapter, we first review several basic concepts for deterministic dynamical systems and measurable dynamical systems, then introduce definitions for stochastic dynamical systems, including canonical sample spaces, cocycles, the multiplicative ergodic theorem, and a stochastic Hartman-Grobman theorem, and finally, we investigate invariant geometric structures called random invariant manifolds. Basic concepts in Sections 6.2, 6.3, and 6.4 for random dynamical systems are from Arnold [9]. In Arnold [9], an abstract stochastic flow or cocycle is called a random dynamical system (RDS). Invariant manifolds are considered in [285, 9], which, however, used cumbersome random norms that depend on the Osledets' structures for the linearized stochastic systems. Our approach in Section 6.5 here is a finite-dimensional adaption of our research on invariant manifolds of infinite dimensional random dynamical systems [79, 80, 47].

In this chapter, we assume that Brownian motion is two-sided (i.e., B_t, $t \in \mathbb{R}$) and that random dynamical systems are two-sided as well.

6.1 Deterministic Dynamical Systems

We now review some basic concepts and results for deterministic dynamical systems [110, 226].

6.1.1 Concepts for Deterministic Dynamical Systems

For a simple linear system $\dot{x} = -x$ with initial condition $x(0) = x_0$, the solution is $x(t) = x_0 e^{-t}$. Each solution is a curve, and it is also called an orbit or trajectory passing through the initial point x_0. The solution is also denoted by $\varphi(t, x_0) \triangleq x_0 e^{-t}$. We can think of $\varphi \colon \mathbb{R} \times \mathbb{R}^1 \to \mathbb{R}^1$ as a mapping. This solution mapping has properties $\varphi(0, x_0) = x_0$ and $\varphi(t + s, x_0) = \varphi(t, \varphi(s, x_0))$. It has an equilibrium state 0, which is an attractor, as all solution orbits (or trajectories) approach it as time $t \to \infty$.

For an n-dimensional linear system $\dot{x} = Ax$ with initial condition $x(0) = x_0$ in $\mathbb{R}^n$, the solution mapping is

$$\varphi(t, x_0) \triangleq e^{At} x_0,$$

where

$$e^{At} = I + \frac{At}{1!} + \frac{A^2 t^2}{2!} + \cdots + \frac{A^k t^k}{k!} + \cdots .$$

Note that the solution mapping φ has properties

$$\varphi(0, x_0) = x_0, \quad \varphi(t + s, x_0) = \varphi(t, \varphi(s, x_0)).$$

This latter one is called the flow property.

For a linear system, its dynamics is determined by eigenvalues and eigenspaces, including stable eigenspaces, unstable eigenspaces, and center eigenspaces.

Consider a nonlinear system of ordinary differential equations in state space $\mathbb{R}^n$:

$$\dot{x} = f(x), \quad x(0) = x_0, \tag{6.1}$$

where $f \colon \mathbb{R}^n \to \mathbb{R}^n$ is often called a vector field and x_0 is an initial point. The vector field f is called locally Lipschitz continuous or satisfies a local Lipschitz condition if, for every open set $U \subset \mathbb{R}^n$,

$$\|f(x) - f(y)\| \leq K \|x - y\|, \quad x, y \in U, \tag{6.2}$$

where the positive Lipschitz constant K depends on U. When f is C^1 smooth, which we will assume in this chapter, it is also locally Lipschitz continuous (by the mean value theorem). The well-known existence and uniqueness theorem

[226, Sections 2.2–2.4] states that the initial value problem (6.1) has a unique local solution $\varphi(t, x_0)$, defined on a maximal time interval $I(x_0)$ containing zero. This is called the maximal interval of existence for the initial point x_0.

For the unique solution $\varphi(t, x_0)$, we introduce a notation $\varphi_t(x_0) \triangleq \varphi(t, x_0)$ for $t \in I(x_0)$. Given a $t \in I(x_0)$, φ_t is a mapping in $\mathbb{R}^n$. For a given x_0, $\varphi_t(x_0)$ is the orbit passing through x_0. Occasionally, we also call either φ_t or φ the solution mapping for system (6.1). By a theorem in [226, Section 2.5], we conclude that if $s \in I(x_0)$ and $t \in I(\varphi_s(x_0))$, then $t + s \in I(x_0)$ and

$$\varphi_t(x_0) = x_0, \ \ \varphi_{t+s}(x_0) = \varphi_t\varphi_s(x_0). \tag{6.3}$$

Definition 6.1 (Dynamical system, or flow) A *dynamical system* or a *flow* on $\mathbb{R}^n$ is a mapping $\varphi\colon \mathbb{R} \times \mathbb{R}^n \to \mathbb{R}^n$ such that

$$\varphi(0, x) = x, \quad \varphi(t + s, x) = \varphi(t, \varphi(s, x)), \quad \text{for all } x \in \mathbb{R}^n, \quad \text{and } \ t \in \mathbb{R}.$$

With the shorthand notation $\varphi_t(x) \triangleq \varphi(x, t)$, the preceding properties become

$$\varphi_0 = \mathrm{Id}, \quad \varphi_{t+s} = \varphi_t\varphi_s,$$

where $\mathrm{Id}\colon \mathbb{R}^n \to \mathbb{R}^n$ is the identity mapping. When this flow is defined by (6.1), we also say (6.1) is a dynamical system.

We should remark that a dynamical system may be defined on a finite-dimensional manifold or a submanifold of Euclidean space $\mathbb{R}^n$. For a dynamical system defined by (6.1), the smoothness of the flow φ defined by its solution mapping depends on the smoothness of f.

We also need a concept to distinguish different dynamical systems. Recall that a homeomorphism is a continuous invertible mapping whose inverse is also continuous.

Definition 6.2 (Equivalence of two dynamical systems) Two dynamical systems φ_t^1 and φ_t^2 on $\mathbb{R}^n$ are topologically equivalent (or conjugate) if there exists a homeomorphism $H\colon \mathbb{R}^n \to \mathbb{R}^n$ such that $H \circ \varphi_t^1 = \varphi_t^2 \circ H$ for all t.

When φ_t^1 and φ_t^2 are topologically equivalent, the homeomorphism H maps orbits of φ_t^1 into orbits of φ_t^2, and vice versa. Thus, topologically equivalent dynamical systems have qualitatively same-phase portraits. Similarly, we can define equivalence of two dynamical systems on different state spaces.

For the dynamical system (6.1), if the solution exist globally (i.e., the maximal interval of existence $I(x_0)$ is the whole $\mathbb{R}$ for each initial point x_0), then it defines a dynamical system as we know from (6.3). If the solution does not exist globally at least for some initial point x_0, then with an appropriate time rescaling, the local solution $\varphi_t(x_0)$ can be defined for all time $t \in \mathbb{R}$. In fact, introducing a new time $\tau = \int_0^t [1 + \|f(x(s))\|]ds$, the original system (6.1) becomes

$$\frac{dx}{d\tau} = \frac{f(x)}{1 + \|f(x)\|}, \quad x(0) = x_0. \tag{6.4}$$

The new vector field $\frac{f(x)}{1+\|f(x)\|}$ is still C^1 (when f is C^1) but is now bounded. It can be shown that the solution, for each initial point x_0, is defined for all time [226]. More importantly, the new dynamical system (6.4) is topologically equivalent ([226, Section 3.1]) to the original system (6.1) on $\mathbb{R}^n$. Therefore, we usually assume that the flow φ_t for system (6.1) is defined for all $t \in \mathbb{R}$.

6.1.2 The Hartman-Grobman Theorem

For a nonlinear system in $\mathbb{R}^n$

$$\dot{x} = f(x), \tag{6.5}$$

a point x_*, in $\mathbb{R}^n$, is called an equilibrium point if $f(x_*) = 0$. The corresponding linearized system at this equilibrium point is $\dot{x} = Df(x_*)\, x, \;\; x \in \mathbb{R}^n$, where $Df(x_*)$ is the Jacobian matrix of f at x_*. If the eigenvalues of $Df(x_*)$ have nonzero real part, then x_* is called a hyperbolic equilibrium point. Otherwise, it is called a nonhyperbolic equilibrium point.

The Hartman-Grobman theorem says that near a hyperbolic equilibrium point x_*, the nonlinear system (6.5) is topologically equivalent to its linearized system at the equilibrium point

$$\dot{x} = Ax \tag{6.6}$$

with $A = Df(x_*)$. So these two systems have qualitatively the same dynamics (i.e., phase portrait) near x_*. The precise statement for this fact is in the following theorem ([59, Section 4.3] or [226, Section 2.8]).

Theorem 6.3 (The Hartman-Grobman theorem) *Let f be C^1 and let φ_t be the flow defined by the nonlinear system* (6.5). *Assume that x_* is a hyperbolic equilibrium point, that is, the Jacobian matrix $A = Df(x_*)$ has no eigenvalue of zero real part. Then there exists a homeomorphism H of an open set U containing x_* onto an open set V containing* 0 *(the equilibrium point of the linearized system* (6.6)*) such that for every $x \in U$, there is an open interval I_x containing zero so that*

$$H \circ \varphi_t(x) = e^{At} H(x), \quad t \in I_x.$$

Therefore, H maps orbits of the nonlinear system (6.5) to the orbits of the linearized system (6.6)) locally, and vice versa, and furthermore preserves the parameterization by time.

6.1.3 Invariant Sets

Invariant sets are essential building blocks for understanding dynamical systems.

Definition 6.4 (Invariant set) A set S in $\mathbb{R}^n$ is an invariant set for a dynamical system (or a flow) φ_t if $\varphi_t(x_0) \in S$, for all $x_0 \in S$ and for all $t \in \mathbb{R}$.

An equilibrium state is an invariant set. An orbit, especially a closed (or periodic) orbit, is an invariant set. A union of orbits is also an invariant set, and it becomes more useful in understanding dynamics when it has a differentiable structure (i.e., when it is a differentiable manifold). In fact, stable and unstable manifolds, which we will discuss later, are such invariant sets.

Let us introduce some concepts [110, 290] that quantify asymptotic behavior for a dynamical system φ_t in the state space $\mathbb{R}^n$.

Definition 6.5 (Attracting set) A set $S \subset \mathbb{R}^n$ is called an attracting set for a dynamical system φ_t if (i) it is a closed, nonempty, invariant set and (ii) there is an open neighborhood U of S such that "all orbits starting inside U will stay inside U and approach S as $t \to +\infty$," that is, $\varphi_t(x) \in U$ for $t \geq 0$ and $\varphi_t(x)$ approaches S (in Hausdorff distance) as $t \to +\infty$, for all $x \in U$.

The basin of attraction for the attracting set S is defined as the set $\bigcup_{t \leq 0} \varphi_t(U)$, that is, it is the union of all those points, so that the orbits starting from such points approach S as $t \to +\infty$.

An attracting set attracts nearby orbits. For example, a stable equilibrium state and a stable limit cycle are attracting sets.

Definition 6.6 (Attractor) An attractor S is an attracting set with a dense orbit.

Definition 6.7 (Sensitive dependence on initial condition) Let S be a bounded and closed invariant set for a dynamical system φ_t. This dynamical system is said to have sensitive dependence on initial conditions on S if there exists a positive number ε such that, for every $x \in S$ and its every neighborhood U, there exists $y \in U$ and $t > 0$ so that $|\varphi_t(x) - \varphi_t(y)| > \varepsilon$.

Thus orbits starting from nearby points in S could have very uncertain destination.

Definition 6.8 (Chaotic attractor and chaotic dynamical system) If a dynamical system φ_t has sensitive dependence on a bounded and closed attractor S, then

S is called a chaotic attractor (also called a strange attractor) and φ_t is called a chaotic dynamical system.

6.1.4 Differentiable Manifolds

We recall some geometric concepts for dynamical systems. For more details, see [36, 204, 1]. A *manifold* M is a set that locally looks like an Euclidean space. Namely, a "patch" of the manifold M looks like a "patch" in $\mathbb{R}^n$. For example, curves in state space $\mathbb{R}^n$ are one-dimensional differentiable manifolds, whereas tori and spheres are two-dimensional differentiable manifolds. A manifold arising from the study of invariant sets for dynamical systems in $\mathbb{R}^n$ can be very complicated. So we give a formal definition of manifolds. For more discussions on differentiable manifolds, see [1, 36]. Recall that a homeomorphism of A to B is continuous, one-to-one onto map $h\colon A \to B$, such that $h^{-1}\colon B \to A$ is also continuous. In this case, we say that A is homeomorphic to B, and vice versa.

Definition 6.9 (Differentiable manifold and Lipschitz manifold) An n-dimensional differentiable manifold M is a connected metric space with an open covering $\{U_\alpha\}$, that is, $M = \bigcup_\alpha U_\alpha$, such that

(i) for every α, U_α is homeomorphic to the open set in $\mathbb{R}^n$
(ii) if $U_\alpha \cap U_\beta \neq \varnothing$ and $h_\alpha\colon U_\alpha \to B$, $h_\beta\colon U_\beta \to B$ are homeomorphisms, then $h_\alpha(U_\alpha \cap U_\beta)$ and $h_\beta(U_\alpha \cap U_\beta)$ are subsets of $\mathbb{R}^n$ and the map

$$h = h_\alpha \circ h_\beta^{-1}\colon h_\beta(U_\alpha \cap U_\beta) \to h_\alpha(U_\alpha \cap U_\beta) \qquad (6.7)$$

is differentiable. Moreover, for every $x \in h_\beta(U_\alpha \cap U_\beta)$, the Jacobian determinant $\det Dh(x) \neq 0$.

When the map (6.7) is C^k with $k \geq 1$ or C^∞, M is called an n-dimensional C^k or C^∞ manifold, respectively. If the map (6.7) is only Lispchitz continuous, then we call M an n-dimensional Lipschitz manifold.

Sometimes, each pair (U_α, h_α) is called a chart, and a collection of charts is called an atlas. The atlas covers the manifold.

Example 6.10 Euclidean space $\mathbb{R}^n$ is an n-dimensional differentiable manifold with a single chart $U_1 = \mathbb{R}^n$ and $h_1 = I$, the identity mapping on $\mathbb{R}^n$.

Example 6.11 The unit circle $M = \{(x, y)\colon x^2 + y^2 = 1\}$ is a one-dimensional differentiable manifold in $\mathbb{R}^2$ with the following atlas of four charts:

U_1 is the upper semicicle; h_1 projects U_1 vertically to the open interval $(-1, 1)$ in $\mathbb{R}^1$

U_2 is the lower semicicle; h_2 projects U_2 vertically to the open interval $(-1, 1)$ in $\mathbb{R}^1$
U_3 is the left semicicle; h_3 projects U_3 horizontally to the open interval $(-1, 1)$ in $\mathbb{R}^1$
U_4 is the right semicicle; h_4 projects U_4 horizontally to the open interval $(-1, 1)$ in $\mathbb{R}^1$

Example 6.12 Define $M \triangleq \{(x, f(x)) \colon x \in \mathbb{R}^1\}$, where f is differentiable. So M is a differentiable curve in $\mathbb{R}^2$. In fact, M is a one-dimensional differentiable manifold with a single chart, $U_1 = M$, and h_1 maps each point $(x, f(x))$ on the curve to a point x in $\mathbb{R}^1$.

More generally, let γ be a Lipschitz continuous or C^k mapping

$$\gamma \colon H^+ \to H^-,$$

where $H^+ \oplus H^- = \mathbb{R}^n$ and $\dim H^+ = m$, and let M be the graph of γ

$$M = \{(x^+, \gamma(x^+)),\ x^+ \in H^+\}.$$

Then M is an m-dimensional Lipschitz or C^k manifold.

Remark 6.13 In the preceding definition and examples, differentiable manifolds are modeled by Euclidean space $\mathbb{R}^n$, that is, these manifolds locally look like a piece of Euclidean space. In fact, a differentiable manifold can also be modeled by a Banach space or a Hilbert space (see definition in [1]), and in this context we usually do not specify the dimension of the manifold, as the dimension of a general Banach space or Hilbert space is infinite.

Let M be a differentiable manifold. A subset S of M is called a submanifold if S itself is a differentiable manifold.

Definition 6.14 (Invariant manifold) If an invariant set M for a dynamical system can be represented as the graph of a Lipschitz continuous mapping

$$\gamma^*(\cdot) \colon H^+ \to H^-, \text{ with direct sum decomposition } \quad H^+ \oplus H^- = \mathbb{R}^n,$$

that is,

$$M = \{(x^+, \gamma^*(x^+)), x^+ \in H^+\},$$

then M is called a Lipschitz invariant manifold. If the mapping γ^* is smooth (e.g., C^k for some positive integer k), we call M a smooth invariant manifold.

In the context of direct sums as introduced in linear algebra, a point in the invariant manifold M would be written as $x^+ + \gamma^*(x^+)$. But in Euclidean space, we usually write it as $(x^+, \gamma^*(x^+))$.

6.1.5 Deterministic Invariant Manifolds

Consider a linear system in $\mathbb{R}^n$

$$\dot{x} = Ax, \quad x(0) = x_0. \tag{6.8}$$

The stable eigenspace E^s is generated by eigenvectors corresponding to eigenvalues with negative real part, the unstable eigenspace E^u is generated by eigenvectors corresponding to eigenvalues with positive real part, while center eigenspace E^c is generated by eigenvectors corresponding to eigenvalues with zero real part (generalized eigenvectors are needed when the geometric multiplicity differs from the algebraic multiplicity for an eigenvalue). These three eigenspaces are invariant subspaces in $\mathbb{R}^n$.

Example 6.15 We consider a linear system

$$\dot{x} = -x,$$

$$\dot{y} = y.$$

The origin (0, 0) is a hyperbolic equilibrium point, with eigenvalues $\lambda_1 = -1$ and $\lambda_2 = 1$, with the corresponding eigenvectors $v_1 = (1, 0)^{\mathrm{T}}$ and $v_2 = (0, 1)^{\mathrm{T}}$, respectively. The stable eigenspace $E^s(0, 0)$ is the x-axis and the unstable eigenspace $E^u(0, 0)$ is the y-axis. See Figure 6.1.

Note that $E^s(0, 0)$ is the linear subspace spanned by the eigenvector v_1 (corresponding to the negative eigenvalue -1), whereas $E^u(0, 0)$ is the linear subspace spanned by the eigenvector v_2 (corresponding to the positive eigenvalue 1).

For nonlinear systems, there are no eigenspaces, but there will be invariant manifolds, i.e., stable manifold W^s, unstable manifold W^u, and center manifold W^c, which may be intuitively regarded as curved versions of stable, unstable, and center eigenspaces (under suitable conditions), respectively. Let us consider a motivating example.

Example 6.16 Consider a nonlinear system

$$\dot{x} = -x - y^2,$$

$$\dot{y} = y + x^2.$$

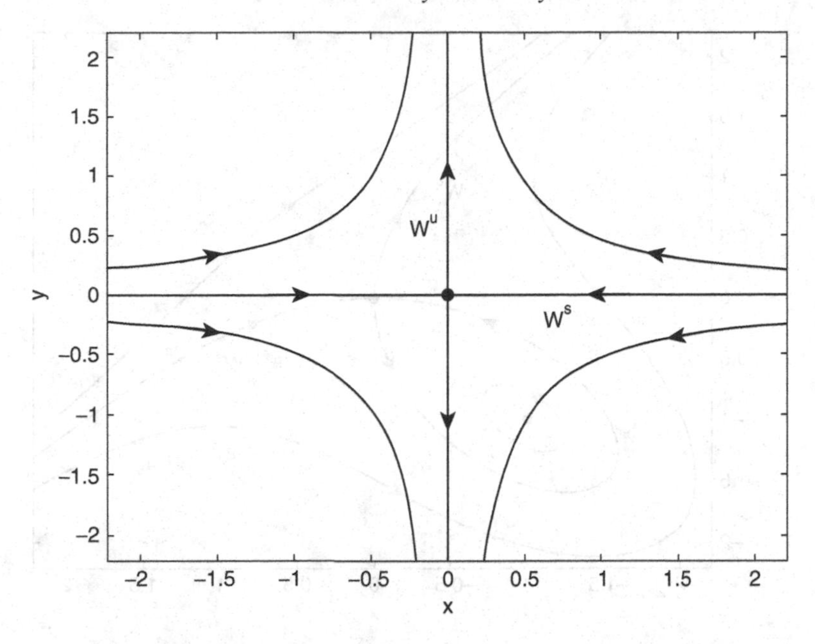

Figure 6.1 Phase portrait for $\dot{x} = -x, \;\; \dot{y} = y$.

This nonlinear system is obtained by perturbing the linear system in Example 6.15 by the nonlinear function $(-y^2, x^2)^{\mathrm{T}}$. We may guess that the stable manifold $W^s(0, 0)$ is a "curved x-axis," whereas the unstable manifold $W^u(0, 0)$ is a "curved y-axis." See Figure 6.2 (generated by Matlab).

Note that $W^s(0, 0)$ is tangent to E^s and $W^u(0, 0)$ is tangent to E^u, both at the equilibrium point (0, 0).

Consider a nonlinear system in $\mathbb{R}^n$

$$\dot{x} = f(x), \;\; x \in \mathbb{R}^n. \tag{6.9}$$

The linearized system at an equilibrium point x_* is

$$\dot{x} = Df(x_*)\, x, \;\; x \in \mathbb{R}^n. \tag{6.10}$$

Let φ_t be the flow associated or generated by the nonlinear system (6.9). The local stable and unstable manifolds of an equilibrium point x_*, in a neighborhood U of x_*, are defined by

$$W^s_{\mathrm{loc}}(x_*) = \left\{ x \in U : \lim_{t \to +\infty} \varphi_t(x) = x_* \right\},$$

$$W^u_{\mathrm{loc}}(x_*) = \left\{ x \in U : \lim_{t \to -\infty} \varphi_t(x) = x_* \right\}.$$

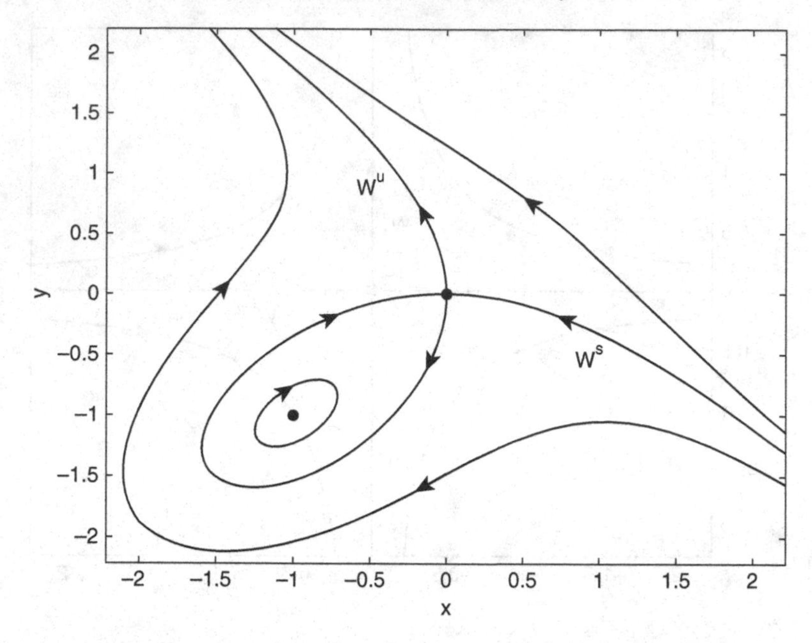

Figure 6.2 Phase portrait for $\dot{x} = -x - y^2$, $\dot{y} = y + x^2$.

We define the global stable manifold W^s and unstable manifold W^u by

$$W^s(x_*) = \bigcup_{t\leq 0} \varphi_t\left(W^s_{\text{loc}}\right), \tag{6.11}$$

$$W^u(x_*) = \bigcup_{t\geq 0} \varphi_t\left(W^u_{\text{loc}}\right). \tag{6.12}$$

A global invariant manifold may not be representable by the graph of a mapping in state space $\mathbb{R}^n$.

Example 6.17 Consider a nonlinear system in $\mathbb{R}^2$

$$\begin{cases} \dot{x} = -x, & x(0) = x_0, \\ \dot{y} = y + x^2, & y(0) = y_0. \end{cases} \tag{6.13}$$

We consider the stable and unstable manifolds for the equilibrium state at the origin, $W^s(0,0)$ and $W^u(0,0)$. The solution mapping, or flow, is $\varphi(t, x_0, y_0) = (x_0e^{-t}, y_0e^t + \frac{1}{3}x_0^2(e^t - e^{-2t}))^{\mathrm{T}}$. Recall that for $(x_0, y_0)^{\mathrm{T}} \in W^s(0,0)$, $\varphi(t, x_0, y_0) \to (0,0)^{\mathrm{T}}$ as $t \to +\infty$, while for $(x_0, y_0)^{\mathrm{T}} \in W^u(0,0)$, $\varphi(t, x_0, y_0) \to (0,0)^{\mathrm{T}}$ as $t \to -\infty$. Thus

$$W^s(0,0) = \left\{(x_0, y_0)^{\mathrm{T}} : \ y_0 = -\frac{1}{3}x_0^2\right\},$$

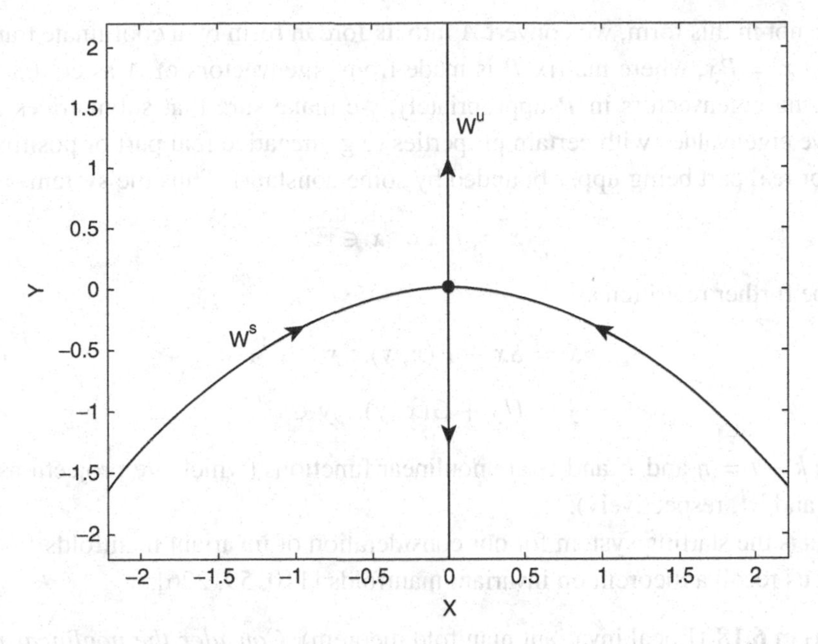

Figure 6.3 Stable and unstable manifolds for $\dot{x} = -x$, $\dot{y} = y + x^2$.

and

$$W^u(0,0) = \{(x_0, y_0)^{\mathrm{T}} : x_0 = 0\}.$$

The stable and unstable manifolds are shown in Figure 6.3.

Pseudo-Stable and Pseudo-Unstable Manifolds

It is convenient to assume that the equilibrium state x_* is 0, that is, $f(0) = 0$. We make a coordinate translation $x \to x - x_*$, if necessary. Using Taylor expansion,

$$\begin{aligned} f(x) &= f(0) + Df(0)x + \text{higher order terms} \\ &= Df(0)x + \text{higher order terms}, \end{aligned}$$

the system $\dot{x} = f(x)$ is then written as

$$\dot{x} = Ax + \tilde{f}(x), \quad x \in \mathbb{R}^n, \tag{6.14}$$

where $A = Df(0)$ and $\tilde{f}(x)$ denotes the (higher-order) nonlinear terms, that is, $\tilde{f}(0) = 0$ and $D\tilde{f}(0) = 0$. We assume that the matrix A is in the following block diagonal form:

$$A = \begin{pmatrix} S & \mathbf{0} \\ \mathbf{0} & U \end{pmatrix}.$$

If A is not in this form, we convert A into its Jordan form by a coordinate transformation $x = Py$, where matrix P is made from eigenvectors of A as columns. By arranging eigenvectors in P appropriately, we make sure that submatrices S and U have eigenvalues with certain properties (e.g., negative real part or positive real part, or real part being upper bounded by some constant). Thus the system

$$\dot{x} = f(x), \quad x \in \mathbb{R}^n$$

may be further rewritten as

$$\dot{x} = Sx + F(x, y), \quad x \in \mathbb{R}^k, \tag{6.15}$$

$$\dot{y} = Uy + G(x, y), \quad y \in \mathbb{R}^l, \tag{6.16}$$

where $k + l = n$ and F and G are nonlinear functions (which are projections of f to $\mathbb{R}^k$ and $\mathbb{R}^l$, respectively).

This is the starting system for our consideration of invariant manifolds.

Let us recall a theorem on invariant manifolds [110, 59, 226].

Theorem 6.18 (Local invariant manifold theorem) *Consider the nonlinear dynamical system*

$$\dot{x} = Sx + F(x, y), \quad x \in \mathbb{R}^k, \tag{6.17}$$

$$\dot{y} = Uy + G(x, y), \quad y \in \mathbb{R}^l. \tag{6.18}$$

Suppose that a and b are real numbers with $a < b$, S is a $k \times k$ matrix with each eigenvalue's real part less than a, and U is an $l \times l$ matrix with each eigenvalue's real part greater than b. If F and G are both C^1 such that $F(0, 0) = G(0, 0) = 0$ and $DF(0, 0) = DG(0, 0) = 0$, then

(i) there exists an open neigborhood Γ_s of 0 in $\mathbb{R}^k$ and a unique C^1 function $\alpha \colon \Gamma_s \subset \mathbb{R}^k \to \mathbb{R}^l$ such that $\alpha(0) = 0$, $D\alpha(0) = 0$ and its graph $W^s_{\text{loc}}(0, 0) = \{(x, y) \in \mathbb{R}^k \times \mathbb{R}^l \colon y = \alpha(x)\}$ is a local invariant manifold;

(ii) there also exists an open neighborhood Γ_u of 0 in $\mathbb{R}^l$ and a unique C^1 function $\beta \colon \Gamma_u \subset \mathbb{R}^l \to \mathbb{R}^k$ such that $\beta(0) = 0$, $D\beta(0) = 0$ and its graph $W^u_{\text{loc}}(0, 0) = \{(x, y) \in \mathbb{R}^k \times \mathbb{R}^l \colon x = \beta(y)\}$ is a local invariant manifold.

Proof We only outline main ideas for the proof. For more details, see [59, Chapter 4], [49, Chapter 3], or [226, Chapter 3].

Under the conditions on the eigenvalues for the matrices S and U, there exists a positive constant K such that for all $t \geq 0$,

$$\|e^{tS}x\| \leq Ke^{at}\|x\|, \quad x \in \mathbb{R}^k, \tag{6.19}$$

$$\|e^{-tU}y\| \leq Ke^{-bt}\|y\|, \quad y \in \mathbb{R}^l. \tag{6.20}$$

This is the so-called pseudo-exponential dichotomy property for the linear system $\dot{x} = Sx,\ \dot{y} = Uy$. When $a < 0$ and $b > 0$ (then, of course, $a < b$), this property is called exponential dichotomy.

Let $(x(t), y(t))$ be on the invariant manifold. Then it satisfies the equations

$$\dot{x} = Sx + F(x, y),\ x \in \mathbb{R}^k,$$
$$\dot{y} = Uy + G(x, y),\ y \in \mathbb{R}^l,$$

with initial conditions $x(0, \xi, \alpha) = \xi,\ y(0, \xi, \alpha) = \alpha(\xi)$. (Strictly speaking, we do not have α yet. It will be determined via a fixed point argument in the later part of this proof.)

Rewriting the two ordinary differential equations in integral forms, we obtain

$$x(t, \xi, \alpha) = e^{tS}\xi + \int_0^t e^{(t-\tau)S} F(x(\tau, \xi, \alpha), \alpha(x(\tau, \xi, \alpha)))d\tau,$$
$$e^{-tU} y(t, \xi, \alpha) - \alpha(\xi) = \int_0^t e^{-\tau U} G(x(\tau, \xi, \alpha), \alpha(x(\tau, \xi, \alpha)))d\tau.$$

Using the assumptions on the matrix U, that is, (6.20), and the nonlinear term $F(x, y)$ and $G(x, y)$, we further have (see [59, Section 4.1])

$$\lim_{t\to\infty} \|e^{-tU} y(t, \xi, \alpha)\| = 0. \tag{6.21}$$

Thus

$$\alpha(\xi) = -\int_0^\infty e^{-\tau U} G(x(\tau, \xi, \alpha), \alpha(x(\tau, \xi, \alpha)))d\tau,$$

which leads to the Lyapunov-Perron operator:

$$\Lambda(\alpha)(\xi) \triangleq -\int_0^\infty e^{-tU} G(x(t, \xi, \alpha), \alpha(x(t, \xi, \alpha)))dt. \tag{6.22}$$

We now only need to show that this Lyapunov-Perron operator Λ has a unique fixed point $\alpha(\xi)$ in an appropriate function space. The graph of $\alpha(\xi)$ will then provide us a local invariant manifold. For more details on choosing a Banach space and applying the contraction mapping principle in this space, see [59, Section 4.1]. □

Remark 6.19 In this theorem, we do not specify the signs of a and b (i.e., we do not assume that a is negative and b is positive). So W^s_{loc} may be called a pseudo-stable manifold and W^u_{loc} a pseudo-unstable manifold. We call $b - a$ the spectral gap. This theorem also says that the pseudo-stable manifold W^s_{loc} is tangent to the pseudo-stable eigenspace $\mathbb{R}^k$, and that the pseudo-unstable manifold W^u_{loc} is tangent to the pseudo-unstable eigenspace $\mathbb{R}^l$, at the equilibrium point $(0, 0)$.

Solutions on Stable and Unstable Manifolds

When $a < 0 < b$, W^s_{loc} and W^u_{loc} are the usual local stable and unstable manifolds. We now formulate the equations to be satisfied by the solutions on the stable and unstable manifolds, respectively.

The Lyapunov-Perron method used to prove the previous theorem does not tell us about what solution orbits look like on stable or unstable manifolds. Let us find a reformulation for these solution orbits for the following system:

$$\dot{x} = Sx + F(x, y), \ x \in \mathbb{R}^k, \tag{6.23}$$

$$\dot{y} = Uy + G(x, y), \ y \in \mathbb{R}^l, \tag{6.24}$$

where S is a $k \times k$ matrix whose eigenvalues have negative real part and U is an $l \times l$ matrix whose eigenvalues have positive real part ($k + l = n$). The stable and unstable eigenspaces are $E^s = \mathbb{R}^k$ and $E^u = \mathbb{R}^l$, respectively. Denote the orthogonal projection into E^s by P^s and into E^u by P^u.

Let us introduce some notations:

$$u = \begin{pmatrix} x \\ y \end{pmatrix}, \tilde{S} = \begin{pmatrix} S & \mathbf{0} \\ \mathbf{0} & \mathbf{0} \end{pmatrix}, \quad \text{and} \quad \tilde{U} = \begin{pmatrix} \mathbf{0} & \mathbf{0} \\ \mathbf{0} & U \end{pmatrix}.$$

The equations (6.23) and (6.24) can be reformulated as

$$u(t) = e^{(\tilde{S}+\tilde{U})(t-t_0)}u(t_0) + \int_{t_0}^{t} e^{(\tilde{S}+\tilde{U})(t-s)} \begin{pmatrix} F(u(s)) \\ G(u(s)) \end{pmatrix} ds. \tag{6.25}$$

First, let us look at solutions on the stable manifold $W^s(0, 0)$, with initial condition at $t_0 = 0$, $u(0) = a$. The stable component is

$$\begin{pmatrix} x(t) \\ 0 \end{pmatrix} = e^{\tilde{S}t}a + \int_0^t e^{\tilde{S}(t-s)} \begin{pmatrix} F(u(s)) \\ 0 \end{pmatrix} ds. \tag{6.26}$$

For the unstable component, take $t_0 \to \infty$ to get

$$\begin{pmatrix} 0 \\ y(t) \end{pmatrix} = \mathbf{0} + \int_\infty^t e^{\tilde{U}(t-s)} \begin{pmatrix} 0 \\ G(u(s)) \end{pmatrix} ds. \tag{6.27}$$

Combining (6.26) and (6.27), we see that solutions in the stable manifold satisfy the following equation:

$$u(t) = e^{\tilde{S}t}a + \int_0^t e^{\tilde{S}(t-s)} \begin{pmatrix} F(u(s)) \\ 0 \end{pmatrix} ds + \int_\infty^t e^{\tilde{U}(t-s)} \begin{pmatrix} 0 \\ G(u(s)) \end{pmatrix} ds. \tag{6.28}$$

Then, let us look at solutions on the unstable manifold $W^u(0, 0)$, with initial condition at $t_0 = 0$, $u(0) = a$. From the reformulated system (6.25), the unstable component is

$$\begin{pmatrix} 0 \\ y(t) \end{pmatrix} = e^{\tilde{U}t}a + \int_0^t e^{\tilde{U}(t-s)} \begin{pmatrix} 0 \\ G(u(s)) \end{pmatrix} ds. \tag{6.29}$$

For the stable component, take $t_0 \to -\infty$ to get

$$\begin{pmatrix} x(t) \\ 0 \end{pmatrix} = 0 + \int_{-\infty}^t e^{\tilde{S}(t-s)} \begin{pmatrix} F(u(s)) \\ 0 \end{pmatrix} ds. \tag{6.30}$$

Combining (6.29) and (6.30), we see that solutions on the unstable manifold satisfy the following equation:

$$u(t) = e^{\tilde{U}t}a + \int_0^t e^{\tilde{U}(t-s)} \begin{pmatrix} 0 \\ G(u(s)) \end{pmatrix} ds + \int_{-\infty}^t e^{\tilde{S}(t-s)} \begin{pmatrix} F(u(s)) \\ 0 \end{pmatrix} ds. \tag{6.31}$$

Stable, Unstable, and Center Manifolds

When the equilibrium point 0 (taking a coordinate translation if necessary) for $\dot{x} = f(x)$ in $\mathbb{R}^n$ is not hyperbolic, the matrix $Df(0)$ has k eigenvalues with negative real part, l eigenvalues with positive real part, m eigenvalues with zero real part, and $k + l + m = n$.

With an appropriate coordinate transformation, the center eigenspace, stable eigenspace, and unstable eigenspace may be assumed to be $E^c = \mathbb{R}^m$, $E^s = \mathbb{R}^k$, and $E^u = \mathbb{R}^l$, respectively. Taking $a > 0$ small enough and applying Theorem 6.18, we get a center-stable manifold W^{cs}. Moreover, taking $b < 0$ small enough, Theorem 6.18 produces a center-unstable manifold W^{cu}. The intersection of these two manifolds can be shown to be a center manifold and is denoted by W^c; see [59, Section 4.1]. At least locally, W^c is the graph of a function $h : \Gamma_c \subset E^c \to E^s \times E^u$ with $h(0) = 0$ and $Dh(0) = 0$.

Theorem 6.18 is for general invariant manifolds. It implies the following theorem specifically for stable, unstable, and center manifolds [226]. The stable, unstable, and center manifolds are the nonlinear counterparts of stable eigenspaces, unstable eigenspaces, and center eigenspaces, respectively.

Theorem 6.20 (Local stable, unstable, and center manifold theorem) *Consider an n-dimensional dynamical system*

$$\dot{x} = f(x),$$

where f is C^r smooth ($r \geq 1$) in a domain containing 0 in $\mathbb{R}^n$. Suppose that 0 is an equilibrium point ($f(0) = 0$) and that $Df(0)$ has k eigenvalues with negative real part, j eigenvalues with positive real part, and $m = n - k - j$ eigenvalues with zero real part. Then,

(i) there exists an m-dimensional C^r local center manifold $W^c(0)$ tangent to the center eigenspace E^c of the linearized system $\dot{x} = Df(0)x$ at 0;
(ii) there exists a k-dimensional C^r local stable manifold $W^s(0)$ tangent to the stable eigenspace E^s of the linearized system $\dot{x} = Df(0)x$ at 0;
(iii) there exists a j-dimensional C^r local unstable manifold $W^u(0)$ tangent to the unstable eigenspace E^u of the linearized system $\dot{x} = Df(0)x$ at 0.

Moreover, $W^c(0)$, $W^s(0)$, and $W^u(0)$ are invariant under the nonlinear flow φ_t. The stable and unstable manifolds are unique, but W^c need not be.

Remark 6.21 To get an unstable manifold, we reverse the time of the original system $t \to -t$. The stable manifold for the new system is the unstable manifold for the original system.

Simulating Nonlinear Dynamics

Matlab codes are widely available for solving and visualizing solution orbits of deterministic dynamical systems. The software package that I use in generating some figures in this book is developed by John C. Polking at Rice University. It has three Matlab codes: `dfield.m` for one-dimensional systems, `pplane.m` for two-dimensional systems, and `odesolve.n` for high-dimensional systems. The `pplane.m` has more functionalities, and it plots stable and unstable manifolds, when they exist. Search for "ODE Software" at http://www.rice.edu for more information.

6.2 Measurable Dynamical Systems

This section is adopted from [9, Appendix A]. Let $(\Omega, \mathcal{F}, \mathbb{P})$ be a probability space. A family of mappings (also called transformations) on the sample space, $\theta_t : \Omega \to \Omega$, $t \in \mathbb{R}$, is called a measurable dynamical system (or a measurable flow) if the following conditions are satisfied:

(i) identity property: $\theta_0 = \mathrm{Id}$;
(ii) flow property: $\theta_{t+s} = \theta_t \circ \theta_s$, for all $t, s \in \mathbb{R}$;
(iii) measurability: $(\omega, t) \to \theta_t\omega$ is measurable.
It is called a measure-preserving dynamical system if, furthermore,
(iv) measure-preserving property: $\mathbb{P}(\theta_t A) = \mathbb{P}(A)$, for every $A \in \mathcal{F}$ and $t \in \mathbb{R}$
In this case, $\mathbb{P}$ is called an invariant measure with respect to the dynamical system θ_t.

Remark 6.22 For every fixed $t \in \mathbb{R}$, a new probability measure $\theta_t\mathbb{P}$ on Ω is defined by $\theta_t\mathbb{P}(A) = \mathbb{P}(\theta_t^{-1}(A))$ for $A \in \mathcal{F}$. The mapping θ_t is measure-preserving if $\theta_t\mathbb{P} = \mathbb{P}$. Because $\theta_t^{-1} = \theta_{-t}$, the measure-preserving property is really $\mathbb{P}(\theta_t A) = \mathbb{P}(A)$. In this case, the probability measure $\mathbb{P}$ is called an *invariant measure* for the dynamical system θ_t.

A measurable function $f: \Omega \to \mathbb{R}^1$ is invariant with respect to a dynamical system θ_t if $f(\theta_t\omega) = f(\omega)$ for all $t \in \mathbb{R}$ and all $\omega \in \Omega$. A set A in Ω is invariant if $\theta_t^{-1}(A) = A$ for all $t \in \mathbb{R}$. Because $\theta_t^{-1} = \theta_{-t}$, this latter definition is just $\theta_t(A) = A$ for all $t \in \mathbb{R}$. The family of all measurable invariant sets forms a sub$-\sigma$-field $\mathcal{I} \subset \mathcal{F}$.

We review a few more concepts ([286] or [70, Chapter 1]). A measure-preserving dynamical system θ_t is called ergodic if

$$\lim_{T\to\infty} \frac{1}{T} \int_0^T \mathbb{P}(\theta_{-t}A \cap B)dt = \mathbb{P}(A)\mathbb{P}(B), \quad \text{for all } A, B \in \mathcal{F}. \tag{6.32}$$

In fact, a measure-preserving dynamical system θ_t is ergodic [70, Theorem 1.2.4] if all invariant sets have probability either 1 or 0. Equivalently, this means that every invariant function is constant, except on a zero probability set, that is, $\mathbb{P}(N) = 0$ for the exceptional set N. In this case, the invariant probability measure $\mathbb{P}$ is called an *ergodic measure*.

A measure-preserving dynamical system θ_t is called strongly mixing if

$$\lim_{t\to\infty} \mathbb{P}(\theta_{-t}A \cap B) = \mathbb{P}(A)\mathbb{P}(B), \quad \text{for all } A, B \in \mathcal{F}. \tag{6.33}$$

Clearly a strongly mixing system is also ergodic.

We state an ergodic theorem [9, p. 538] for a measurable dynamical system θ_t on a probability space $(\Omega, \mathcal{F}, \mathbb{P})$. Let $f: \Omega \to \mathbb{R}^1$ be a measurable function (think of it as an "observable"). We introduce the "average," $\bar{f}$, of f as

$$\bar{f}(\omega) = \lim_{n\to\infty} \frac{1}{n} \int_0^n f(\theta_t\omega)dt = \lim_{n\to\infty} \frac{1}{n} \int_{-n}^0 f(\theta_t\omega)dt \tag{6.34}$$

whenever both limits exist and are equal for some $\omega \in \Omega$. For a general measurable dynamical system θ_t, this average may not exist for all sample $\omega \in \Omega$. Let us quantify the set of all such samples. Define

$$\begin{aligned}\Omega_f(\omega) = \Bigg\{ \omega\colon\ & f(\theta_t\omega) \text{ is locally integrable in } t, \\ & \lim_{n\to\pm\infty} \frac{1}{n} \int_0^1 |f(\theta_{n+t}\omega)|dt = 0, \text{ and } \lim_{n\to\infty} \frac{1}{n} \int_0^n f(\theta_t\omega)dt \\ & = \lim_{n\to\infty} \frac{1}{n} \int_{-n}^0 f(\theta_t\omega)dt = \bar{f}(\omega) \text{ exist} \Bigg\}.\end{aligned} \tag{6.35}$$

Theorem 6.23 *Let θ_t be a measurable dynamical system on a probability space $(\Omega, \mathcal{F}, \mathbb{P})$ and $f: \Omega \to \mathbb{R}^1$ be a measurable function. Then Ω_f is an invariant set, $\bar{f}$ is an invariant function, and for all $\omega \in \Omega_f$,*

$$\lim_{t\to\infty} \frac{1}{t} \int_0^t f(\theta_s\omega)ds = \lim_{t\to\infty} \frac{1}{t} \int_{-t}^0 f(\theta_s\omega)ds = \bar{f}(\omega). \tag{6.36}$$

Corollary 6.24 (Birkhoff-Khintchine theorem) *When θ_t is additionally measure preserving (i.e., $\mathbb{P}$ is an invariant measure), then*

(i) $\mathbb{P}(\Omega_f) = 1$

(ii) if $f \in L^p(\Omega)$ for some $1 \le p < \infty$, then $\bar{f} \in L^p(\Omega)$, and the preceding convergence to $\bar{f}$ also holds in $L^p(\Omega)$

6.3 Random Dynamical Systems

In this section, we review basic concepts for "random flows" (cocycles) or random dynamical systems, following [9, Chapter 1]. First, we introduce canonical sample spaces for SDEs and the Wiener shift defined on such a sample space. Then, we define random dynamical systems, as generated by solution mappings of SDEs. Finally, we discuss structural stability and stationary orbits for random dynamical systems.

6.3.1 Canonical Sample Spaces for SDEs

Consider an SDE system in $\mathbb{R}^n$

$$dX_t = b(X_t)dt + \sigma(X_t)dB_t. \tag{6.37}$$

In the definition of Brownian motion B_t, the probability space $(\Omega, \mathcal{F}, \mathbb{P})$ is arbitrary. Now we introduce a specific or natural probability space to facilitate dynamical systems research. We will call it canonical probability space for this SDE system.

We may think the scalar Brownian motion $B_t(\omega)$ as a random variable taking values in $C(\mathbb{R}, \mathbb{R}^1)$, that is, a measurable mapping

$$B: \Omega \to C(\mathbb{R}, \mathbb{R}^1), \qquad \omega \to \omega(t), t \in \mathbb{R}, \tag{6.38}$$

where $C(\mathbb{R}, \mathbb{R}^1)$ is the space of all continuous, real-valued functions $\omega(t)$ on $\mathbb{R}$ with metric

$$\rho(\omega_1, \omega_2) = \sum_{n=1}^{\infty} \frac{1}{2^n} \frac{\max_{-n\le t\le n} |\omega_1(t) - \omega_2(t)|}{1 + \max_{-n\le t\le n} |\omega_1(t) - \omega_2(t)|}. \tag{6.39}$$

This metric induces the topology of uniform convergence on compact time intervals.

This space is a natural or canonical sample space for Brownian motion. It is similar to the situation of dice tossing, where we take six face values, 1, 2, 3, 4, 5, and 6, as samples in the *canonical* sample space $\Omega = \{1, 2, 3, 4, 5, 6\}$. When we "toss" a Brownian motion B_t, we see continuous (but nowhere differentiable) curves as "face values," or samples.

Given this metric, the σ-field on $C(\mathbb{R}, \mathbb{R}^1)$ is generated by all Borel sets of $C(\mathbb{R}, \mathbb{R}^1)$. We denote it by $\mathcal{B}(C(\mathbb{R}, \mathbb{R}^1))$. Thus, the random variable B induces a probability measure (i.e., the law of B) on $(C(\mathbb{R}, \mathbb{R}^1), \mathcal{B}(C(\mathbb{R}, \mathbb{R}^1)))$, and we denote it as P_B. This measure is called the Wiener measure.

Let us observe [237] that for each t, $B_t : \Omega \to \mathbb{R}^1$ is a scalar random variable with distribution determined by $\mathbb{P}(B_{t_1} \in A_1, \ldots, B_{t_k} \in A_k)$, for all $t_1, \ldots, t_k$ in $\mathbb{R}$ and all Borel sets $A_1, \ldots, A_k$ in $\mathbb{R}^1$. Conversely, we can define a new stochastic process $\tilde{B}_t$ on $C(\mathbb{R}, \mathbb{R}^1)$. Namely, for each t, we define $\tilde{B}_t$ by

$$\tilde{B}_t : C(\mathbb{R}, \mathbb{R}^1) \to \mathbb{R}^1, \quad \tilde{B}_t(\omega) = \omega(t). \tag{6.40}$$

Note that

$$\mathbb{P}(B_{t_1} \in A_1, \ldots, B_{t_k} \in A_k) = \mathbb{P}_B(\tilde{B}_{t_1} \in A_1, \ldots, \tilde{B}_{t_k} \in A_k). \tag{6.41}$$

This says that the two stochastic processes B_t and $\tilde{B}_t$, although defined on *different* probability spaces and both taking real values, actually have the *same distribution.* We call $\tilde{B}_t$ a version of B_t, and vice versa. Thus we regard $\tilde{B}_t$ as a scalar Brownian motion on the probability space $(C(\mathbb{R}, \mathbb{R}^1), \mathcal{B}(C(\mathbb{R}, \mathbb{R}^1)), \mathbb{P}_B)$. In fact, we call $\tilde{B}_t$ a canonical version of Brownian motion B_t.

Similarly, for a Brownian motion B_t in $\mathbb{R}^n$, we think of it as a random variable taking values in $C(\mathbb{R}, \mathbb{R}^n)$, with the Borel σ-field $\mathcal{B}(C(\mathbb{R}, \mathbb{R}^n))$ and Wiener measure P_B as earlier. That is, the Brownian motion is a measurable mapping

$$B : \Omega \to C(\mathbb{R}, \mathbb{R}^n), \qquad \omega \to \omega(t), t \in \mathbb{R}. \tag{6.42}$$

We also define its canonical version $\tilde{B}_t$ on $C(\mathbb{R}, \mathbb{R}^n)$:

$$\tilde{B}_t : C(\mathbb{R}, \mathbb{R}^n) \to \mathbb{R}^n, \quad \tilde{B}_t(\omega) = \omega(t). \tag{6.43}$$

See Figure 6.4.

From now on, $(C(\mathbb{R}, \mathbb{R}^n), \mathcal{B}(C(\mathbb{R}, \mathbb{R}^n)), \mathbb{P}_B)$ is called the *canonical probability space* for Brownian motion B_t or for an SDE system (6.37) driven by B_t. For simplicity, we still denote this canonical space by our original notation $(\Omega, \mathcal{F}, \mathbb{P})$, and we identify B_t and $\tilde{B}_t$. Note that

$$\tilde{B}_t(\omega) = \omega(t), \quad \omega \in \Omega, \quad t \in \mathbb{R}. \tag{6.44}$$

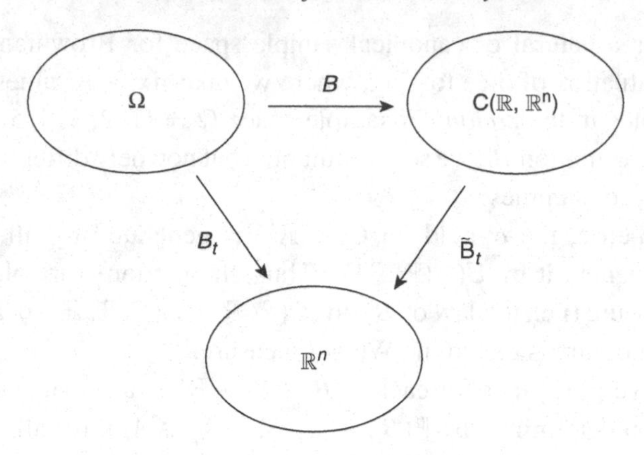

Figure 6.4 A canonical version of Brownian motion B_t.

There is another way to visualize elements in $\mathcal{B}(C(\mathbb{R}, \mathbb{R}^n))$, in terms of *cylinder sets*. A *cylinder set* of $C(\mathbb{R}, \mathbb{R}^n)$ is a subset of the form

$$A = \{\omega : \omega(t_1) \in B_1, \ldots \omega(t_n) \in B_n\},$$

where t_is are time instants with $-\infty < t_1 < \cdots < t_n < \infty$ and B_is are Borel sets of $\mathbb{R}^n$. It is known [147, Chapter 2] that the smallest σ-field generated by all cylinder sets equals $\mathcal{B}(C(\mathbb{R}, \mathbb{R}^n))$.

6.3.2 Wiener Shift

From the previous subsection, the *canonical probability space* for an SDE system (6.37) in $\mathbb{R}^n$ is $(\Omega, \mathcal{F}, \mathbb{P}) = (C(\mathbb{R}, \mathbb{R}^n), \mathcal{B}(C(\mathbb{R}, \mathbb{R}^n)), \mathbb{P}_B)$. The canonical sample space is now very concrete, as the samples are curves (i.e., paths of Brownian motion). To understand better this new sample space and its samples, we could do some differentiation, as we often do in analysis. But unfortunately, Brownian paths are not differentiable. So, instead, we examine differences or increments such as $\omega(t+s) - \omega(t)$. This leads to an important concept, the Wiener shift, in the canonical sample space. We now introduce an interesting "flow," the Wiener shift, in the canonical sample space. The Wiener shift enriches the structures of this canonical probability space but will also play a significant role in describing cocycle properties for SDEs.

The Wiener shift θ_t is defined as a mapping in the canonical sample space Ω, for each fixed $t \in \mathbb{R}$,

$$\theta_t : \Omega \to \Omega$$

$$\omega \mapsto \breve{\omega} \quad \text{such that} \quad \theta_t\omega(s) = \breve{\omega}(s) \triangleq \omega(t+s) - \omega(t), \quad s, t \in \mathbb{R}. \qquad (6.45)$$

By a simple calculation, we see that $\theta_0 = \mathrm{Id}$ (the identity mapping in Ω) and $\theta_{s+t} = \theta_s\theta_t$. Moreover, $(\omega, t) \to \theta_t\omega$ is continuous and therefore measurable. Hence, the Wiener shift is a measurable dynamical system (or a flow) in Ω. The equation (6.45) means that

$$B_s(\theta_t\omega) = B_{t+s}(\omega) - B_t(\omega). \tag{6.46}$$

When s is infinitesimally small, the right-hand side is $dB_t(\omega)$. Thus, θ_t is closely related to the noise in the stochastic system (6.37) and is often called the *driving flow*.

In fact, θ_t is a homeomorphism for each t and $(t, \omega) \rightarrowtail \theta(t)\omega$ is continuous, hence measurable. The Wiener measure $\mathbb{P}$ is invariant and ergodic [9, Appendix A] under the Wiener shift θ_t.

In summary, θ_t has the following properties:

- $\theta_0 = \mathrm{Id}$,
- $\theta_t\theta_s = \theta_{t+s}$, for all $s, t \in \mathbb{R}$,
- The mapping $(t, \omega) \mapsto \theta_t\omega$ is measurable and $\theta_t\mathbb{P} = \mathbb{P}$, for all $t \in \mathbb{R}$.

That is, the Wiener shift $\theta_t \colon \Omega \to \Omega$ is a measurable, measure preserving, and ergodic dynamical system. It is closely related to the Brownian motion B_t or the noise "dB_t" by its definition. We see in the next subsection that it is embedded in the solution mappings for SDEs and gives them the so-called cocycle property.

6.3.3 Cocycles and Random Dynamical Systems

This subsection is adopted from [9, Chapter 2].

We first recall an important concept. As defined in Section 2.8, a filtration is an increasing family of information accumulations, called σ-fields, $\mathcal{F}_t$. For each t, σ-field $\mathcal{F}_t$ is a collection of events in sample space Ω. One might observe the Brownian motion B_t over time t and use $\mathcal{F}_t$ to represent the information accumulated up to and including time t. More formally, on $(\Omega, \mathcal{F})$, a filtration is a family of σ-fields $\mathcal{F}_s \colon 0 \le s \le t$ with $\mathcal{F}_s$ contained in $\mathcal{F}$ for each s and $\mathcal{F}_{t_1} \subset \mathcal{F}_{t_2}$ for $t_1 \le t_2$.

To understand an SDE system (6.37) from a dynamical systems point of view, the natural filtration is defined as a two-parameter family of σ-fields generated by increments of B_t:

$$\mathcal{F}_s^t = \sigma(B(\tau_1) - B(\tau_2) \colon s \le \tau_1, \tau_2 \le t), \quad s, t \in \mathbb{R}.$$

This represents the information accumulated from time s up to and including time t. This two-parameter filtration allows us to define forward as well as backward stochastic integrals, and thus we can solve a stochastic differential equation from an initial time forward as well as backward in time. For the rest of this chapter, we assume that this procedure has been taken care of and omit it here. For more details, see [9, Chapter 2].

A continuous random dynamical system (RDS) in the state space X, with the time set $\mathbb{T}$ (in this book, X is usually $\mathbb{R}^n$, and the time set $\mathbb{T}$ is either the real number set $\mathbb{R}$ or natural number set $\mathbb{N}$) and the underlying probability space $(\Omega, \mathcal{F}, \mathbb{P})$, consists of two ingredients [9, Chapter 1]:

(i) a model of the noise, namely, a driving flow $(\theta_t)_{t\in\mathbb{T}}$ on the sample space Ω, such that $(t, \omega) \mapsto \theta_t\omega$ is a measurable flow that leaves $\mathbb{P}$ invariant, that is, $\theta_t\mathbb{P} = \mathbb{P}$ for all $t \in \mathbb{T}$.

(ii) a model of the system evolution influenced by noise, namely, a cocycle φ over θ_t, that is, a measurable mapping $\varphi\colon \mathbb{T} \times \Omega \times X \to X$, $(t, \omega, x) \mapsto \varphi(t, \omega, x)$, such that $(t, x) \mapsto \varphi(t, \omega, x)$ is continuous for all $\omega \in \Omega$ and the family $\varphi(t, \omega, \cdot) = \varphi(t, \omega)\colon X \to X$ of random mappings has the cocycle property

$$\varphi(0, \omega) = Id_X, \tag{6.47}$$

$$\varphi(t + s, \omega) = \varphi(t, \theta_s\omega) \circ \varphi(s, \omega), \tag{6.48}$$

for all $t, s \in \mathbb{T}$, $x \in X$, and $\omega \in \Omega$. Here Id_X is the identity mapping on the state space X. We usually say φ is a RDS, over θ_t, or often simply say φ is a cocycle. The cocycle property (6.48) is also written as follows:

$$\varphi(t + s, \omega, x) = \varphi(t, \theta_s\omega, \varphi(s, \omega, x)), \tag{6.49}$$

for all $t, s \in \mathbb{T}$, $x \in X$, and $\omega \in \Omega$. The cocycle property is shown in Figure 6.5.

The cocycle property is a stochastic flow property for SDEs considered in this book. It is a generalization of the well-known flow property for ordinary differential equations.

In fact, the cocycle defined here is a perfect cocycle, because the relation (6.48) holds identically for all t and all ω. If it only holds almost surely for fixed s and/or fixed t, where the exceptional set depends on s and/or t, then it is called a crude cocycle. However, under quite general conditions for SDEs, it is possible to make a crude cocycle a prefect cocycle [9, Section 1.3], that is, to construct a perfect cocycle ψ that is indistinguishable from the original crude cocycle φ. Two cocycles φ and ψ are indistinguishable if $\{\omega\colon \varphi(t, \omega) \neq \psi(t, \omega) \text{ for some } t \in \mathbb{T}\}$ has probability zero. We only deal with the perfected cocycles, assuming that the perfection procedure has been conducted.

It follows from (6.48) that $\varphi(t, \omega)$ is a homeomorphism of X and

$$\varphi(t, \omega)^{-1} = \varphi(-t, \theta_t\omega).$$

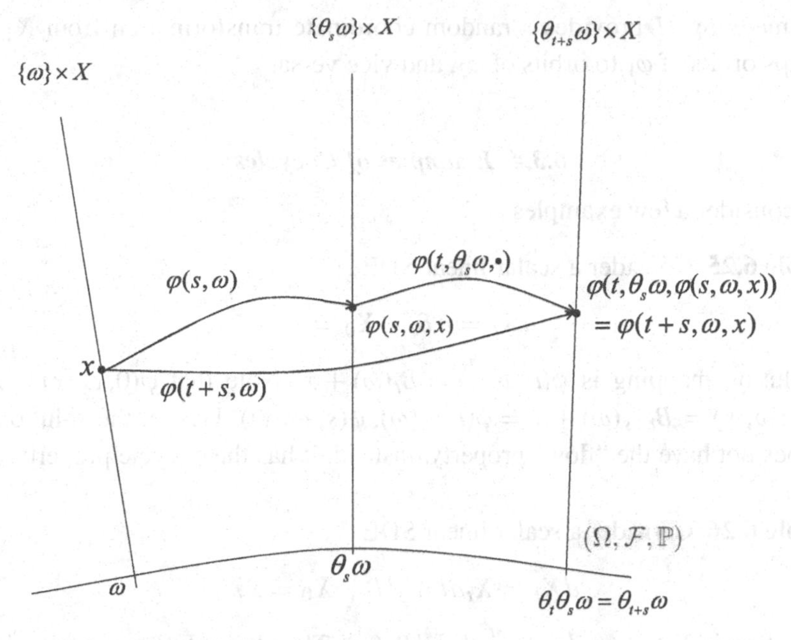

Figure 6.5 Visualizing the cocycle property.

When φ is a discrete random dynamical system and φ is the time-one map of φ, that is, $\varphi(\omega) = \varphi(1, \omega)\colon X \to X$, then we call φ the random homeomorphism determined by φ. Conversely, if φ is a random homeomorphism, then it generates via iteration a discrete RDS $\varphi(n, \omega, x)$. So we identify the discrete RDS φ as a random homeomorphism.

The concept of topological equivalence (or conjugacy) of RDS is adapted from the deterministic case as in Definition 6.2. We follow [9, p. 369] for the following definition.

Let φ_1 and φ_2 be two RDS over the same driving flow θ, but with state spaces X_1 and X_2, respectively. The RDS φ_1 and φ_2 are said to be *topologically equivalent* or conjugate if there exists a mapping $H\colon \Omega \times X_1 \to X_2$ with the following properties:

(i) The mapping $x \to H(\omega, x)$ is a homeomorphism from X_1 onto X_2 for every $\omega \in \Omega$.
(ii) Both mappings $\omega \to H(\omega, x_1)$ and $\omega \to H^{-1}(\omega, x_2)$ are measurable for $x_1 \in X_1$ and $x_2 \in X_2$.
(iii) The cocycles φ_1 and φ_2 satisfy the following condition:

$$\varphi_2(t, \omega, H(\omega, x)) = H(\theta_t\omega, \varphi_1(t, \omega, x)) \text{ for } x \in X_1 \text{ and } \omega \in \Omega. \quad (6.50)$$

The mapping H provides a random coordinate transformation from X_1 to X_2 and maps orbits of φ_1 to orbits of φ_2, and vice versa.

6.3.4 Examples of Cocycles

Let us consider a few examples.

Example 6.25 Consider a scalar linear SDE

$$dX_t = dB_t, \quad X_0 = x. \tag{6.51}$$

The solution mapping is $\varphi(t, \omega, x) = B_t(\omega) + x$. Note that $\varphi(0, \omega, x) = x$, and $\varphi(t + s, \omega, x) = B_{t+s}(\omega) + x = \varphi(t, \theta_s(\omega), \varphi(s, \omega, x))$. Hence, the solution mapping does not have the "flow" property. Instead, it has the cocycle property.

Example 6.26 Consider a scalar linear SDE

$$dX_t = X_t dt + dB_t, \; X_0 = x. \tag{6.52}$$

The solution is $X_t(\omega) = e^t x + \int_0^t e^{t-\tau} dB_\tau(\omega)$. Thus the solution mapping is

$$\varphi(t, \omega, x) \triangleq e^t x + \int_0^t e^{t-\tau} dB_\tau(\omega), \; x \in \mathbb{R}^1, \; t \in \mathbb{R}.$$

Note that

$$\varphi(0, \omega, x) = x. \tag{6.53}$$

Now let us show that

$$\varphi(t + s, \omega, x) = \varphi(t, \theta_s\omega, \varphi(s, \omega, x)). \tag{6.54}$$

Indeed, on one hand,

$$\varphi(t + s, \omega, x) = e^{t+s} x + \int_0^{t+s} e^{t+s-\tau} dB_\tau(\omega).$$

On the other hand,

$$\begin{aligned}
\varphi(t, \theta_s\omega, \varphi(s, \omega, x)) &= e^t \varphi(s, \omega, x) + \int_0^t e^{t-\tau} dB_\tau(\theta_s\omega) \\
&= e^t \left[e^s x + \int_0^s e^{s-\tau} dB_\tau \right] + \int_0^t e^{t-\tau} dB_\tau(\theta_s\omega).
\end{aligned}$$

Hence, we only need to prove the following *claim*:

$$\int_0^t e^{t-\tau} dB_\tau(\theta_s\omega) = \int_s^{t+s} e^{t+s-\tau} dB_\tau(\omega). \tag{6.55}$$

To this end, we prove that both sides of (6.55) are equal, for each fixed t and s. On $[0, t]$, consider a sequence of partitions with the maximal subinterval length δ^n:

$$0 = \tau_0^n < \cdots < \tau_j^n < \tau_{j+1}^n < \cdots < \tau_n^n = t.$$

On $[s, t+s]$, this sequence of partitions become

$$s = s + \tau_0^n < \cdots < s + \tau_j^n < s + \tau_{j+1}^n < \cdots < s + \tau_n^n = t + s.$$

Noticing that $dB_\tau(\theta_s\omega) = d(B_{s+\tau} - B_s)$, the left-hand side is

$$\lim_{\delta^n \to 0} \text{ in m.s.} \sum_j e^{t-\tau_j^n} \left(B_{s+\tau_{j+1}^n} - B_{s+\tau_j^n}\right), \tag{6.56}$$

and the right-hand side is

$$\begin{aligned} &\lim_{\delta^n \to 0} \text{ in m.s.} \sum_j e^{t+s-(s+\tau_j^n)} \left(B_{s+\tau_{j+1}^n} - B_{s+\tau_j^n}\right) \\ &= \lim_{\delta^n \to 0} \text{ in m.s.} \sum_j e^{t-\tau_j^n} \left(B_{s+\tau_{j+1}^n} - B_{s+\tau_j^n}\right). \end{aligned} \tag{6.57}$$

Hence, the claim (6.55) is proved. Therefore, the solution mapping $\varphi(t, \omega, x)$ defines a crude cocycle

$$\varphi(t + s, \omega, x) = \varphi(t, \theta_s\omega, \varphi(s, \omega, x)). \tag{6.58}$$

With a perfection procedure in [9, Section 1.3], we get a perfect cocycle.

Example 6.27 Consider the following linear scalar SDE with multiplicative noise:

$$dX_t = rX_t dt + \alpha X_t dB_t, \quad X_0 = x, \tag{6.59}$$

where r and α are real constants. By Example 4.20, the solution mapping is

$$\varphi(t, \omega, x) = x \exp\left(\left(r - \frac{1}{2}\alpha^2\right) t + \alpha B_t(\omega)\right).$$

Clearly $\varphi(0, \omega, x) = x$. Furthermore,

$$\varphi(t + s, \omega, x) = x \exp\left(\left(r - \frac{1}{2}\alpha^2\right)(t + s) + \alpha B_{t+s}(\omega)\right),$$

and

$$\begin{aligned}
&\varphi(t, \theta_s\omega, \varphi(s, \omega, x)) \\
&= \varphi(s, \omega, x) \exp\left(\left(r - \frac{1}{2}\alpha^2\right) t + \alpha B_t(\theta_s\omega)\right) \\
&= x \exp\left(\left(r - \frac{1}{2}\alpha^2\right) s + \alpha B_s(\omega)\right) \exp\left(\left(r - \frac{1}{2}\alpha^2\right) t + \alpha B_t(\theta_s\omega)\right) \\
&= x \exp\left(\left(r - \frac{1}{2}\alpha^2\right)(t+s) + \alpha B_{t+s}(\omega)\right),
\end{aligned}$$

where we have used the fact that $B_t(\theta_s\omega) = B_{t+s}(\omega) - B_s(\omega)$. Hence $\varphi(t + s, \omega, x) = \varphi(t, \theta_s\omega, \varphi(s, \omega, x))$; that is, $\varphi(t, \omega, x)$ is a cocycle.

More generally, consider an SDE system in $\mathbb{R}^n$

$$dX_t = b(X_t)dt + \sigma(X_t)dB_t, \; X_0 = x, \tag{6.60}$$

where B_t is a two-sided Brownian motion. We denote the solution as $\varphi(t, \omega, x)$. For each fixed t, ω, it is a mapping from $\mathbb{R}^n \to \mathbb{R}^n$, and it is also called the solution mapping. Under very general conditions, this SDE generates a random dynamical system [9, Chapter 2]. To this end, we need to introduce backward stochastic integration, and we refer the reader to [9, Section 2.3]. However, we summarize the result in the following.

For a natural number k and a real number $\delta \in [0, 1]$, the Banach space $C_b^{m,\delta}$ has already been introduced in Section 4.6. We present the following theorem about generation of cocycles for nonlinear SDE systems [9, Section 2.3] or [163].

Theorem 6.28 *Assume that $b \in C_b^{k,\delta}$, $\sigma \in C_b^{k+1,\delta}$, and $\sum_{i=1}^n b_i \frac{\partial b_i}{\partial x_i} \in C_b^{k,\delta}$ for some natural number $k \geq 1$ and $\delta > 0$. Then the solution mapping $\varphi(t, x, \omega)$ for $dX_t = b(X_t)dt + \sigma(X_t)dB_t$, $X_0 = x \in \mathbb{R}^n$, generates a unique global C^k cocycle over the Wiener shift θ_t.*

6.3.5 Structural Stability and Stationary Orbits

This section is adopted from our earlier works [79, 80].

A random dynamical system $\varphi(t, \omega)$ generated by an SDE is called structurally stable if, for any sufficiently small perturbation of the SDE, the perturbed random dynamical system $\Phi(t, \omega)$ is topologically equivalent to $\varphi(t, \omega)$. Namely, there exists a random homeomorphism $H(\omega)$ so that $\Phi(t, \omega) \circ H(\omega) = H(\theta_t\omega) \circ \varphi(t, \omega)$.

Definition 6.29 (Stationary orbit) A random variable $y(\omega)$ is called a stationary orbit (or random fixed point) for a random dynamical system φ if

$$\varphi(t, \omega, y(\omega)) = y(\theta_t \omega), \text{ a.s., for all } t.$$

Let us consider some examples.

Example 6.30 Stationary orbit for a Langevin equation
Consider an SDE

$$dX_t = -X_t dt + dB_t, \quad X_0 = x. \tag{6.61}$$

This SDE defines a random dynamical system

$$\varphi(t, \omega, x) = e^{-t} x + \int_0^t e^{-(t-s)} dB_s. \tag{6.62}$$

By examining the relation $\varphi(t, \omega, y(\omega)) = y(\theta_t \omega)$ in this specific case, a stationary orbit of this random dynamical system is guessed out to be

$$y(\omega) = \int_{-\infty}^0 e^s dB_s(\omega). \tag{6.63}$$

Indeed, it follows from (6.62) and (6.63) that

$$\begin{aligned} \varphi(t, \omega, y(\omega)) &= e^{-t} y(\omega) + \int_0^t e^{-(t-s)} dB_s(\omega) \\ &= e^{-t} \int_{-\infty}^0 e^s dB_s(\omega) + \int_0^t e^{-(t-s)} dB_s(\omega) \\ &= \int_{-\infty}^0 e^{-(t-s)} dB_s(\omega) + \int_0^t e^{-(t-s)} dB_s(\omega) \\ &= \int_{-\infty}^t e^{-(t-s)} dB_s(\omega). \end{aligned} \tag{6.64}$$

By (6.63), we also see that

$$\begin{aligned} y(\theta_t \omega) &= \int_{-\infty}^0 e^s dB_s(\theta_t \omega) \\ &= \int_{-\infty}^0 e^s dB_{s+t}(\omega) \\ &= \int_{-\infty}^t e^{-(t-s)} dB_s(\omega). \end{aligned} \tag{6.65}$$

Thus $\varphi(t, \omega, y(\omega)) = y(\theta_t \omega)$; that is, $y(\omega) = \int_{-\infty}^0 e^s dB_s(\omega)$ is a stationary orbit for the random dynamical system (6.61).

Example 6.31 Stationary orbit for a nonlinear SDE
Consider the stochastic equation

$$dX_t = \left(\frac{3}{2}X_t - X_t^3\right) dt + X_t dB_t, \ X_0 = x. \tag{6.66}$$

The random dynamical system generated by this SDE is

$$\varphi(t, \omega, x) = \frac{xe^{t+B_t(\omega)}}{(1 + 2x^2 \int_0^t e^{2(s+B_s(\omega))} ds)^{\frac{1}{2}}}. \tag{6.67}$$

By examining the relation $\varphi(t, \omega, y(\omega)) = y(\theta_t \omega)$ in this specific case, it appears that a stationary orbit of this random dynamical system is

$$y(\omega) = \left(\int_{-\infty}^{0} e^{2s+2B_s(\omega)} ds\right)^{-\frac{1}{2}}. \tag{6.68}$$

We can verify that this is indeed a stationary orbit for this random dynamical system.

Unlike deterministic cases, it is difficult to find the stationary orbits of stochastic systems explicitly. The existence of the stationary orbits for SDEs is a subtle problem. There is no method yet that applies to general SDEs.

6.4 Linear Stochastic Dynamics

In this section we consider linear stochastic dynamical systems in a probability space $(\Omega, \mathcal{F}, \mathbb{P})$. A fundamental result about linear stochastic dynamical systems is the Oseledets' multiplicative ergodic theorem [9, Chapter 3].

6.4.1 Oseledets' Multiplicative Ergodic Theorem and Lyapunov Exponents

The multiplicative ergodic theorem (MET) is a spectral theory for linear cocycles or random matrices. It provides us stochastic counterparts for eigenspaces and eigenvalues, that is, random invariant subspaces and Lyapunov exponents for stochastic linear systems.

We look at a few motivating examples about eigenvalues and eigenspaces for deterministic linear dynamical systems. Recall that the eigenspace E_λ for an eigenvalue λ is spanned by the corresponding eigenvectors or generalized eigenvectors. When the algebraic multiplicity of each eigenvalue is equal to its geometric multiplicity (i.e., the dimension of its eigenspace), the matrix is diagonalizable. Otherwise, it is not diagonalizable. The state space $\mathbb{R}^n$ is decomposed as a direct sum of eigenspaces.

Example 6.32 Consider a linear system in $\mathbb{R}^2$:

$$\dot{x}_1 = 2x_1 + 3x_2, \tag{6.69}$$

$$\dot{x}_2 = -4x_2. \tag{6.70}$$

The eigenvalues are $\lambda_1 = 2 > 0$ and $\lambda_2 = -4 < 0$ with the corresponding eigenvectors $V_1 = (1, 0)^T$ and $V_2 = (1, -2)^T$. The flow is

$$\Phi(t, x) \triangleq \left(x_1 + \frac{x_2}{2}\right) e^{2t}(1, 0)^T - \frac{x_2}{2} e^{-4t}(1, -2)^T, \quad t \in \mathbb{R}.$$

In this and the following examples in this chapter, we also denote $(x_1, x_2, x_3)^T$ as an initial point (a common practice in dynamical systems). The unstable eigenspace E_1 is the x-axis and the stable eigenspace $E_2 = \{y = -2x\}$. Moreover, $\mathbb{R}^2 = E_1 \bigoplus E_2$. Also note that

$$\lim_{t\to\infty} \frac{1}{t} \ln \|\Phi(t, x)\| = \lambda_1, \quad \text{if and only if } x \in E_1 \setminus \{0\}, \tag{6.71}$$

$$\lim_{t\to\infty} \frac{1}{t} \ln \|\Phi(t, x)\| = \lambda_2, \quad \text{if and only if } x \in E_2 \setminus \{0\}. \tag{6.72}$$

Both E_1 and E_2 are invariant subspaces under the flow.

More generally for a linear system $\dot{X} = AX,\ X(0) = x$ in $\mathbb{R}^n$, where A is diagonalizable, we have a similar result. The linear flow is

$$\Phi(t, x) = e^{tA}x, \quad t \in \mathbb{R}.$$

The eigenvalue λ and its corresponding eigenspace E_λ are determined as follows:

$$\lambda = \lim_{t\to\infty} \frac{1}{t} \ln \|\Phi(t, x)\| \text{ if and only if } x \in E_\lambda \setminus \{0\}. \tag{6.73}$$

A linear system with real matrix A is not diagonalizable when the algebraic multiplicity of at least one eigenvalue λ is bigger than its geometric multiplicity (i.e., the dimension of its eigenspace). In such a case, we need to consider generalized eigenvectors. Let λ be an eigenvalue, of algebraic multiplicity k, for A. Its generalized eigenvectors v are nonzero vectors satisfying $(A - \lambda I)^k v = 0$. The set of all generalized eigenvectors for a given eigenvalue λ spans a linear space and is still called the eigenspace for λ.

Example 6.33 Consider a linear system in $\mathbb{R}^3$:

$$\dot{x}_1 = 2x_1 + x_2, \tag{6.74}$$

$$\dot{x}_2 = 2x_2, \tag{6.75}$$

$$\dot{x}_3 = -4x_3. \tag{6.76}$$

The eigenvalues are $\lambda_1 = 2$ of algebraic multiplicity 2 and $\lambda_2 = -4$ of algebraic multiplicity 1. The eigenspace for $\lambda_1 = 2$ is spanned by $V_1 = (1, 0, 0)^T$ and is thus one-dimensional, that is, the geometric multiplicity is 1. Thus we need to consider generalized eigenvectors: $(A - \lambda_1 I)^2 v = 0$. The corresponding eigenspace E_1 is spanned by $V_1 = (1, 0, 0)^T$ and $V_2 = (0, 1, 0)^T$, that is, it is the x_1x_2-plane. The eigenspace E_2 is spanned by $V_3 = (0, 0, 1)^T$, that is, it is the x_3-axis. Note that $\mathbb{R}^3 = E_1 \bigoplus E_2$. The flow is

$$\Phi(t, x) \triangleq (x_1 e^{2t} + x_2 t e^{2t}, x_2 e^{2t}, x_3 e^{-4t})^T, \quad t \in \mathbb{R}.$$

Note that

$$\lim_{t\to\infty} \frac{1}{t} \ln \|\Phi(t, x)\| = \lambda_1, \quad \text{if and only if } x \in E_1 \setminus \{0\}, \tag{6.77}$$

$$\lim_{t\to\infty} \frac{1}{t} \ln \|\Phi(t, x)\| = \lambda_2, \quad \text{if and only if } x \in E_2 \setminus \{0\}. \tag{6.78}$$

Both E_1 and E_2 are invariant subspaces under the flow.

Example 6.34 Consider a linear system in $\mathbb{R}^3$:

$$\dot{x}_1 = x_1, \tag{6.79}$$

$$\dot{x}_2 = 4x_2 - 5x_3, \tag{6.80}$$

$$\dot{x}_3 = 5x_2 + 4x_3. \tag{6.81}$$

The eigenvalues are $\lambda_1 = 1$, $\lambda_2 = 4 + 5i$, and $\lambda_3 = 4 - 5i$. The eigenspace E_1 for $\lambda_1 = 1$ is spanned by $V_1 = (1, 0, 0)^T$ and is the x_1-axis. The real and imaginary parts for the eigenvectors for λ_2 (and λ_3) span a linear space E_2, which is the x_2x_3-space. Moreover, $\mathbb{R}^3 = E_1 \bigoplus E_2$. The flow is [226, p. 41]

$$\Phi(t, x) \triangleq (x_1 e^t, e^{4t}(x_2 \cos 5t - x_3 \sin 5t), e^{4t}(x_2 \sin 5t + x_3 \cos 5t))^T, \quad t \in \mathbb{R}.$$

Both E_1 and E_2 are invariant subspaces under the flow. Note that

$$\lim_{t\to\infty} \frac{1}{t} \ln \|\Phi(t, x)\| = \lambda_1 = 1, \quad \text{if and only if } x \in E_1 \setminus \{0\}, \tag{6.82}$$

$$\lim_{t\to\infty} \frac{1}{t} \ln \|\Phi(t, x)\| = \Re\lambda_2 = \Re\lambda_3 = 4, \quad \text{if and only if } x \in E_2 \setminus \{0\}. \tag{6.83}$$

Now consider a linear random dynamical system $\Phi(t, \omega)$ in $\mathbb{R}^n$, which may be defined by the solution mapping of a system of linear stochastic differential equations.

The Lyapunov exponent is able to determine the rate of exponential growth of $\|\Phi(t, \omega)v\|$ along an orbit passing v but is not able to detect the presence of subexponential growth. If $\lambda < 0$, then the orbits of x and of a nearby point y will converge; if $\lambda > 0$, then the orbits of x and y will diverge (at least initially); if $\lambda = 0$, then generally we conclude nothing.

The existence of Lyapunov exponents is assured by the multiplicative ergodic theorem [9, Chapter 3]. This Oseledets' multiplicative ergodic theorem provides the theoretical background for computation of Lyapunov exponents of a linear stochastic dynamical system.

Theorem 6.35 (Multiplicative ergodic theorem) *Let $\Phi(t, \omega)$ be a linear random dynamical system (i.e., a linear cocycle) in $\mathbb{R}^n$, for $t \in \mathbb{R}$, on a probability space $(\Omega, \mathcal{F}, \mathbb{P})$, over a measurable driving flow θ_t. Assume that the following integrability conditions are satisfied:*

$$\sup_{0 \le t \le 1} \ln^+ \|\Phi(t, \omega)\| \in L^1(\Omega), \quad \sup_{0 \le t \le 1} \ln^+ \|\Phi(-t, \omega)\| \in L^1(\Omega), \tag{6.84}$$

where $\ln^+(z) \triangleq \max\{\ln(z), 0\}$, denoting the nonnegative part of the natural logarithm. Then there exists an invariant set $\tilde{\Omega} \in \mathcal{F}$ (i.e., $\theta_t^{-1}\tilde{\Omega} = \tilde{\Omega}$ for all $t \in \mathbb{R}$) of full probability measure, such that for every $\omega \in \tilde{\Omega}$,

1. *the 'asymptotic geometric mean' $\lim_{t \to \pm\infty} [\Phi(t, \omega)^{\mathrm{T}} \, \Phi(t, \omega)]^{\frac{1}{2t}} = \bar{\Phi}(\omega)$ exists, and $\bar{\Phi}(\omega)$ is a nonnegative definite $n \times n$ matrix*
2. *the matrix $\bar{\Phi}(\omega)$ has distinct eigenvalues $e^{\lambda_{p(\omega)}} < \cdots < e^{\lambda_1(\omega)}$, $\omega \in \tilde{\Omega}$, with corresponding eigenspaces $E_{p(\omega)}(\omega), \ldots, E_1(\omega)$ of dimensions $d_i(\omega) = dim E_i(\omega)$, $i = 1, \ldots, p(\omega)$. These eigenspaces are such that $\mathbb{R}^n = E_1(\omega) \bigoplus E_2(\omega) \bigoplus \cdots \bigoplus E_{p(\omega)}(\omega)$. Moreover, $p, \lambda_i's, d_i's$ are invariant under the driving flow θ_t in the following sense:*

$$p(\theta_t \omega) = p(\omega),$$

$$\lambda_i(\theta_t \omega) = \lambda_i(\omega),$$

and

$$d_i(\theta_t \omega) = d_i(\omega),$$

for $i = 1, \ldots, p$, $t \in \mathbb{R}$, and $\omega \in \tilde{\Omega}$
3. *each $E_i(\omega)$ is invariant for the linear random dynamical system Φ in the following sense: $\Phi(t, \omega)E_i(\omega) = E_i(\theta_t \omega)$, for all $\omega \in \tilde{\Omega}$ and all $t \in \mathbb{R}$*
4. *$\lim_{t \to \pm\infty} \frac{1}{t} \ln \|\Phi(t, \omega)x\| = \lambda_i$ if and only if $x \in E_i(\omega) \setminus \{0\}$, for all $\omega \in \tilde{\Omega}$ and $i = 1, 2, \ldots, p$*

We call $E_i(\omega)$s Oseledets spaces corresponding to Lyapunov exponents λ_i with multiplicities d_i. The decomposition $\mathbb{R}^n = E_1(\omega) \bigoplus E_2(\omega) \bigoplus \cdots \bigoplus E_{p(\omega)}(\omega)$ is called an Oseledets splitting. Moreover, $\{\lambda_1, \ldots, \lambda_p; d_1, \ldots, d_p\}$ is called the

Lyapunov spectrum. When all Lyapunov exponents are nonzero, we call the linear stochastic system $\Phi(t, \omega)$ hyperbolic.

A proof of this theorem is in [9, Sections 3.3 and 3.4]. The original proof is based on the triangularization of a linear cocycle and the classical ergodic theorem for the triangular cocycle. Another class of proofs involves the subadditive ergodic theorem.

Remark 6.36

(i) In a linear SDE seeting in $\mathbb{R}^n$, we work in the canonical sample space $\Omega = C(\mathbb{R}, \mathbb{R}^n)$ with the Wiener measure. The Wiener shift θ_t is an ergodic measurable driving flow. Thus, for a linear random dynamical system $\Phi(t, \omega)$ defined by the solution mapping of such a linear SDE system, the invariant quantities $p, \lambda_i's, d_i's$ are actually constant almost surely on $\tilde{\Omega}$ (see [9, Section 3.4.1]).

(ii) The integrability conditions mean

$$\mathbb{E} \sup_{0 \le t \le 1} \ln^+ \|\Phi(t, \omega)\| < \infty, \quad \mathbb{E} \sup_{0 \le t \le 1} \ln^+ \|\Phi(-t, \omega)\| < \infty, \quad (6.85)$$

where $\|\cdot\|$ is for a matrix norm. Because all matrix norms are equivalent, we could pick any one norm to check these integrability conditions. Common matrix norms include Euclidean norm (square root of the sum of squared entries), maximal absolute value of eigenvalues, and operator norm (as often used in functional analysis). There are two methods to verify these conditions: a probabilistic method (see Example 6.40) and an analytical method (see Problem 6.18).

(iii) For the symmetric nonnegative definite matrix $\Phi^T\Phi$, its fractional power $[\Phi(t, \omega)^T \, \Phi(t, \omega)]^{\frac{1}{2t}}$, for each t and ω, is defined as in linear algebra. Indeed, let K be an $n \times n$ symmetric nonnegative definite matrix, with eigenvalues $\mu_1, \ldots, \mu_n$ (which are, of course, nonnegative) whose corresponding eigenvectors form an orthonormal basis $\{e_1, \ldots, e_n\}$ for $\mathbb{R}^n$. Then the fractional power of K of order $\gamma \in (0, 1)$ is defined by $K^\gamma x = \sum_i \mu_i^\gamma < x, e_i > e_i$, for $x \in \mathbb{R}^n$. In particular, if K is diagonal with diagonal entries $a_1, \ldots, a_n$ (which are nonnegative), then K^γ is a diagonal matrix with diagonal entries $a_1^\gamma, \ldots, a_n^\gamma$.

(iv) Recall that $(ab)^{\frac{1}{2}}$ is the geometric mean for two positive numbers a and b. Hence, the matrix $\bar{\Phi}(\omega)$ may be regarded as an asymptotic geometric mean for the linear cocycle $\Phi(t, \omega)$ and its transpose.

(v) For a linear deterministic system $\dot{X} = AX$ with a real symmetric $n \times n$ matrix A, the flow (or solution mapping) is e^{At}. The eigenvalues of A are real and there exists an orthonormal basis for $\mathbb{R}^n$ formed by eigenvectors. The asymptotic geometric mean is $\lim_{t\to\infty}[(e^{At})^{\mathrm{T}} e^{At}]^{\frac{1}{2t}} = e^A$. Thus the Lyapunov exponents are the (real parts) of the distinct eigenvalues of A, and the Oseledets spaces are the corresponding eigenspaces for e^A (which are eigenspaces for A as well). In fact, the multiplicative ergodic theorem also applies when A is a normal matrix (i.e., $A^{\mathrm{T}}A = AA^{\mathrm{T}}$).

(vi) There are two methods to determine Lyapunov exponents and their corresponding Oseledets spaces. The first method follows Part 1 of the theorem and requires finding out the asymptotic geometric mean matrix $\bar{\Phi}(\omega)$. If this matrix is unavailable, this method will not apply. The second method follows Part 4 of the theorem. It indicates that λ_i is a Lyapunov exponent, precisely when the nonzero initial point x is in the corresponding Oseledets space $E_i(\omega)$, for $i = 1, \dots, p$. This offers a way to identify each Oseledets space $E_i(\omega)$.

To check the integrability conditions, the following Burkholder-Davis-Gundy (B-D-G) inequality [61, Corollary 3.4] and Doob inequality [151, p. 201] are useful. We do not need the concept of martingales in this book, except for these two inequalities. A stochastic process M_t is called a martingale with respect to a filtration $\mathcal{F}_t$ if it is integrable and satisfies the condition $\mathbb{E}\{M_t|\mathcal{F}_s\} = M_s$, a.s. for every $t > s$. The examples of martingales include Brownian motion and Itô integrals. A scalar process M_t is called a submartingale if it satisfies the condition $\mathbb{E}\{M_t|\mathcal{F}_s\} \geq M_s$, a.s. for every $t > s$. Recall that if M_t is a martingale and φ is a convex function, then $\varphi(M_t)$ is a submartingale [165, p. 51].

Theorem 6.37 (B-D-G inequality)

(i) Let M_t be a scalar-continuous martingale, defined for $t \in [0, \tau]$ with a given positive τ, such that $M_0 = 0$ and $\mathbb{E}|M_\tau|^p < \infty$ for some $p > 0$. Then there exist two positive constants c_p and C_p such that

$$c_p \mathbb{E}[M]_\tau^{\frac{p}{2}} \leq \mathbb{E}\left\{\sup_{0\leq t\leq\tau} |M_t|^p\right\} \leq C_p\, \mathbb{E}[M]_\tau^{\frac{p}{2}}, \tag{6.86}$$

where $[M]_\tau$ is the quadratic variation over $[0, \tau]$.

(ii) Let $\sigma(t)$ be an $n \times m$ stochastic matrix defined for $t \in [0, \tau]$ and $B(t)$ be an m-dimensional Brownian motion. Then, for every $p > 0$, there exists a positive constant K_p such that

$$\mathbb{E}\left\{\sup_{0\leq t\leq\tau}\left\|\int_0^t \sigma(s)dB(s)\right\|^p\right\} \leq K_p\, \mathbb{E}\left\{\int_0^\tau Tr[\sigma(s)\sigma^{\mathrm{T}}(s)]ds\right\}^{\frac{p}{2}}, \tag{6.87}$$

provided that the right-hand side is finite. Here superscript T denotes the transpose of a matrix.

Theorem 6.38 (Doob inequality) *Let M_t be a scalar martingale or positive submartingale defined for $t \in [0, \tau]$ with a given positive τ. Then*

$$\mathbb{E}\left\{ \sup_{0\le t\le \tau} (M_t)^2 \right\} \le 4\, \mathbb{E}(M_\tau)^2. \tag{6.88}$$

We consider a couple of examples.

Example 6.39 Consider a scalar linear SDE

$$dX_t = 3X_t dt + 2X_t dB_t, \;\; X(0) = x \in \mathbb{R}^1. \tag{6.89}$$

The solution mapping for this linear stochastic system is $\Phi(t, \omega)x = e^{t+2B_t}x$. Note that

$$\begin{aligned}
\mathbb{E} \sup_{0\le t\le 1} \ln^+ \|\Phi(t, \omega)\| &= \mathbb{E} \sup_{0\le t\le 1} \max\{\ln e^{t+2B_t}, 0\} \\
&= \mathbb{E} \sup_{0\le t\le 1} \max\{t + 2B_t, 0\} \\
&\le 1 + 2\mathbb{E} \sup_{0\le t\le 1} |B_t| < \infty,
\end{aligned}$$

where in the final step, we have used the B-D-G inequality. Note that B_t is a martingale as seen in [151, Section 3.3]. Similarly,

$$\mathbb{E} \sup_{0\le t\le 1} \ln^+ \|\Phi(-t, \omega)\| < \infty.$$

So the integrability conditions hold and the multiplicative ergodic theorem applies. Thus, the Lyapunov exponent is

$$\begin{aligned}
\lambda &= \lim_{t\to\infty} \frac{1}{t} \ln \|\Phi(t, \omega)\| \\
&= \lim_{t\to\infty} \frac{t + 2B_t}{t} = 1 + \lim_{t\to\infty} \frac{2B_t}{t} = 1.
\end{aligned} \tag{6.90}$$

The only Oseledets space is the whole space $\mathbb{R}^1$.

Example 6.40 Consider a linear stochastic system in $\mathbb{R}^2$:

$$dX_1 = 2X_1 dt + X_1 dB_t^1, \quad X_1(0) = x_1, \tag{6.91}$$

$$dX_2 = -4X_2 dt + 6X_2 dB_t^2, \;\; X_2(0) = x_2. \tag{6.92}$$

The solution mapping or the linear cocycle is

$$\Phi(t, \omega, x) = \begin{pmatrix} x_1 e^{\frac{3}{2}t + B_t^1} \\ x_2 e^{-22t + 6B_t^2} \end{pmatrix} := \Phi(t, \omega)x,$$

where

$$\Phi(t,\omega) = \begin{pmatrix} e^{\frac{3}{2}t+B_t^1} & 0 \\ 0 & e^{-22t+6B_t^2} \end{pmatrix}.$$

Let us check the integrability conditions in the multiplicative ergodic theorem. Note that

$$\begin{aligned} \|\Phi(t,\omega)\| &= \sup_{\|x\|=1} \|\Phi(t,\omega)x\| \\ &= \sup_{x_1^2+x_2^2=1} \sqrt{e^{3t+2B_t^1}x_1^2 + e^{-44t+12B_t^2}x_2^2} \\ &\le e^{\frac{3}{2}t+B_t^1} + e^{-22t+6B_t^2}. \end{aligned}$$

Recall that $\ln x \le x$ and thus $\ln^+(x) \le x$ for positive x. Thus,

$$\begin{aligned} &\mathbb{E} \sup_{0\le t\le 1} \ln^+ \|\Phi(t,\omega)\| \\ &\le \mathbb{E} \sup_{0\le t\le 1} \ln^+ \left(e^{\frac{3}{2}t+B_t^1} + e^{-22t+6B_t^2}\right) \\ &\le \mathbb{E} \sup_{0\le t\le 1} \left[e^{\frac{3}{2}t+B_t^1} + e^{-22t+6B_t^2}\right] \\ &\le \mathbb{E} \sup_{0\le t\le 1} e^{\frac{3}{2}t+B_t^1} + \mathbb{E} \sup_{0\le t\le 1} e^{-22t+6B_t^2} \\ &:= A + B. \end{aligned}$$

For the term A,

$$\begin{aligned} A &= \mathbb{E} \sup_{0\le t\le 1} e^{\frac{3}{2}t+B_t^1} \\ &= \mathbb{E} \sup_{0\le t\le 1} \left(e^{B_t^1/2-\frac{1}{8}t}\right)^2 e^{\frac{7}{4}t} \\ &\le e^{\frac{7}{4}} \mathbb{E} \sup_{0\le t\le 1} \left|\left(e^{B_t^1/2-\frac{1}{8}t}\right)^2\right| < \infty, \end{aligned}$$

where in the final step, we have used the Doob inequality, together with the fact that $e^{cB_t-\frac{c^2}{2}t}$ is a martingale, for every constant c and Brownian motion B_t

[151, Section 3.3]. Likewise, for the term B,

$$B = \mathbb{E} \sup_{0 \le t \le 1} e^{-22t + 6B_t^2}$$
$$= \mathbb{E} \sup_{0 \le t \le 1} \left(e^{3B_t^2 - \frac{9}{2}t}\right)^2 e^{-13t}$$
$$\le e^0 \mathbb{E} \sup_{0 \le t \le 1} \left| \left(e^{3B_t^2 - \frac{9}{2}t}\right)^2 \right| < \infty.$$

Therefore,

$$\mathbb{E} \sup_{0 \le t \le 1} \ln^+ \|\Phi(t, \omega)\| < \infty.$$

Similarly,

$$\mathbb{E} \sup_{0 \le t \le 1} \ln^+ \|\Phi(-t, \omega)\| < \infty.$$

So the integrability conditions hold and the multiplicative ergodic theorem applies. The above method for verifying the integrability conditions involves probabilistic techniques (B-D-G inequality and Doob inequality). There is also an analytical method to check these integrability conditions (see Problem 6.18) at the end of this chapter.

We calculate the following symmetric matrix:

$$\Phi^{\mathrm{T}}(t, \omega)\Phi(t, \omega) = \begin{pmatrix} e^{3t + 2B_t^1} & 0 \\ 0 & e^{-44t + 12B_t^2} \end{pmatrix}.$$

Thus

$$\lim_{t \to \infty} (\Phi^{\mathrm{T}}(t, \omega)\Phi(t, \omega))^{\frac{1}{2t}} = \bar{\Phi}(\omega) = \begin{pmatrix} e^{\frac{3}{2}} & 0 \\ 0 & e^{-22} \end{pmatrix}.$$

The matrix $\bar{\Phi}(\omega)$ may be regarded as an asymptotic geometric mean for the linear cocycle $\Phi(t, \omega)$. The eigenvalues for $\bar{\Phi}(\omega)$ are $e^{\lambda_2} = e^{-22} < e^{\lambda_1} = e^{\frac{3}{2}}$, with the corresponding eigenspaces or Oseledets spaces $E_2 = \{x_1 = 0\}$ and $E_1 = \{x_2 = 0\}$, respectively. Note that $\mathbb{R}^2 = E_1 \bigoplus E_2$. Finally, the Lyapunov exponents are

$$\lim_{t \to \infty} \frac{1}{t} \ln \|\Phi(t, \omega)x\| = \lambda_1 = \frac{3}{2}, \quad \text{if and only if } x \in E_1 \setminus \{0\}, \tag{6.93}$$

$$\lim_{t \to \infty} \frac{1}{t} \ln \|\Phi(t, \omega)x\| = \lambda_2 = -22, \quad \text{if and only if } x \in E_2 \setminus \{0\}. \tag{6.94}$$

6.4.2 A Stochastic Hartman-Grobman Theorem

We start with two motivating examples.

Example 6.41 Consider a linear system in $\mathbb{R}^2$:

$$\dot{x}_1 = -x_1 - 3x_2, \tag{6.95}$$

$$\dot{x}_2 = -3x_1 - x_2. \tag{6.96}$$

By a coordinate transform H, this system can be converted to the following diagonal system (Jordan form):

$$\dot{y}_1 = 2y_1, \tag{6.97}$$

$$\dot{y}_2 = -4y_2. \tag{6.98}$$

This transform H and its inverse H^{-1} are both continuous, and H maps orbits in one system to the other, and vice versa. Hence, these two systems are topologically equivalent (or conjugate), that is, they have the qualitatively similar phase portraits.

Example 6.42 As we have seen in Section 6.1.2, for a nonlinear system $\dot{x} = f(x)$ and its linearized system at an equilibrium point x_0, $\dot{x} = Df(x_0)x$, there exists a local coordinate transform H that locally maps orbits of one system to the orbits of the other, and vice versa. This is the deterministic Hartman-Grobman theorem.

In this subsection, we discuss a stochastic Hartman-Grobman theorem.

Recall that we have defined the function space $C_b^{k,\delta}$ earlier. For a natural number $k \geq 1$ and a positive real number $\delta > 0$, we define the function space $C_b^{k,\delta}$ as the set of functions $f\colon \mathbb{R}^n \to \mathbb{R}^n$ that are sublinear (i.e., $\|f(x)\| \leq M(1 + \|x\|)$ for some positive constant M), all derivatives up to order k exist and are bounded, and furthermore, the kth-order derivatives are Hölder continuous with exponent δ.

We present the following Hartman-Grobman theorem [133]. Other versions of this theorem, involving random norms, are in [285, 9, 63].

Theorem 6.43 (A stochastic Hartman-Grobman theorem [133, Theorem 5.1]) *Let $\delta > 0$ and assume that $f_0 \in C_b^{2,\delta}$, $f_1, \ldots, f_m \in C_b^{3,\delta}$, and $\sum_{i=1}^m \sum_{j=1}^d f_i^j \frac{\partial f_i}{\partial x_j} \in C_b^{3,\delta}$. Suppose further that 0 is a fixed point for the vector fields $f_0, \ldots, f_m$, and set $A_i = \frac{\partial}{\partial x} f_i(0)$, $0 \leq i \leq m$.*

Let Ψ be the cocycle generated by the following nonlinear SDE system:

$$dx_t = f_0(x_t)dt + \sum_{i=1}^m f_i(x_t) \circ dB_t^i, \; t \in \mathbb{R}. \tag{6.99}$$

Let Φ be the cocycle generated by the linearized SDE system

$$dy_t = A_0 y_t dt + \sum_{i=1}^{m} A_i y_t \circ dB_t^i, \quad t \in \mathbb{R}, \tag{6.100}$$

and suppose that Φ is hyperbolic, that is, all Lyapunov exponents are nonzero.

Then, there exists a measurable mapping $h : \Omega \times \mathbb{R}^n \to \mathbb{R}^n$ such that

(i) the mapping $h(\omega, \cdot) : \mathbb{R}^n \to \mathbb{R}^n$ is a homeomorphism of R^n, such that $h(\omega, 0) = 0$, $\omega \in \Omega$; and

(ii) the topological equivalence relation

$$\Phi_t(\omega) \circ h(\omega, x) = h(\theta_t \omega, \cdot) \circ \Psi_t(\omega)$$

holds in a random time interval $\tau_-(\omega, x) \le t \le \tau_+(\omega, x)$ for $\omega \in \Omega$.

Remark 6.44 Because the random time interval is space dependent, this topological equivalence holds locally in space. Therefore, the conditions for the drift and diffusion terms and their derivatives need only to be in $C_b^{2,\delta}$ or $C_b^{3,\delta}$ locally.

6.5 *Random Invariant Manifolds

This section may be omitted in the first reading.

Invariant manifolds provide geometric structures that describe dynamical behavior of nonlinear systems. Dynamical reductions to attracting invariant manifolds or dynamical restrictions to other (not necessarily attracting) invariant manifolds are often sought to gain understanding of nonlinear dynamics.

There have been recent works on invariant manifolds for stochastic differential equations by [48, 285, 9, 38, 39, 201], among others. These authors use the cocycle property for the solution operator of the stochastic differential equations, the Oseledets' multiplicative ergodic theorem, and a less physical but technically convenient random norm to prove the existence of invariant manifolds. The construction of a random norm needs the knowledge of Oseledets spaces (a kind of eigenspace in random linear algebra) as well as Lyapunov exponents; both are hardly available [9, pp. 191, 379]. We use a different norm that is easier to utilize.

Earlier approaches to deriving dynamical reductions on stochastic center-like manifolds by series expansions are considered by [153, 251, 296].

In this section, we establish random invariant manifolds via a Lyapunov-Perron approach as in [47, 80], adapted to the finite-dimensional Euclidean space setting here.

In the framework of cocycles [9], the suitable concept for invariance of a random set is that each orbit starting inside it stays inside it sample-wisely, modulo the change of sample due to noise (via Wiener shift).

6.5.1 Definition of Random Invariant Manifolds

Just as invariant sets and invariant manifolds are building blocks for deterministic dynamical systems, random invariant sets and random invariant manifolds are also basic geometric objects to help understand stochastic dynamics [9].

Definition 6.45 (Random set) A collection $M = M(\omega)$, $\omega \in \Omega$, is called a random set for a random dynamical system φ in $\mathbb{R}^n$ if

(i) $M(\omega)$ is a nonempty closed set contained in $\mathbb{R}^n$, for every $\omega \in \Omega$;
(ii) $V_x : \Omega \to \mathbb{R}^1$, defined by $V_x(\omega) \triangleq \inf_{y \in M(\omega)} d(x, y)$, is a scalar random variable for every $x \in \mathbb{R}^n$.

Definition 6.46 (Random invariant set) A random set M is called an invariant set for a random dynamical system φ if

$$\varphi(t, \omega, M(\omega)) = M(\theta_t \omega), \quad t \in \mathbb{R}, \quad \text{and} \quad \omega \in \Omega.$$

Definition 6.47 (Random invariant manifold) If a random invariant set M for a random dynamical system φ can be represented by a graph of a Lipschitz mapping

$$\gamma^*(\omega, \cdot) \colon \; H^+ \to H^-, \text{ with direct sum decomposition } \quad H^+ \oplus H^- = \mathbb{R}^n$$

such that

$$M(\omega) = \{(x^+, \gamma^*(\omega, x^+)), \; x^+ \in H^+\},$$

then M is called a Lipschitz continuous invariant manifold. If the mapping γ^* is C^k for $k \geq 1$, that is, it has up to kth-order continuous Fréchet derivatives, then M is called a C^k invariant manifold.

6.5.2 Converting SDEs to RDEs

We would like to examine random invariant manifolds for an SDE system in $\mathbb{R}^n$:

$$dX_t = b(X_t)dt + \sigma(X_t)dB_t. \tag{6.101}$$

However, the current theory on random invariant manifolds is for random dynamical systems or cocycles ([47, 285], or [9, Chapter 7]). To this end, we better convert an SDE system to a system of random differential equations (or RDEs), which are differential equations with random coefficients and which are easily seen to generate random dynamical systems [9, Section 2.2]. This conversion from SDEs to RDEs is often possible but not always straightforward [132, 133, 129]. With the inverse conversion, the random invariant manifolds for RDEs become those for the original SDEs. We present an example at the end of this section (Example 6.55).

In this subsection, we present several examples of converting certain SDEs to RDEs.

Example 6.48 (*A nonlinear SDE with linear multiplicative noise*) Consider the following scalar SDE:

$$dX_t = b(t, X_t)dt + c(t)X_t dB_t, \quad X_0 = x, \tag{6.102}$$

where b and c are given scalar-continuous deterministic functions.

To motivate our approach, let us recall how we solve a deterministic nonhomogeneous differential equation,

$$\dot{x} = c(t)x + g(t), \quad x(0) = x_0, \tag{6.103}$$

with c and g given. The general solution (also called the fundamental solution) to the corresponding homogeneous equation $\dot{x} = c(t)x$ is $Ce^{\int_0^t c(s)ds}$, with C the constant of integration. Taking $C = 1$, we get $e^{\int_0^t c(s)ds}$. Define an integrating factor to be the reciprocal of this function, $F(t) = e^{-\int_0^t c(s)ds}$. Multiply the integrating factor F with each side of $\dot{x} = c(t)x + g(t)$ to get $F\dot{x} = c(t)F(t)x + F(t)g(t)$. Thus

$$\frac{d}{dt}(F(t)x(t)) = F(t)g(t).$$

Integrating from 0 to t, we obtain the solution $x(t) = F^{-1}(t)[x_0 + \int_0^t F(s)g(s)ds]$ to the differential equation (6.103).

This has inspired the use of a stochastic integrating factor in solving the SDE (6.102), as in [213, Exercise 5.16]. A solution to the corresponding SDE with no drift,

$$dX_t = c(t)X_t dB_t,$$

is $\exp(-\frac{1}{2}\int_0^t c^2(s)ds + \int_0^t c(s)dB_s)$. Its reciprocal is taken as a stochastic integrating factor

$$F_t(\omega) = \exp\left(\frac{1}{2}\int_0^t c^2(s)ds - \int_0^t c(s)dB_s\right).$$

Multiplying each side of the SDE (6.102) by this stochastic integrating factor and using the stochastic product rule, we obtain

$$d(F_t X_t) = F_t b(t, X_t)dt.$$

Now define $Y_t(\omega) = F_t(\omega)X_t(\omega)$ and thus $X_t = F_t^{-1}Y_t$. Then the SDE (6.102) on X_t is converted to the following RDE on Y_t:

$$\frac{dY_t}{dt} = F_t b\big(t, F_t^{-1}Y_t\big), \ Y_0 = x. \tag{6.104}$$

From the solution Y_t for this RDE, we get the solution for the original SDE (6.102), via $X_t = F_t^{-1}Y_t$.

Example 6.49 (*A general nonlinear SDE*) In fact, the method of stochastic integrating factors in the previous example works for more general nonlinear SDEs. This is shown to the author by Jia-an Yan.

Consider a nonlinear scalar SDE

$$dY_t = \mu(t, Y_t)dt + \sigma(t, Y_t)dB_t, \tag{6.105}$$

where B_t is a scalar Brownian motion. Assume that $\sigma(t, x) > 0$ and $\frac{1}{\sigma(t,\cdot)} \in L^1(\mathbb{R})$. Also assume that $\sigma(t, x)$ is continuously differentiable in t and x. Define

$$g(t, x) = \exp\left\{\int_0^x \sigma^{-1}(t, y)dy\right\}, \quad x \in \mathbb{R}^1.$$

For each $t \geq 0$, denote by $h(t, x)$ the inverse function of $g(t, x)$. Then $h(t, x)$ is defined on $(0, \infty)$, but take values in $\mathbb{R}^1$. Note that

$$g(t, h(t, x)) = x, \quad g_x(t, x)\sigma(t, x) = g(t, x), \quad x > 0,$$
$$g_{xx}(t, x)\sigma^2(t, x) + g(t, x)\sigma_x(t, x) = g(t, x), \quad x > 0.$$

Let $X_t = g(t, Y_t)$. Then X_t takes values in $(0, \infty)$ and $Y_t = h(t, X_t)$.

By Itô's formula, we conclude that

$$\begin{aligned} dX_t &= g_t(t, Y_t)dt + g_x(t, Y_t)[\mu(t, Y_t)dt + \sigma(t, Y_t)dB_t] \qquad (6.106)\\ &\quad + \frac{1}{2}g_{xx}(t, Y_t)\sigma^2(t, Y_t)dt \\ &= b(t, X_t)dt + X_t dB_t, \qquad (6.107)\end{aligned}$$

where

$$\begin{aligned} b(t, x) &= g_t(t, h(t, x)) + g_x(t, h(t, x))\mu(t, h(t, x)) \\ &\quad + \frac{1}{2}g_{xx}(t, h(t, x))\sigma^2(t, h(t, x)) \\ &= g_t(t, h(t, x)) + \frac{x\mu(t, h(t, x))}{\sigma(t, h(t, x))} \\ &\quad + \frac{1}{2}x[1 - \sigma_x(t, h(t, x))], \quad x > 0.\end{aligned}$$

As seen in the previous example, this SDE (6.106) contains a linear multiplicative noise term and can thus be converted to a RDE. Thus, we can convert the nonlinear SDE (6.105) to a RDE.

For a more recent development in converting SDEs to RDEs, see [132, 133], and discussions in [9, Section 2.3] about random dynamical systems (i.e., cocycles) generated by RDEs and SDEs.

Example 6.50 (A nonlinear SDE system with additive noise) Consider the following SDE system in $\mathbb{R}^n$ with additive noise:

$$dX_t = (AX_t + F(X_t))dt + KdB_t, \tag{6.108}$$

where A is an $n \times n$ matrix, F is a nonlinear vector function taking values in $\mathbb{R}^n$, K is an $m \times n$ constant matrix, and B_t is Brownian motion in $\mathbb{R}^n$. Let η_t be the stationary orbit of the corresponding linear SDE system (a Langevin SDE system)

$$d\eta_t = A\eta_t dt + KdB_t.$$

See Example 6.30 for the existence of this stationary orbit. Define $Y_t = X_t - \eta_t$. Then Y_t solves the following RDE system:

$$\frac{d}{dt}Y_t = (AY_t + F(Y_t + \eta_t)). \tag{6.109}$$

6.5.3 Local Random Pseudo-Stable and Pseudo-Unstable Manifolds

This subsection is adopted from our earlier work [47].

We would like to examine random stable and unstable manifolds for an SDE system in $\mathbb{R}^n$:

$$dX_t = b(X_t)dt + \sigma(X_t)dB_t. \tag{6.110}$$

As we said ealier, the current theory on random invariant manifolds is for random dynamical systems or cocycles ([47], [285], or [9, Chapter 7]). To this end, we need to convert an SDE system to a RDE system (random differential equations), which is known to generate a random dynamical system, and then develop a random invariant manifold framework for RDEs.

Therefore, assuming that the conversion from SDEs to RDEs is achieved, we start with a RDE system in Euclidean space $\mathbb{R}^n$:

$$\frac{du}{dt} = A(\theta_t\omega)u + F(\theta_t\omega, u), \tag{6.111}$$

with the random linear operator A and nonlinear part F. We assume that the corresponding linear equation

$$\frac{du}{dt} = A(\theta_t \omega)u \tag{6.112}$$

generates a linear random dynamical system $U(t, \omega)$ on $\mathbb{R}^n$ for $t \in \mathbb{R}$. We first introduce a weak hyperbolicity or a weak exponential dichotomy condition on the linear random dynamics.

Recall that a positive random variable $\xi_t : \Omega \to (0, \infty)$ is called tempered, with respect to θ_t, if the invariant set for which

$$\lim_{t \to \pm\infty} \frac{1}{t} \ln \xi(\theta_t \omega) = 0$$

has probability 1. In fact, the only alternative for a positive random variable (if not tempered) is that the limit equal ∞, which may be regarded as a degenerated case [9, Proposition 4.1.3]. A positive random variable ξ is called tempered from above if

$$\lim_{t \to \pm\infty} \frac{1}{|t|} \ln^+ \xi(\theta_t \omega) = 0, \text{ a.s.}$$

It is called tempered from below if $\frac{1}{\xi}$ is tempered from above. We comment that temperedness is of fundamental importance in stochastic dynamical systems theory, although we will not be able to appreciate this fully in this introductory book.

Definition 6.51 (Nonuniform pseudo-hyperbolicity) A linear random system $U(t, \omega)$ (or $u = 0$) is said to be **nonuniformly pseudo-hyperbolic** if there exists a θ_t-invariant set $\tilde{\Omega} \subset \Omega$ of full measure, such that for each $\omega \in \tilde{\Omega}$, the state space $\mathbb{R}^n$ decomposes into

$$\mathbb{R}^n = E^s(\omega) \oplus E^u(\omega),$$

where $E^s(\omega)$, $E^u(\omega)$ are closed subspaces satisfying the following conditions:

(i) This decomposition is invariant under $U(t, \omega)$:

$$U(t, \omega)E^s(\omega) \subset E^s(\theta_t \omega),$$
$$U(t, \omega)E^u(\omega) \subset E^u(\theta_t \omega),$$

and $U(t, \omega)|_{E^u(\omega)}$ is an isomorphism from $E^u(\omega)$ to $E^u(\theta_t \omega)$.

(ii) There are θ-invariant random variables $\alpha(\omega)$, $\beta(\omega)$ such that $\alpha(\omega) > \beta(\omega)$, and a tempered random variable $K(\omega)$: $\tilde{\Omega} \to [1, \infty)$ such that

$$||U(t, \omega)P^s(\omega)|| \le K(\omega)e^{\beta(\omega)t} \quad \text{for } t \ge 0, \tag{6.113}$$
$$||U(t, \omega)P^u(\omega)|| \le K(\omega)e^{\alpha(\omega)t} \quad \text{for } t \le 0, \tag{6.114}$$

where $P^s(\omega)$ and $P^u(\omega)$ are the measurable projections associated with the decomposition. For our special setting of *ergodicity*, we can assume that α, β are constant on a $\{\theta_t\}_{t\in\mathbb{R}}$-invariant set of full measure.

In the following we also use the notations

$$U^u(t,\omega) = U(t,\omega)|_{E^u(\omega)} \quad \text{and} \quad U^s(t,\omega) = U(t,\omega)|_{E^s(\omega)}.$$

One can verify that the cocycle property holds for both $U^u(t,\omega)$ and $U^s(t,\omega)$ with $t \in \mathbb{R}$.

Remark 6.52 As ω varies, $\beta(\omega)$ may be arbitrarily small and $K(\omega)$ may be arbitrarily large. However, along each orbit $\{\theta_t\omega\}$, $\alpha(\omega)$ and $\beta(\omega)$ are constant and $K(\omega)$ can increase only at a subexponential rate. Thus, the linear system $U(t,\omega)$ is nonuniformly hyperbolic in the sense of Barreira-Pesin [15].

We make the following two hypotheses.

Hypothesis A (Linear part):
The linear cocycle $U(t,\omega)$ *is nonuniformly pseudo-hyperbolic.*

Hypothesis B (Nonlinear part):
There exists a ball, $\mathcal{N}(\omega) = \{u \in \mathbb{R}^n : \|u\| < \rho(\omega)\}$, *with radius* ρ *tempered from below and* $\rho(\theta_t\omega)$ *locally integrable in* t, *such that* $F(\omega,\cdot)\colon \mathcal{N}(\omega) \to \mathbb{R}^n$ *is Lipschitz continuous and satisfies the following conditions:*

$$F(\omega, 0) = 0$$

and

$$\|F(\omega,u) - F(\omega,v)\| \le Ł(\omega)\left(\|u\|^{\varepsilon} + \|v\|^{\varepsilon}\right)\|u-v\|, \quad \omega \in \Omega,\ u, v \in \mathcal{N}(\omega).$$

Here $\varepsilon \in (0,1]$, *and the random Lipschitz constant* $Ł(\omega)$ *is a random variable tempered from above such that* $Ł(\theta_t\omega)$ *is locally integrable in* t.

We state the following local random invariant manifold theorem [47].

Theorem 6.53 (Local random invariant manifold theorem) *Under the hypotheses A and B, local random invariant manifolds exist for the stochastic system (6.111). More precisely,*

(i) there exists a Lipschitz pseudo-stable manifold, and it is represented as

$$M^s(\omega) = \{(q, h^s(q,\omega)) : q \in \Gamma_s(\omega) \subset E^s(\omega)\},$$

where $\Gamma_s(\omega)$ *is a random neighborhood of* 0 *in* E^s *and* $h^s(\cdot,\omega)\colon E^s(\omega) \to E^u(\omega)$ *is Lipschitz continuous such that* $h^s(0,\omega) = 0$;

(ii) there exists a Lipschitz pseudo-unstable manifold for the stochastic system (6.111), and it is represented as

$$M^u(\omega) = \{(p, h^u(p, \omega)) \colon p \in \Gamma_u(\omega) \subset E^u(\omega)\},$$

where $\Gamma_u(\omega)$ is a random neighborhood of 0 in E^u, and $h^u(\cdot, \omega) \colon E^u(\omega) \to E^s(\omega)$ is Lipschitz continuous such that $h^u(0, \omega) = 0$.

Remark 6.54 If F is continuously differentiable in the state variable, then h^s, h^u are also continuously differentiable in the state variable.

6.5.4 Local Random Stable, Unstable, and Center Manifolds

Hypothesis A, about the nonuniform pseudo-hyperbolicity for the linear system, is quite difficult to verify. However, when the linear part satisfies the multiplicative ergodic theorem, we can verify this hypothesis using Lyapunov exponents. Consider a nonlinear random differential equation in $\mathbb{R}^n$

$$\frac{du}{dt} = A(\theta_t\omega)u + F(\theta_t\omega, u), \; u(0) = u_0. \tag{6.115}$$

Assume that the nonlinear part F satisfies Hypothesis B. When the multiplicative ergodic theorem, Theorem 6.35, holds for the corresponding linear system $\frac{du}{dt} = A(\theta_t\omega)u$, the nonuniform pseudo-hyperbolicity we introduced earlier, Definition 6.51, automatically follows [47]. Indeed, we can split the Lyapunov spectrum $\{\lambda_1, \ldots, \lambda_p\}$ between any two Lyapunov exponents. Thus we obtain various pseudo-stable and pseudo-unstable manifolds.

In particular, when we split between the zero and positive Lyapunov exponents (if any), we obtain the unstable manifold $M^u(\omega)$ and the center-stable manifold $M^{cs}(\omega)$, in the usual sense that is familiar to us in deterministic dynamical systems. Similarly, the usual stable manifold $M^s(\omega)$ and the center-unstable manifold $M^{cu}(\omega)$ follow with splitting between the zero and negative Lyapunov exponents (if any).

Let us discuss this more precisely. For the linear RDE system in $\mathbb{R}^n$,

$$\frac{du}{dt} = A(\theta_t\omega)u, \tag{6.116}$$

the fundamental matrix defines the linear cocycle $\Phi(t, \omega) \triangleq e^{\int_0^t A(\theta_\tau\omega)\,d\tau}$. Assume that $\Phi(t, \omega)$ satisfies the integrability conditions in Theorem 6.35. Then the multiplicative ergodic theorem (MET), Theorem 6.35, yields the Lyapunov exponents $\lambda_1 > \lambda_2 > \cdots > \lambda_r$, with $r \leq n$, together with r Oseledets subspaces $E_i(\omega)$, so that $\mathbb{R}^n$ decomposes into the direct sum

$$\mathbb{R}^n = E_1(\omega) \oplus E_2(\omega) \oplus \ldots E_r(\omega). \tag{6.117}$$

All the subspaces $E_i(\omega)$ are measurable and are random invariant subspaces of $\Phi(t, \omega)$, that is, for $i = 1, 2, \ldots, r$,

$$\Phi(t, \omega)E_i(\omega) = E_i(\theta_t\omega), \quad \text{for all } t \in \mathbb{R}, \ \mathbb{P}-\text{a.s.} \tag{6.118}$$

We define stable, center, and unstable subspaces as follows:

$$E^s(\omega) \triangleq \oplus_{\lambda_i<0} E_i(\omega), \tag{6.119}$$

$$E^c(\omega) \triangleq \oplus_{\lambda_i=0} E_i(\omega), \tag{6.120}$$

$$E^u(\omega) \triangleq \oplus_{\lambda_i>0} E_i(\omega). \tag{6.121}$$

Therefore,

$$\mathbb{R}^n = E^s(\omega) \oplus E^c(\omega) \oplus E^u(\omega),$$

and the linear cocycle $\Phi(t, \omega)$ can be decomposed as

$$\Phi(t, \omega) = (\Phi_s(t, \omega), \Phi_c(t, \omega), \Phi_u(t, \omega)).$$

The solution mapping for the nonlinear RDE (6.115), that is, the nonlinear cocycle φ, is

$$\varphi(t, \omega, u_0) = \Phi(t, \omega)u_0 + \int_0^t \Phi(t-\tau, \omega)F(\theta_\tau\omega, u(\tau))\, d\tau. \tag{6.122}$$

Let F^s, F^c, F^u be the projections of F to the stable, center, and unstable subspaces E^s, E^c, and E^u, respectively. Similarly, φ^s, φ^c, and φ^u are the projections of the nonlinear cocycle φ to these subspaces.

The random center manifold

$$M^c(\omega) = \{(x^c, h^c(\omega, x^c)) \colon x^c \in E^c\} \tag{6.123}$$

is the graph of a measurable mapping $h^c(\omega, \cdot)\colon E^c \to E^s \oplus E^u$ such that $h^c(\omega, 0) = 0$ and $Dh^c(\omega, 0) = 0$ (the tangency condition). It is called a Lipschitz center manifold if the mapping $h^c(\omega, \cdot)\colon E^c \to E^s \oplus E^u$ is Lipschitz and the tangency condition is absent. The mapping h^c satisfies the following Lyapunov-Perron integral equation [38, 39]:

$$\begin{aligned} h^c(\omega, x^c) = &\int_{-\infty}^0 \Phi_s^{-1}(\tau, \omega)F^s(\theta_\tau\omega, (\varphi^c(\tau, \omega, x^c, h^c(\omega, x^c)), h^c(\theta_\tau\omega, \varphi^c \\ &\times (\tau, \omega, x^c, h^c(\omega, x^c)))))\, d\tau \\ &+ \int_{\infty}^0 \Phi_u^{-1}(\tau, \omega)F^u(\theta_\tau\omega, (\varphi^c(\tau, \omega, x^c, h^c(\omega, x^c)), h^c(\theta_\tau\omega, \varphi^c \\ &\times (\tau, \omega, x^c, h^c(\omega, x^c)))))\, d\tau, \end{aligned} \tag{6.124}$$

for $x^c \in E^c$.

When the center subspace E^c is absent, then

$$\mathbb{R}^n = E^s(\omega) \oplus E^u(\omega),$$

and $\Phi(t, \omega)$ can be decomposed as

$$\Phi(t, \omega) = (\Phi_s(t, \omega), \Phi_u(t, \omega)).$$

In this case, the random stable manifold

$$M^s(\omega) = \{(x^s, h^s(x^s, \omega)): \ x^s \in E^s(\omega)\} \tag{6.125}$$

is the graph of a mapping h^s satisfying the following Lyapunov-Perron integral equation:

$$\begin{aligned} h^s(x^s, \omega) = \int_{\infty}^{0} &\Phi_u(-\tau, \theta_\tau\omega) F^u(\theta_\tau\omega, (\varphi^s(\tau, \omega, x^s, h^s(\omega, x^s)), h^s(\theta_\tau\omega, \varphi^s \\ &\times (\tau, \omega, x^s, h^s(\omega, x^s))))) \, d\tau, \end{aligned} \tag{6.126}$$

for $x^s \in E^s$. Similarly, the random unstable manifold

$$M^u(\omega) = \{(x^u, h^u(x^u, \omega)): \ x^u \in E^u(\omega)\} \tag{6.127}$$

is the graph of a mapping h^u satisfying the following Lyapunov-Perron integral equation:

$$\begin{aligned} h^u(x^u, \omega) = \int_{-\infty}^{0} &\Phi_s(-\tau, \theta_\tau\omega) F^s(\theta_\tau\omega, (\varphi^u(\tau, \omega, x^u, h^u(\omega, x^u)), h^u(\theta_\tau\omega, \varphi^u \\ &\times (\tau, \omega, x^u, h^u(\omega, x^u))))) \, d\tau, \end{aligned} \tag{6.128}$$

for $x^u \in E^u$.

Let us look at an example.

Example 6.55 Consider a nonlinear stochastic system in $\mathbb{R}^2$:

$$dX_1 = 2X_1 dt + X_2^2 dt + X_1 dB_t^1, \tag{6.129}$$

$$dX_2 = -4X_2 dt + 6X_2 dB_t^2. \tag{6.130}$$

The origin is an equilibrium state. As in Example 6.48, we first convert this system of SDEs to a system of RDEs. Define new system state variables

$$Y_1 = F_1 X_1, \quad Y_2 = F_2 X_2, \tag{6.131}$$

with $F_1 = e^{\frac{1}{2}t - B_t^1}$ and $F_2 = e^{18t - 6B_t^2}$ as stochastic integrating factors. The new RDE system is

$$\dot{Y}_1 = 2Y_1 + \frac{F_1}{F_2^2} Y_2^2, \tag{6.132}$$

$$\dot{Y}_2 = -4Y_2. \tag{6.133}$$

As in Example 6.40, the multiplicative ergodic theorem holds for the corresponding linear cocycle (which is actually a deterministic linear system). The Lyapunov exponents are $\lambda_1 = 2, \lambda_2 = -4$, with the corresponding Oseledets spaces $E_1 = \{y_2 = 0\}$ and $E_2 = \{y_1 = 0\}$, respectively. Note that $\mathbb{R}^2 = E_1 \bigoplus E_2$. The nonlinear term $\frac{F_1}{F_2^2} Y_2^2$ is locally Lipachitz continuous. By the discussions in this subsection, the local random stable and unstable manifolds for the RDE system (6.132) and (6.133) exist. By the inverse transform of (6.131), the local random stable and unstable manifolds for the original SDE system (6.129) and (6.130) exist.

Remark 6.56 There are other "invariant structures," such as invariant measures, that may also help us understand stochastic dynamics.

6.6 Problems

6.1
Consider a linear system

$$\dot{x} = -x, \ \ \dot{y} = 2z, \ \ \dot{z} = 3z, \ \ (x(0), y(0), z(0)) = (x_0, y_0, z_0).$$

Solve the equations and find the flow generated by this system $\varphi(t, x_0, y_0, z_0) = (x(t, x_0, y_0, z_0), y(t, x_0, y_0, z_0))^{\mathrm{T}}$.

What are eigenvalues and the corresponding stable, center, and unstable eigenspaces E^s, E^c, E^u for this linear system? What are stable, center, and unstable manifolds $W^s(0,0,0)$, $W^c(0,0,0)$, $W^u(0,0,0)$?

6.2
Consider a linear system

$$\dot{x} = y, \ \ \dot{y} = -x, \ \ \dot{z} = -z, \ \ (x(0), y(0), z(0)) = (x_0, y_0, z_0).$$

Solve the equations and find the flow generated by this system $\varphi(t, x_0, y_0, z_0) = (x(t, x_0, y_0, z_0), y(t, x_0, y_0, z_0))^{\mathrm{T}}$.

What are eigenvalues and the corresponding stable, center, and unstable eigenspaces E^s, E^c, E^u for this linear system? What are stable, center, and unstable manifolds $W^s(0,0,0)$, $W^c(0,0,0)$, $W^u(0,0,0)$?

6.3
Consider a linear system

$$\dot{x}_1 = -3x_1 + x_2,$$
$$\dot{x}_2 = -4x_2,$$
$$\dot{x}_3 = 5x_3 - x_4,$$
$$\dot{x}_4 = 6x_4.$$

Does this system satisfy the exponential dichotomy? What are the dichotomy constants?

6.4
Consider a linear system

$$\dot{x}_1 = -2x_1 + 5x_2,$$
$$\dot{x}_2 = -3x_2,$$
$$\dot{x}_3 = x_4 + x_5,$$
$$\dot{x}_4 = 4x_4 + x_5,$$
$$\dot{x}_5 = 7x_5.$$

Does this system satisfy a pseudo-exponential dichotomy? What are the associated dichotomy constants K, a, and b?

6.5 Stable and unstable manifolds
Consider a nonlinear system

$$\dot{x} = -x + y^2, \ \dot{y} = y, \ (x(0), y(0)) = (x_0, y_0).$$

Solve the equations and find the flow generated by this system $\varphi(t, x_0, y_0) = (x(t, x_0, y_0), y(t, x_0, y_0))^{\mathrm{T}}$.

Find the stable and unstable manifolds (you may use a method of your choice) at the equilibrium point (0, 0) and sketch the phase portrait.

6.6 Center manifold
Consider a nonlinear system

$$\dot{x} = -x - y^2, \dot{y} = 2xy.$$

(a) Find the equilibrium point, find the linearized system at the equilibrium point, and sketch the phase portrait for the linearized system (indicating eigenspaces E^s, E^c, E^u, if any). Is the equilibrium point hyperbolic or nonhyperbolic?
(b) Does the center manifold theorem apply here? If yes, find a local center manifold W^c_{loc} (approximately) at this equilibrium point. What is the dynamical system

restricted on this center manifold? This provides a reduced dynamical system (i.e., dimension is reduced). Sketch the phase portrait for this reduced dynamical system. Is the equilibrium point for the reduced system stable, asymptotically stable, or unstable? Is the equilibrium point for the original system stable, asymptotically stable, or unstable?

(c) Sketch the (local) phase portrait of the original nonlinear system.

6.7 Integral reformulation of a nonautonomous system
Consider a nonautonomous system in $\mathbb{R}^n$

$$\dot{x} = Ax + f(x, t),$$

where A is a real $n \times n$ matrix. Let P be an orthogonal projection operator from $\mathbb{R}^n$ to a k-dimensional subspace, that is, it is a bounded linear operator satisfying $P^2 = P$. Denote $Q = I - P$, which is an orthogonal projection operator from $\mathbb{R}^n$ to a $(n-k)$-dimensional subspace. Verify that the equation may be reformulated as the following integral equation:

$$x(t) = \int_{-\infty}^{t} e^{(t-s)A} Pf(x(s), s)ds + \int_{\infty}^{t} e^{(t-s)A} Qf(x(s), s)ds.$$

In particular, the autonomous system in $\mathbb{R}^n$,

$$\dot{x} = Ax + f(x),$$

may be reformulated as the following integral equation:

$$x(t) = \int_{-\infty}^{t} e^{(t-s)A} Pf(x(s))ds + \int_{\infty}^{t} e^{(t-s)A} Qf(x(s))ds.$$

6.8 An example of cocycle
Consider a nonlinear SDE $dX_t = X_t^2 dt + X_t dB_t$ with deterministic initial data $X_0 = x$.

(a) Solve this SDE. Does the solution explode (becoming ∞) at some random time for some x? Does the solution exist for $x = 0$?
(b) The solution for the original SDE depends on time, sample, and initial position: t, ω, and x. So let us denote it as $\varphi(t, \omega, x)$. Calculate $\varphi(0, \omega, x)$ and $\varphi(t+s, \omega, x)$ with $s < t$. Is there any relation between $\varphi(t+s, \omega, x)$ and $\varphi(t, \omega, \varphi(s, \omega, x))$?

6.9 Another example of cocycle
Consider a nonlinear SDE $dX_t = X_t^{\gamma} dt + X_t dB_t$ with deterministic initial data $X_0 = x > 0$. Here $\gamma \in (0, 1)$.

(a) Solve this SDE.
(b) The solution for this SDE depends on time, sample, and initial position: t, ω, and x. So let us denote it as $\varphi(t, \omega, x)$. Does $\varphi(t, \omega, x)$ satisfy the cocycle property for $x > 0$? Prove or disprove it.
(c) Is there a stationary orbit for this system?

6.10 A cocycle defined by an SDE system

Consider a system of SDEs $dX_t = -X_t dt + 2dB_t^1$, $dY_t = -3Y_t dt + 4dB_t^2$ with initial data $(X_0, Y_0) = (x, y) \in \mathbb{R}^2$. Find the solution and show that it defines a random dynamical system.

6.11

First find the solution $\varphi(t, \omega, x)$ for the scalar stochastic differential equation

$$dX_t = (\alpha X_t - X_t^2)dt + \sigma X_t \circ dB_t, \ X_0 = x,$$

where α and σ are real constants, and then show that $\varphi(t, \omega, x)$ is a cocycle. Are there any stationary orbits for this system?

6.12

First find the solution $\varphi(t, \omega, x)$ for the scalar stochastic differential equation

$$dX_t = (\alpha X_t - X_t^3)dt + \sigma X_t \circ dB_t, \ X_0 = x,$$

where α and σ are real constants, and then show that $\varphi(t, \omega, x)$ is a cocycle. Are there any stationary orbits for this system?

6.13

First find the solution $\varphi(t, \omega, x)$ for the scalar stochastic differential equation

$$dX_t = (\alpha - X_t^2)dt + \sigma X_t \circ dB_t, \ X_0 = x,$$

where α and σ are real constants, and then show that $\varphi(t, \omega, x)$ is a cocycle. Are there any stationary orbits for this system?

6.14

Consider a nonlinear scalar SDE

$$dX_t = (-X_t + X_t^2)dt + \varepsilon dB_t, \ X_0 = x,$$

with small parameter $0 < \varepsilon \ll 1$. Sketch the phase portrait for the unperturbed system ($\varepsilon = 0$). It has an attractor $\{0\}$ and a repeller $\{1\}$. Numerically generate solution paths $X_t(\omega)$ starting from various initial points X_0. What do you see as the impact of small noise on the attractor and the repeller?

6.15
Consider a simple pendulum under small external random forcing $\ddot{X} + \sin(X) = \varepsilon dB_t$. It may be written as a system of two nonlinear SDEs

$$dX_t = Y_t dt, \quad dY_t = -\sin(X_t)dt + \varepsilon dB_t, \quad (X_0, Y_0) = (x, y),$$

with small parameter $0 < \varepsilon \ll 1$. Sketch or numerically generate the phase portrait for the unperturbed system ($\varepsilon = 0$), in a domain surrounding $X \in (-\pi, \pi)$ and $Y \in (-2, 2)$. It has a center $(0, 0)$ and two saddles $(\pm\pi, 0)$. Numerically generate solution paths $(X_t(\omega), Y_t(\omega))$ starting from various initial points (X_0, Y_0). What do you see as the impact of small noise on the motion of the pendulum?

6.16 Stochastic Duffing–van der Pol system
Consider the stochastic system

$$\begin{aligned} dX_t &= Y_t dt, \\ dY_t &= \left(aX_t + bY_t - X_t^3 - X_t^2 Y_t\right)dt + c\, X_t\, dB_t, \end{aligned}$$

where B_t is a standard scalar Brownian motion and a, b, c are real parameters.

(a) Apply Itô's formula in differential form to the energy $E = \frac{1}{2}(X_t^2 + Y_t^2)$.
(b) Write down the Fokker-Planck equation for this system.

6.17 Lyapunov exponents
Consider a linear scalar stochastic differential equation

$$dX_t = rX_t dt + \alpha X_t dB_t, \; X_0 = x,$$

where r and α are real constants. The solution is $X_t = x\exp((r - \frac{1}{2}\alpha^2)t + \alpha B_t)$. Find out the Lyapunov exponent and the Oseledets space for this random system.

6.18 Integrability conditions in MET
This is an analytical method to verify the integrability conditions in Example 6.40, to apply the multiplicative ergodic theorem. Note that

$$\mathbb{E} \sup_{0 \le t \le 1} e^{\frac{3}{2}t + B_t^1} = \int_{-\infty}^{\infty} \sup_{0 \le t \le 1} e^{\frac{3}{2}t + z} \frac{1}{\sqrt{2\pi t}} e^{-\frac{z^2}{2t}} dz.$$

Show that the integral $\int_{-\infty}^{\infty} [\cdots] dz$ is finite.

6.19 A linear cocycle defined by an SDE system
Consider a system of SDEs

$$\begin{aligned} dX_t &= X_t dt + Y_t dB_t^1, \; X_0 = x, \\ dY_t &= -4Y_t dt + 0 dB_t^2, \; Y_0 = y. \end{aligned}$$

Find the linear cocycle $\Phi(t, \omega)$ generated by this linear system. Check whether the multiplicative ergodic theorem applies, and then try to find out the Oseledets spaces, Lyapunov exponents, and Oseledets splitting.

6.20 Another linear cocycle defined by an SDE system

Consider a system of SDEs

$$dX_t = 2X_t dt + 8Y_t dB_t^1, \; X_0 = x,$$
$$dY_t = 0dt - Y_t dB_t^2, \; Y_0 = y.$$

Find the linear cocycle $\Phi(t, \omega)$ generated by this linear system. Check whether the multiplicative ergodic theorem applies, and then try to find out the Oseledets spaces, Lyapunov exponents, and Oseledets splitting.

6.21 A linear cocycle defined by a RDE system

Consider a system of RDEs (differential equations with random coefficients)

$$\dot{X} = a(\theta_t \omega)X, \; X_0 = x,$$
$$\dot{Y} = b(\theta_t \omega)Y, \; Y_0 = y,$$

where the driving system $\theta_t \colon \Omega \to \Omega$ is assumed to be ergodic. Find the linear cocycle $\Phi(t, \omega)$ generated by this linear system. Impose appropriate conditions on a and b so that the multiplicative ergodic theorem applies, then try to find out the Oseledets spaces, Lyapunov exponents, and Oseledets splitting.

6.22 A nonlinear cocycle defined by an SDE system

Consider a system of nonlinear SDEs

$$dX_t = 2X_t dt + Y_t^2 dB_t^1, \; X_0 = x,$$
$$dY_t = 0dt - Y_t dB_t^2, \; Y_0 = y.$$

Find the nonlinear cocycle $\varphi(t, \omega, x, y)$ generated by this system. Linearize, that is, take the Jacobian of $\varphi(t, \omega, x, y)$ at the equilibrium point (0, 0) to obtain a linear cocycle $\Phi(t, \omega)(x, y)^{\mathrm{T}} := D\varphi(t, \omega, 0, 0)(x, y)^{\mathrm{T}}$. Is the linear cocycle $\Phi(t, \omega)$ just the cocycle generated by the original SDE system linearized at the equilibrium point (0, 0), that is, by the linear system

$$dX_t = 2X_t dt + 0dB_t^1, \; X_0 = x,$$
$$dY_t = 0dt - Y_t dB_t^2, \; Y_0 = y?$$

Check whether the Hartman-Grobman theorem applies and then explain what the theorem implies.

6.23
Consider a system of nonlinear SDEs

$$dX_t = (3X_t + Y_t^2)dt + (2Y_t + X_t^3)dB_t^1, \; X_0 = x \in \mathbb{R}^1,$$
$$dY_t = (X_t \sin X_t - 6Y_t)dt + Y_t^2 dB_t^2, \; Y_0 = y \in \mathbb{R}^1.$$

(a) What is the linearized SDE system at the equilibrium point (0, 0)? Does the multiplicative ergodic theorem apply to this linear system? If so, what are the Lyapunov exponents, Oseledets spaces, and Osledets splitting? Show details of your work.
(b) Does the Hartman-Grobman theorem apply to this nonlinear system? Show details of your work and then explain what the theorem implies.

6.24
Consider a system of nonlinear SDEs

$$dX_t = (9X_t + Y_t^2)dt + (2Y_t + X_t^3)dB_t^1, \; X_0 = x \in \mathbb{R}^1,$$
$$dY_t = (\sin(X_t^2) - 5Y_t)dt + Y_t^3 dB_t^2, \; Y_0 = y \in \mathbb{R}^1.$$

(a) What is the linearized SDE system at the equilibrium point (0, 0)? Does the multiplicative ergodic theorem apply to this linear system? If so, what are the Lyapunov exponents, Oseledets spaces, and Oseledets splitting? Show details of your work.
(b) Does the stochastic Hartman-Grobman theorem apply to this nonlinear system? Show details of your work and then explain what the theorem implies.

6.25
Consider a nonlinear SDE system

$$dX_t = -X_t dt + \varepsilon X_t dB_t, \quad x \in \mathbb{R}^1, \tag{6.134}$$
$$dY_t = (Y_t + X_t^2)dt + \varepsilon Y_t dB_t, \quad y \in \mathbb{R}^1, \tag{6.135}$$

where ε is a positive parameter and B_t is a scalar Brownian motion. Check if the random stable and unstable manifolds exist.

6.26 Impact of noise on a stable manifold
Consider a nonlinear SDE system

$$dX_t = (-X_t + Y_t^2)dt, \; dY_t = Y_t dt + \varepsilon dB_t, \; (X_0, Y_0) = (x, y) \in \mathbb{R}^2,$$

where ε is a small nonnegative real parameter. The corresponding deterministic system ($\varepsilon = 0$) is known to have an invariant stable manifold $W^s(0,0) = \{(x,y)^{\mathrm{T}} : y = 0\}$.

(a) Solve the two SDEs and denote the solutions as X_t^ε and Y_t^ε.
(b) What can you say about the impact of small noise on the deterministic invariant stable manifold $W^s(0, 0)$?

6.27 Impact of noise on an unstable manifold
Find an approximate random unstable manifold for the system in Example 6.55 when the noise terms are multiplied by a sufficiently small positive parameter ε.

7
Dynamical Systems Driven by Non-Gaussian Lévy Motions

In this chapter, we consider stochastic differential equations with non-Gaussian processes. In fact, as there are so many non-Gaussian processes, we will focus on α-stable Lévy motions and dynamical systems driven by these Lévy motions.

A matrix K is called normal if $KK^* = K^*K$, where K^* is the conjugate transpose of K. Normal matrices are precisely those matrices that are unitarily similar to a diagonal matrix (i.e., diagonalizable by a unitary transformation). This is a very special class of matrices. All other matrices are not diagonalizable in this fashion and are nonnormal. Likewise, as we have learned in Chapter 2, a normal (or Gaussian) random variable has a specific distribution and is indeed very special. All other random variables are nonnormal (or non-Gaussian).

Brownian motion is defined in terms of normal random variables (i.e., Gaussian random variables), whereas a α-stable Lévy motion is defined via stable random variables (which are non-Gaussian). Brownian motion is a Gaussian process with independent and stationary increments. Lévy motions, especially the α-stable Lévy motions, are non-Gaussian processes with independent and stationary increments that mimic many fluctuating processes in complex systems in physics, geophysics, biophysics, chemistry, engineering, and other disciplines.

Although Brownian motion has been widely used in describing fluctuations in mathematical modeling of complex systems under uncertainty, many complex phenomena involve non-Gaussian Lévy motions, especially α-stable Lévy motions.

Levy motions are thought to be appropriate models for non-Gaussian fluctuations [244, 27]. A Lévy motion $L(t)$ has independent and stationary increments, that is, increments $\Delta L(t) = L(t + \Delta t) - L(t)$ are stationary (therefore ΔL has no statistical dependence on t) and independent for any nonoverlapping time lags Δt. Moreover, its sample paths are only continuous in probability, namely, $\mathbb{P}(|L(t) - L(t_0)| \geq \delta) \to 0$ as $t \to t_0$ for any positive δ. This continuity is weaker than the usual continuity in time. This generalizes Brownian motion $B(t)$, which satisfies all these three conditions, and *additionally*, (1) almost every sample path of Brownian motion is continuous in time in the usual sense and (2) increments for Brownian motion have normal distribution.

Lévy motions arise as models for fluctuations in various systems. For instance, in the turbulent motions of rotating annular fluid flows [289], a passive tracer particle may experience a series of "pauses," when the particle is trapped by a vortex for a random time period, and "flights" (or "jumps"), when the particle is ejected to the jet flow, before being trapped in another vortex after a random time period. The duration of flight events is found to follow an α-stable distribution. The α-stable Lévy motions have also been found to be reasonable models for the following phenomena or systems: Lagrangian drifts in certain oceanic fluid flows and anomalous nonlocal surface diffusions [294], a class of biological evolution [142], random search [283], and finance [208, 66], among others [258, 243].

This chapter is a brief introduction to non-Gaussian stochastic dynamical systems. We first present an example of geophysical systems modeled by a differential equation with α-stable Lévy motion (Section 7.1). Then, we introduce Lévy motions (Section 7.2), and particularly, α-stable Lévy motions (Section 7.3). Following are discussions on stochastic differential equations with non-Gaussian Lévy motions (Section 7.4). Finally, we investigate mean exit time (Section 7.5), escape probability (Section 7.6), and Fokker-Planck equations (Section 7.7) for these non-Gaussian stochastic dynamical systems.

7.1 Modeling via Stochastic Differential Equations with Lévy Motions

Stochastic differential equations with α-stable Lévy motions emerge as mathematical models for various complex systems. Let us look at an example of SDEs with α-stable Lévy motions arising in climate dynamics. Other stochastic models for climate dynamics are discussed in, for example, [131, 134].

Example 7.1 In the Greenland ice core project, a large number of ice cores were drilled. Analysis of isotopes and other atmospheric chemicals in the cores have revealed dramatic climatic variations during the last ice age. Particularly, the calcium (Ca) signal for the ice core carries climatic information, as its logarithm, ln(Ca), is negatively correlated with temperature. Taking $X = \ln(\mathrm{Ca})$ as the climate state and analyzing the ice core records, Ditlevsen [73] found that X is modeled by the following SDE:

$$dX_t = -U'(X_t)dt + \sigma_1 dY_t + \sigma_2 dL_t^\alpha, \tag{7.1}$$

where U is a potential function, σ_1, σ_2 are noise intensities, Y_t satisfies a stochastic differential equation with Brownian motion,

$$dY_t = -Y_t dt + \sqrt{1 + Y_t^2} dB_t,$$

and L_t^{α} is an α-stable Lévy motion with $\alpha = 1.75$. Imkeller, Pavlyukevich, and colleagues [135, 136] further analyzed dynamical behaviors (e.g., exit time estimates for certain climate events) for this model.

The study of non-Gaussian dynamical systems is still in its infancy [164, 7]. But some interesting works are emerging [233, 97]. More recently, partial differential equations with α-stable Lévy motions have also been actively investigated [227, 71].

In this chapter, we consider the following stochastic differential equation (SDE) with a Lévy motion L_t as well as a Brownian motion B_t, in $\mathbb{R}^n$:

$$dX_t = f(X_t)dt + \sigma(X_t)dB_t + dL_t,$$

where f is a vector field (or drift), B_t is a Brownian motion, L_t is a Lévy motion (especially, an α-stable Lévy motion), and σ is an $n \times n$ matrix function. In engineering and applied sciences literature, this SDE is also written as

$$\dot{x} = f(x) + \sigma(x)\dot{B}_t + \dot{L}_t.$$

7.2 Lévy Motions

Recall that a Brownian motion B_t in $\mathbb{R}^n$ is a Gaussian process with the following properties:

(i) $B_0 = 0$, a.s.
(ii) B_t has independent increments: for $t_1 < t_2 < \cdots < t_{n-1} < t_n$, the random variables $B_{t_2} - B_{t_1}, \ldots,$ and $B_{t_n} - B_{t_{n-1}}$ are independent.
(iii) B_t has stationary increments: $B_t - B_s$ and B_{t-s} have the same distribution $N(0, (t-s)I)$ for $s < t$, where I is the $n \times n$ identity matrix.
(iv) B_t has continuous sample paths, a.s.

Lévy motions [7, p. 43] are defined similarly.

Definition 7.2 A Lévy motion, or a Lévy process, L_t or $L(t)$, is a stochastic process satisfying the following conditions:

(i) $L_0 = 0$, a.s.
(ii) Independent increments: for $t_1 < t_2 < \cdots < t_{n-1} < t_n$, the random variables $L_{t_2} - L_{t_1}, \ldots, L_{t_n} - L_{t_{n-1}}$ are independent.
(iii) Stationary increments: $L_t - L_s$ and L_{t-s} have the same distribution.
(iv) Stochastically continuous sample paths (i.e., sample paths are continuous in probability): for all $\delta > 0$ and all $s \geq 0$,

$$\mathbb{P}(|L_t - L_s| > \delta) \to 0$$

as $t \to s$.

Remark 7.3 The continuity in probability is weaker than the usual continuity. Owing to (i)–(iii), the condition (iv) is equivalent to the following condition: for all $\delta > 0$,

$$\mathbb{P}(|L_t| > \delta) \to 0$$

as $t \to 0$.

Brownian motion B_t, as a Gaussian stochastic process, is a special Lévy motion. In general, a Lévy motion is a non-Gaussian stochastic process.

Remark 7.4 For a stochastically continuous process, there exists a modification (i.e., a version) whose paths are continuous from the right and have left limits ("cadlag") at every time [7, Chapter 2]. We are going to take this modification for the Lévy motion. Therefore, the paths of a Lévy motion are cadlag. Note that a cadlag function can only have (at most) a countable number of jumps (see Section 7.2.1), and the jumps are their only possible discontinuities in time.

7.2.1 Functions that have One-Side Limits

Let f be a real valued function on $\mathbb{R}^1$. The set of points at which f is not continuous but with right limit is countable. Define

$$S \triangleq \{x \in \mathbb{R}^1 : f \text{ is discountinous at } x \text{ but the right limit } \lim_{y \to x+} f(y) \text{ exists}\}. \tag{7.2}$$

Then S is countable [155, 310]. Thus the discontinuous points for a right-continuous real function are countable. Recall that a set is countable if either it is finite or it has a one-to-one correspondence with the set $\mathbb{N}$ of the natural numbers.

In fact, for each natural number n, define

$$E_n \triangleq \{x \in \mathbb{R}^1 : \text{ There exists } \delta > 0 \text{ such that } |f(x') - f(x'')| < \frac{1}{n}$$
$$\text{whenever } x', x'' \in (x - \delta, x + \delta)\}.$$

Then $\cap_{n=1}^{\infty} E_n$ is the set of points of continuity for f. Now we only need to show that $S \setminus E_n$, $n = 1, 2, \ldots$, are countable, as this will imply that S is countable.

To this end, for any fixed n, let $x \in S \setminus E_n$. Then there exists a $\delta > 0$ such that

$$|f(x') - f(x + 0)| < \frac{1}{2n} \text{ whenever } x' \in (x, x + \delta).$$

Hence $|f(x') - f(x'')| < \frac{1}{n}$ whenever $x', x'' \in (x, x + \delta)$. So $(x, x + \delta) \subset E_n$, that is, every point x in $S \setminus E_n$ is the left end point of an open interval $I_x = (x, x + \delta)$, and I_x does not intersect with $S \setminus E_n$. For two distinct points x_1, x_2 in $S \setminus E_n$, I_{x_1}

and I_{x_2} do not intersect. Thus the family of intervals $\{I_x : x \in S \setminus E_n\}$ is countable, and so $S \setminus E_n$ is countable. This verifies that the discontinuous points for a right-continuous real function are countable.

It is also true that the set of points at which f is not continuous but with left limit is countable.

This also implies that the points of discontinuity for a real cadlag function (i.e., right continuous with left limit at every point) are countable. Note that a right-continuous function also has right limit at every point.

Moreover, let f be a real function that is either right continuous or left continuous at every point. Then its discontinuous points are countable.

If f is an increasing function, then both left and right limits exist at every point [6, p. 95]. Thus the discontinuous points for an increasing function are countable.

A function is of finite variation on $[0, T]$ if and only if it can be expressed as the difference of two increasing functions. Thus the discontinuous points for a function of finite variation are countable.

7.2.2 Lévy-Itô Decomposition

Let L_t be a Lévy motion in $\mathbb{R}^n$. Define the jump process $\Delta L(t)$ by

$$\Delta L(t) \triangleq L_t - L_{t-}, \quad t \geq 0, \tag{7.3}$$

where L_{t-} is the left limit of L_t at time t.

Let us count the jumps of specified size. For a Borel set $S \in \mathcal{B}(\mathbb{R}^n \setminus \{0\})$ and for $t > 0$, define

$$N(t, S)(\omega) \triangleq \#\{0 \leq s < t\colon \Delta L(s)(\omega) \in S\} \tag{7.4}$$

whenever $\omega \in \Omega_0$ and $N(t, S)(\omega) \triangleq 0$ whenever $\omega \in \Omega_0^c$.

Define a Borel measure ν on $\mathcal{B}(\mathbb{R}^n \setminus \{0\})$ by

$$\nu(S) \triangleq \mathbb{E}N(1, S)(\omega). \tag{7.5}$$

Furthermore, define the compensated Poisson random measure by

$$\tilde{N}(t, S) \triangleq N(t, S) - t\,\nu(S). \tag{7.6}$$

We recall the Lévy-Itô decomposition theorem [7, p. 126].

Theorem 7.5 (Lévy-Itô decomposition) *If L_t is a Lévy motion in $\mathbb{R}^n$, then there exist a vector $b \in \mathbb{R}^n$, a covariance matrix Q, and an independent Poisson random* *measure* N *on* $\mathbb{R}^+ \times (\mathbb{R}^n \setminus \{0\})$ *such that, for each* $t \geq 0$,

$$L_t = bt + B_t^Q + \int_{\|y\|<1} y\tilde{N}(t, dy) + \int_{\|y\|\geq 1} yN(t, dy), \tag{7.7}$$

where $N(dt, dx)$ *is the Poisson random measure (quantifying the number of jumps of* L_t*),* $\tilde{N}(dt, dx) \triangleq N(dt, dx) - \nu(dx)dt$ *is the compensated Poisson random measure,* $\nu(S) \triangleq \mathbb{E}N(1, S)$ *is the jump measure, and* B_t^Q *is an independent* n*-dimensional Brownian motion with covariance matrix* Q.

A Brownian motion with covariance matrix Q has been introduced in Definition 3.11. It means that $B_t^Q = \sigma B_t$, where σ is an $n \times m$ real nonzero matrix and B_t is an m-dimensional standard Brownian motion, such that $Q = \sigma\sigma^{\mathrm{T}}$. For this reason, we occasionally write $\sigma = Q^{\frac{1}{2}}$. Therefore,

$$B_t^Q = Q^{\frac{1}{2}} B_t. \tag{7.8}$$

The number 1 in $\|y\| < 1$ and $\|y\| \geq 1$ allows us to specify relatively "small" and "large" jumps, respectively [7, p. 364]. It may be replaced by an arbitrary positive number c. The standard or usual Brownian motion B_t in $\mathbb{R}^n$ has the identity matrix I as covariance matrix (i.e., $Q = I$).

The triplet (b, Q, ν) is called the generating triplet for the Lévy motion L_t.

7.2.3 Lévy-Khintchine Formula

The Lévy-Khintchine formula specifies the expression for the characteristic function of a Lévy motion [7, p. 45]. Recall that the indicator function, I_s, for a set S is defined as

$$I_S(y) = \begin{cases} 1, & \text{if } y \in S, \\ 0, & \text{if } y \notin S. \end{cases}$$

Theorem 7.6 (Lévy-Khintchine formula) *If* L_t *is a Lévy motion in* $\mathbb{R}^n$*, then its characteristic function is*

$$\Phi_t(u) \triangleq \mathbb{E}e^{i\langle u, L_t\rangle} = e^{t\eta(u)} \qquad \textit{for each } t \geq 0,\, u \in \mathbb{R}^n,$$

where

$$\eta(u) = ib \cdot u - \frac{1}{2}u \cdot Qu + \int_{\mathbb{R}^n\setminus\{0\}} [e^{iu\cdot y} - 1 - iI_{\{\|y\|<1\}}\, u \cdot y]\, \nu(dy) \tag{7.9}$$

for a vector $b \in \mathbb{R}^n$*, a nonnegative definite symmetric* $n \times n$ *matrix* Q*, and a Borel measure* ν *on* $\mathbb{R}^n \setminus \{0\}$ *for which* $\int_{\mathbb{R}^n\setminus\{0\}}(\|y\|^2 \wedge 1)\nu(dy) < \infty$*. Conversely, given a* *mapping of the form* (7.9)*, there exists a Lévy motion with characteristic function* $\Phi_t(u) = e^{t\eta(u)}$*. It is called the Lévy motion with triplet* (b, Q, ν)*.*

Both a dot and $\langle \cdot, \cdot \rangle$ in this theorem denote the scalar product in $\mathbb{R}^n$. We use both notations in this book.

In this theorem, (b, Q, ν) is the triplet, or generating triplet, for the Lévy motion L_t. The vector b is usually called the drift vector, Q is called the covariance matrix or diffusion matrix, and the Borel measure ν is called the *jump measure*, for L_t.

Each term appearing in the Lévy-Khintchine formula has a probabilistic significance, as emphasized in [237]. Every Lévy motion is obtained as a sum of independent processes with three types of triplets $(b, 0, 0)$, $(0, Q, 0)$, and $(0, 0, \nu)$.

Generator of a Lévy Motion

The generator A of a Lévy motion with triplet (b, Q, ν) is

$$A\varphi = b \cdot \nabla\varphi + \frac{1}{2}\mathrm{Tr}(QH(\varphi)) + \int_{\mathbb{R}^n\setminus\{0\}} [\varphi(x+y) - \varphi(x) - I_{\{\|y\|<1\}}\, y \cdot \nabla\varphi(x)]\, \nu(dy) \qquad (7.10)$$

for φ in the domain of definition of A.

7.2.4 Basic Properties of Lévy Motions

We discuss a few basic features of Lévy motions.

Transformations of Lévy Motions

If L_t is a Lévy motion with the generating triplet (b, Q, ν), then $-L_t$ is also a Lévy motion [7, p. 45] with the generating triplet $(-b, Q, \tilde{\nu})$, where the new jump measure is defined by $\tilde{\nu}(S) \triangleq \nu(-S)$ for every $S \in \mathcal{B}(\mathbb{R}^n)$. Here $-S \triangleq \{-x : x \in S\}$.

Moreover, for each real constant c, $L_t + tc$ is a Lévy motion with triplet $(b + c, Q, \nu)$. The sum of two independent Lévy motions is still a Lévy motion.

Moments of Lévy Motions

Recall from [186], for $p > 0$, $\mathbb{E}\|X_1\|_p < +\infty$ if and only if $\mathbb{E}\|X_t\|_p < +\infty$ for every $t \geq 0$. Furthermore, $\mathbb{E}\|X_1\|_p < +\infty$ if and only if $\int_{\|x\|>1} \|x\|^p d\nu(x) < +\infty$.

Variations of Lévy Motions

See Section 4.2 about functions of finite variation.

A stochastic process has finite variation if its paths have finite variation for almost all samples $\omega \in \Omega$. We say a stochastic process has finite variation if it has finite variation on every bounded and closed time interval.

By [7, Theorem 2.4.25, p. 129], we have the following result about finite variations of Lévy motions.

Theorem 7.7 (Finite variations) *A Lévy motion L_t with triplet (b, Q, ν) has finite variation if and only if the diffusion matrix $Q = 0$ and $\int_{\|x\|<1} \|x\|\nu(dx) < \infty$.*

Remark 7.8

(i) This theorem implies that Brownian motion $B(t)$ has infinite variation, because its diffusion matrix (i.e., the identity matrix) is nonzero.
(ii) For other properties of Lévy motions, see [210, 209].

A Geometric View of Jump Measure

We examine Lévy jump measures in a geometric way. For simplicity, we only consider the one-dimensional case.

A scalar Lévy motion is characterized by a drift parameter b, a nonnegative variance parameter a, and a nonnegative Borel measure ν, defined on $(\mathbb{R}^1, \mathcal{B}(\mathbb{R}^1))$ and concentrated on $\mathbb{R}^1 \setminus \{0\}$, that satisfies

$$\int_{\mathbb{R}^1\setminus\{0\}} (y^2 \wedge 1)\ \nu(dy) < \infty, \tag{7.11}$$

or equivalently,

$$\int_{\mathbb{R}^1\setminus\{0\}} \frac{y^2}{1+y^2}\ \nu(dy) < \infty. \tag{7.12}$$

Here $y^2 \wedge 1 \triangleq \min\{y^2, 1\}$. This measure ν is the so-called the Lévy jump measure of the Lévy motion $L(t)$. We also call (b, a, ν) the *generating triplet*. Note that $\nu\{(-\varepsilon, \varepsilon)^c\} < \infty$.

The integrands in the definition of Lévy jump measure $\nu(dy)$, $y^2 \wedge 1$ and $\frac{y^2}{1+y^2}$, are plotted in Figure 7.1.

Let L_t be a Lévy motion with the generating triplet (b, a, ν). It is known that a scalar Lévy motion is completely determined by the Lévy-Khintchine formula (see [7]). This says that for every one-dimensional Lévy motion L_t, there exists a real

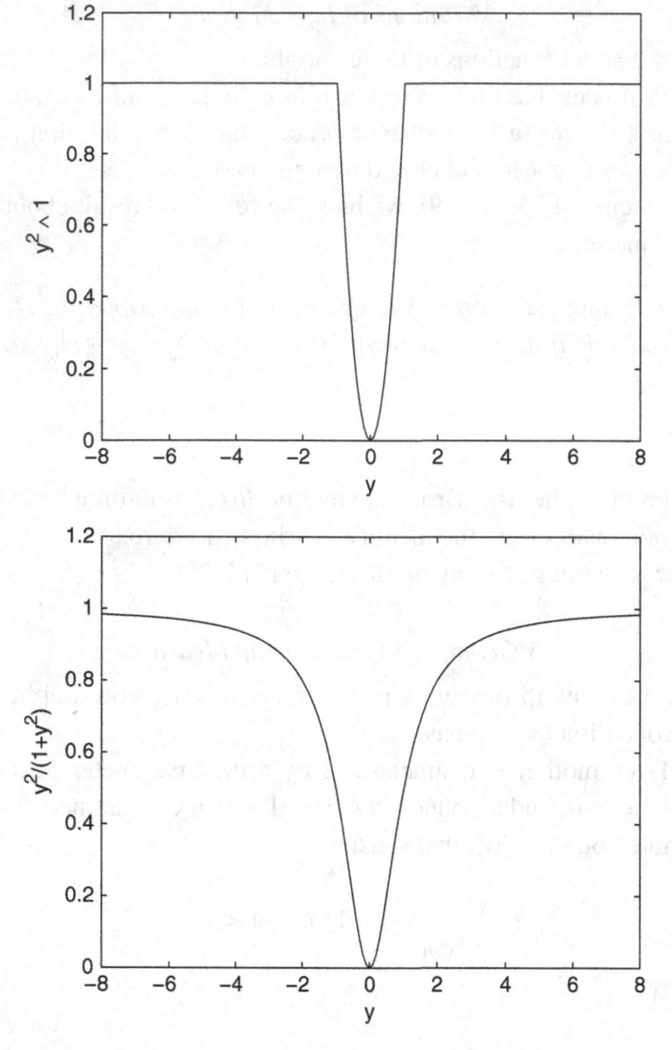

Figure 7.1 The integrands in the definition of Lévy jump measure $\nu(dy)$: $y^2 \wedge 1$ and $\frac{y^2}{1+y^2}$.

number b, a nonnegative number a, and a Borel measure ν such that

$$\mathbb{E}e^{iuL_t} = \exp\left\{ibut - at\frac{u^2}{2} + t\int_{\mathbb{R}^1\setminus\{0\}} (e^{iuy} - 1 - iuyI_{\{|y|<1\}})\nu(dy)\right\}, \quad u \in \mathbb{R}^1, \tag{7.13}$$

where I_S is the indicator function of the set S, that is, it takes value 1 on this set and takes zero value otherwise.

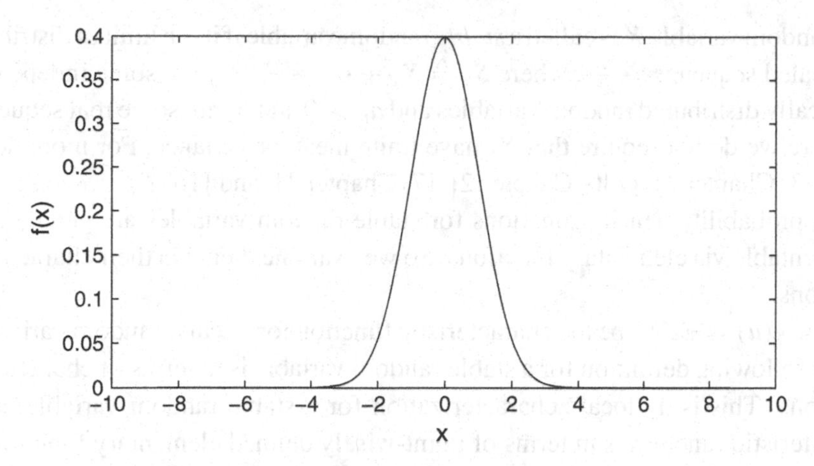

Figure 7.2 Bell shape: the probability density function for the standard Gaussian random variable $X \sim \mathcal{N}(0, 1)$.

7.3 The α-Stable Lévy Motions

We now consider a special but important class of Lévy motions, the α-stable Lévy motions ([140, p. 30] and [243, p. 113]).

7.3.1 Stable Random Variables

In this subsection, we review basic concepts for stable random variables. They will be used to define (non-Gaussian) α-stable Lévy motions L_t^α, much like normal random variables are used to define (Gaussian) Brownian motion B_t.

Gaussian Random Variables as Limits

Let $X_1, X_2, \ldots$ be a sequence of independent, identically distributed random variables with finite mean γ and finite variance σ^2. Denote $S_n \triangleq X_1 + \cdots + X_n$. By the central limit theorem in [13], $\frac{S_n - n\gamma}{\sigma\sqrt{n}}$ converges in distribution to a standard normal random variable $X \sim \mathcal{N}(0, 1)$. A normal random variable is also called a Gaussian random variable.

Figure 7.2 shows the probability density function $f(x)$ for the standard Gaussian random variable $X \sim \mathcal{N}(0, 1)$.

All other random variables are called non-Gaussian random variables. But a special class of non-Gaussian random variables, stable random variables, stands out.

Stable Random Variables as Limits

We first consider scalar and then vector stable random variables as special non-Gaussian random variables.

A random variable X is called a *stable* random variable if it is a limit in distribution of a scaled sequence $\frac{S_n - b_n}{a_n}$, where $S_n \triangleq X_1 + \cdots + X_n$, X_i are some independent, identically distributed random variables and $a_n > 0$ and b_n are some real sequences. But here we do not require that X_i have finite mean or variance. For more details, see [243, Chapter 1], [140, Chapter 2], [7, Chapter 1], and [167].

The probability density functions for stable random variables are generally not representable via elementary functions. So we examine them via their characteristic functions.

Let $\Phi_X(u) \triangleq \mathbb{E}e^{iuX}$ be the characteristic function for a scalar random variable X.

The following definition for a stable random variable is in terms of characteristic functions. This is a "local" characterization for a stable random variable, as the characteristic function is in terms of point-wisely defined elementary functions.

Definition 7.9 A scalar random variable X is stable if there exist four real parameters, that is, a stability parameter $\alpha \in (0, 2]$, a scaling parameter $\sigma > 0$, a symmetry parameter $\beta \in [-1, 1]$, and a shift parameter $\gamma \in \mathbb{R}^1$, such that its characteristic function $\Phi_X(u)$ has the following representation

(i) $0 < \alpha < 1$:

$$\Phi_X(u) = \exp\left\{i\gamma u - \sigma|u|^\alpha \left[1 - i\beta \text{sign}(u) \tan\frac{\pi\alpha}{2}\right]\right\};$$

(ii) $\alpha = 1$:

$$\Phi_X(u) = \exp\left\{i\gamma u - \sigma|u| \left[1 + i\beta\frac{2}{\pi}\text{sign}(u) \ln|u|\right]\right\};$$

(iii) $1 < \alpha < 2$:

$$\Phi_X(u) = \exp\left\{i\gamma u - \sigma|u|^\alpha \left[1 - i\beta \text{sign}(u) \tan\frac{\pi\alpha}{2}\right]\right\};$$

(iv) $\alpha = 2$:

$$\Phi_X(u) = \exp\left\{i\gamma u - \frac{1}{2}\sigma^2 u^2\right\},$$

where

$$\text{sign}(u) = \begin{cases} 1, & u > 0, \\ 0, & u = 0, \\ -1, & u < 0. \end{cases}$$

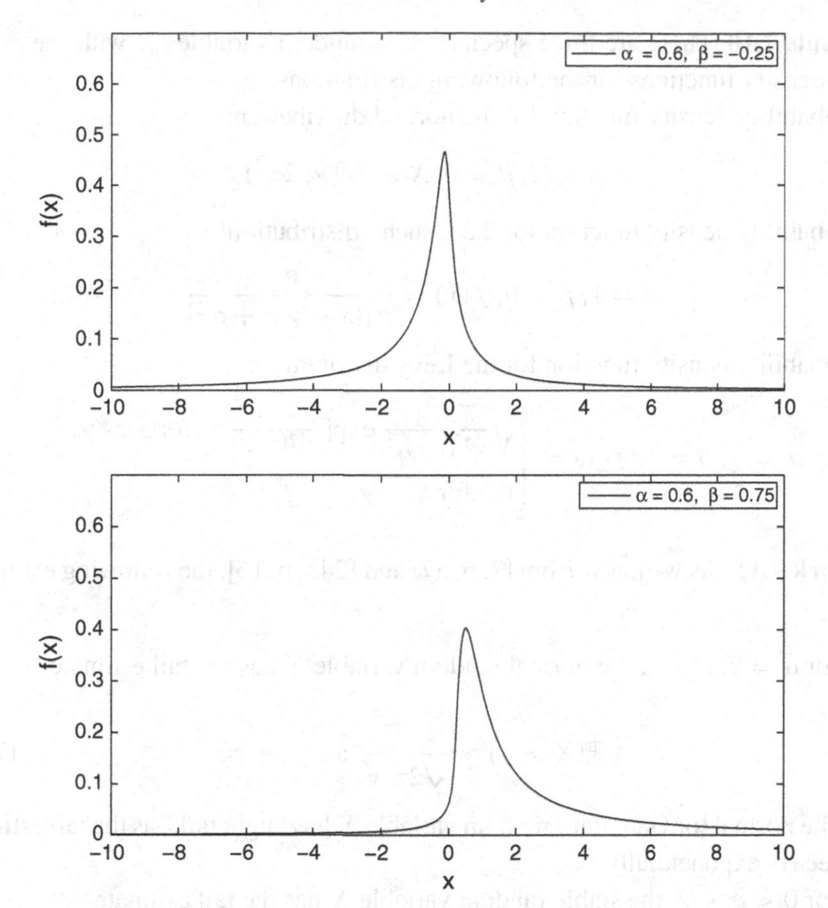

Figure 7.3 The probability density functions for the α-stable random variable $X \sim S_\alpha(\sigma, \beta, \gamma)$: $\sigma = 1, \gamma = 0$.

Note that $\Phi_X(u) = \exp\{i\gamma u - \frac{1}{2}\sigma^2 u^2\}$ is the characteristic function for a Gaussian random variable. So when $\alpha = 2$, the stable random variable is just the Gaussian random variable.

The distribution for a stable random variable is denoted as $S_\alpha(\sigma, \beta, \gamma)$. Usually, α is called the index of stability (or non-Gaussianity index), σ the scale parameter, β the skewness parameter, and γ the shift parameter. The symbol $S_\alpha(\sigma, \beta, \gamma)$ refers to either the distribution function or the probability density function for a stable random variable. To indicate the importance of the index of stability, α, we often call such a random variable the α-stable random variable.

Note that $S_2(\sigma, 0, \gamma) = \mathcal{N}(\gamma, 2\sigma^2)$, as seen in [243, pp. 7–10].

Figure 7.3 shows the probability density functions for various $\alpha, \beta, \sigma, \gamma$ values. Probability density functions for stable random variables are generated by a Matlab code of Mark Veillette.

Example 7.10 There are three special stable random variables X with the probability density functions for the following distributions.

Probability density function for the normal distribution:

$$\alpha = 2, \beta = 0, X \sim \mathcal{N}(\gamma, 2\sigma^2)$$

Probability density function for the Cauchy distribution:

$$\alpha = 1, \beta = 0, f(x) = \frac{\sigma}{\pi[(x-\gamma)^2 + \sigma^2]}$$

Probability density function for the Lévy distribution:

$$\alpha = \frac{1}{2}, \beta = 1, f(x) = \begin{cases} \sqrt{\frac{\sigma}{2\pi}} \frac{1}{(x-\gamma)^{\frac{3}{2}}} \exp[-\frac{\sigma}{2(x-\gamma)}], & \text{for } x > \gamma, \\ 0, \ \text{for } x \leq \gamma. \end{cases}$$

Remark 7.11 As we know from [7, p. 37] and [243, p. 16], the following estimates hold:

(i) For $\alpha = 2$, that is, the normal random variable X has the tail estimate

$$\mathbb{P}(X > y) \sim \frac{e^{-\frac{y^2}{2}}}{\sqrt{2\pi}\, y} \quad \text{as} \quad y \to \infty. \tag{7.14}$$

The normal (or Gaussian) random variable X has "light tail," as the tail estimate decays exponentially

(ii) For $0 < \alpha < 2$, the stable random variable X has the tail estimate

$$\lim_{y \to \infty} y^\alpha \mathbb{P}(X > y) = C_\alpha \frac{1+\beta}{2} \sigma^\alpha, \tag{7.15}$$

$$\lim_{y \to \infty} y^\alpha \mathbb{P}(X < -y) = C_\alpha \frac{1-\beta}{2} \sigma^\alpha, \tag{7.16}$$

where C_α is a positive constant; a stable random variable X (with $0 < \alpha < 2$) has "heavy tail," as the tail estimate decays polynomially.

We now discuss more detailed properties for stable random variables. See Remark 14.6 in [244, p. 80].

Basic Properties of α-Stable Random Variables

We recall some properties of stable random variables [243, Chapter 1].

Theorem 7.12

(i) If $X \sim S_\alpha(\sigma, \beta, \gamma)$ and a is a real constant, then $X + a \sim S_\alpha(\sigma, \beta, \gamma + a)$.

(ii) If X_1 and X_2 are independent stable random variables with $X_1 \sim S_\alpha(\sigma_1, \beta_1, \gamma_1)$ and $X_2 \sim S_\alpha(\sigma_2, \beta_2, \gamma_2)$, then

$$X_1 + X_2 \sim S_\alpha(\sigma, \beta, \gamma), \tag{7.17}$$

with $\sigma = (\sigma_1^\alpha + \sigma_2^\alpha)^{\frac{1}{\alpha}}$, $\beta = \frac{\beta_1\sigma_1^\alpha + \beta_2\sigma_2^\alpha}{\sigma_1^\alpha + \sigma_2^\alpha}$, and $\gamma = \gamma_1 + \gamma_2$.

(iii) If $X \sim S_\alpha(\sigma, \beta, \gamma)$ and k is a real constant, then

$$kX \sim \begin{cases} S_\alpha(|k|\sigma, sign(k)\beta, k\gamma), & for \ \alpha \neq 1, \\ S_1(|k|\sigma, sign(k)\beta, k\gamma - \frac{2}{\pi}k(\log|k|)\sigma\beta), & for \ \alpha = 1. \end{cases} \tag{7.18}$$

In particular, if $X \sim S_\alpha(1, 0, 0)$ and k is a real constant, then

$$kX \sim S_\alpha(|k|, 0, 0), \quad for\ \alpha \in (0, 2). \tag{7.19}$$

(iv) If $X \sim S_\alpha(\sigma, \beta, 0)$, then $-X \sim S_\alpha(\sigma, -\beta, 0)$, for $0 < \alpha < 2$.

(v) If X_1, X_2 are independent stable random variables with the same distribution $S_\alpha(\sigma, \beta, \gamma)$ for $\alpha \neq 1$ and A, B are positive constants, then

$$AX_1 + BX_2 \sim S_\alpha(\sigma(A^\alpha + B^\alpha)^{\frac{1}{\alpha}}, \beta, \gamma(A + B)). \tag{7.20}$$

In particular, for $\alpha \neq 1$, $AX_1 \sim S_\alpha(\sigma A, \beta, \gamma A)$.

Proof See [243, Chapter 1]. □

Hence, if $f_\alpha(x, \sigma, \beta, \gamma)$ is the probability density function of the stable random variable $X \sim S_\alpha(\sigma, \beta, \gamma)$, then $f_\alpha(x, \sigma, \beta, \gamma + a)$ is the probability density function of $X + a$ (for every real constant a) and $f_\alpha(x, \sigma A, \beta, \gamma A)$ is the probability density function of AX (for every positive constant A and $\alpha \neq 1$).

Symmetric Scalar α-Stable Random Variables

Definition 7.13 $X \sim S_\alpha(\sigma, \beta, \gamma)$ is called a symmetric α-stable random variable if $\beta = 0$ and $\gamma = 0$, that is, $X \sim S_\alpha(\sigma, 0, 0)$. This distribution is often denoted by $S\alpha S$. When $\sigma = 1$, it is called a standard symmetric α-stable random variable, and we denote this by $X \sim S_\alpha(1, 0, 0)$.

Figure 7.4 shows the probability density functions of the standard symmetric α-stable random variable $X \sim S_\alpha(1, 0, 0)$ for various α values.

Remark 7.14 If $X \sim S_\alpha(1, 0, 0)$ and A is a positive constant, then $AX \sim S_\alpha(A, 0, 0)$. Also note that if the probability density function for the standard symmetric α-stable random variable $X \sim S_\alpha(1, 0, 0)$ is $f_\alpha(x)$, then AX has the probability density function $\frac{1}{A} f_\alpha(\frac{x}{A})$. This comes from the fact that $\mathbb{P}(AX \leq x) = \mathbb{P}(X \leq \frac{x}{A}) = \int_{-\infty}^{\frac{x}{A}} f_\alpha(\xi) d\xi$ and $\frac{d}{dx}\mathbb{P}(AX \leq x) = \frac{1}{A} f_\alpha(\frac{x}{A})$.

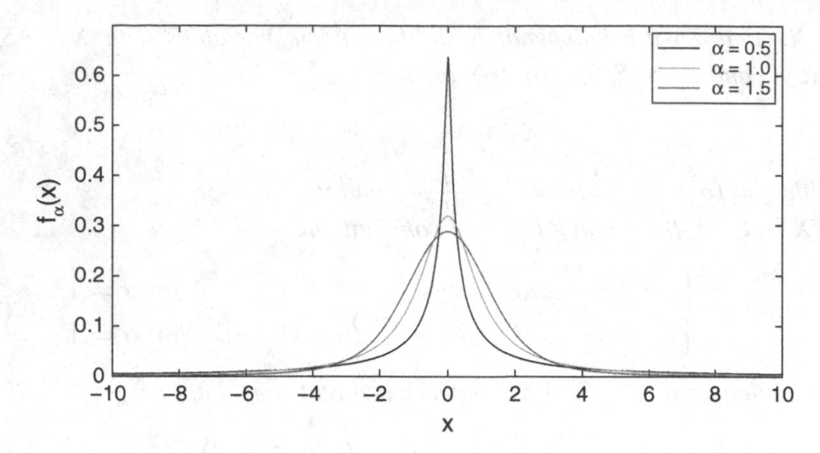

Figure 7.4 The probability density function for the standard symmetric α-stable random variable $X \sim S_\alpha(1,0,0)$: $\alpha = 0.5$ (with highest peak or in black), $\alpha = 1.0$ (with second highest peak or in light gray), and $\alpha = 1.5$ (with lowest peak or in dark gray).

Namely, if $f_\alpha(x)$ is the probability density function corresponding to $S_\alpha(1,0,0)$, then $\frac{1}{A} f_\alpha(\frac{x}{A})$ is the probability density function corresponding to $S_\alpha(A,0,0)$.

Remark 7.15 The probability density function $f_\alpha(x)$ for the standard symmetric α-stable random variable $X \sim S_\alpha(1,0,0)$ can be represented as infinite series ([256] or [140, p. 48]):

$$f_\alpha(x) = \begin{cases} \frac{1}{\pi x} \sum_{k=1}^{\infty} \frac{(-1)^{k-1}}{k!} \Gamma(\alpha k + 1)|x|^{-\alpha k} \sin\left(\frac{k\alpha\pi}{2}\right), & x \neq 0, 0 < \alpha < 1, \\ \frac{1}{\pi} \int_0^\infty e^{-u^\alpha} du, & x = 0, 0 < \alpha < 1, \\ \frac{1}{\pi(1+x^2)}, & \alpha = 1, \\ \frac{1}{\pi\alpha} \sum_{k=0}^{\infty} \frac{(-1)^k}{(2k)!} \Gamma\left(\frac{2k+1}{\alpha}\right) x^{2k}, & 1 < \alpha < 2. \end{cases} \tag{7.21}$$

A symmetric scalar α-stable random variable X has distribution $S_\alpha(\sigma,0,0)$, that is, $\beta = \gamma = 0$, with the characteristic function

$$\Phi_X(u) = \begin{cases} e^{-\sigma^\alpha |u|^\alpha}, & 0 < \alpha < 2, \\ e^{-\frac{1}{2}\sigma^2 |u|^2}, & \alpha = 2. \end{cases} \tag{7.22}$$

Note that a symmetric α-stable random variable $X \sim S_\alpha(\sigma,0,0)$ has the following moment properties [140, p. 24]: for $\alpha \in (0,2)$,

$$\mathbb{E}X = \begin{cases} \text{does not exist}, & \alpha \in (0,1], \\ 0, & \alpha \in (1,2], \end{cases} \tag{7.23}$$

$$\mathbb{E}|X| \begin{cases} = \infty, & \alpha \in (0, 1], \\ < \infty, & \alpha \in (1, 2], \end{cases} \tag{7.24}$$

$$\mathbb{E}|X|^p \begin{cases} < \infty, & \text{for } p \in (0, \alpha), \\ = \infty, & \text{for } p \in [\alpha, 2), \end{cases} \tag{7.25}$$

$$\mathbb{E}|X|^2 = \infty. \tag{7.26}$$

Therefore, $\mathbb{E}|X| < \infty$ if and only if $\alpha \in (1, 2]$, and $\mathbb{E}|X|^2 < \infty$ if and only if $\alpha = 2$ (i.e., X is a Gaussian random variable).

Remark 7.16 We now examine whether a symmetric scalar α-stable Lévy motion L_t^α has finite variation, by Theorem 7.7. For L_t^α, the jump measure is $\nu_\alpha(dx) = c_\alpha \frac{1}{|x|^{1+\alpha}} dx$. Note that $\int_{|x|<1} |x| \nu_\alpha(dx) = \int_{-1}^{1} |x|/|x|^{1+\alpha} dx = \int_{-1}^{1} |x|^{-\alpha} dx = \frac{2}{1-\alpha}$ for $\alpha \in (0, 1)$. When $\alpha \in [1, 2)$, $\int_{|x|<1} |x| \nu_\alpha(dx) = \infty$. Therefore, when $\alpha \in (0, 1)$, the α-stable Lévy motion L_t^α has finite variation.

Another Definition for Stable Random Variables

There is another definition for a stable random variable. It is a "global" characterization of a stable random variable, as the characteristic function is in terms of an integral in the whole Euclidean space, where the stable random variable takes values.

Definition 7.17 A scalar random variable X is stable if its characteristic function takes the following form:

(i) $0 < \alpha < 1$:

$$\Phi_X(u) = \exp\left\{ \int_{\mathbb{R}^1 \setminus \{0\}} (e^{ixu} - 1)\nu(dx) + i\gamma_0 u \right\};$$

(ii) $\alpha = 1$:

$$\Phi_X(u) = \exp\left\{ \int_{\mathbb{R}^1 \setminus \{0\}} (e^{ixu} - 1 - ixu I_{\{|x|<1\}}(x))\nu(dx) + i\gamma_* u \right\};$$

(iii) $1 < \alpha < 2$:

$$\Phi_X(u) = \exp\left\{ \int_{\mathbb{R}^1 \setminus \{0\}} (e^{ixu} - 1 - ixu)\nu(dx) + i\gamma_1 u \right\};$$

(iv) $\alpha = 2$:

$$\Phi_X(u) = \exp\left\{ i\gamma u - \frac{1}{2}\sigma^2 u^2 \right\},$$

where $\gamma_0, \gamma_*, \gamma_1$ are real constants and $\nu(dx) = \frac{c_1}{|x|^{1+\alpha}} I_{(0,\infty)}(x)dx + \frac{c_2}{|x|^{1+\alpha}} I_{(-\infty,0)}(x)dx$, with nonnegative constants c_1, c_2 satisfying $c_1 + c_2 > 0$.

It can be shown that the two definitions, Definitions 7.9 and 7.17, are equivalent (Problem 7.1).

Stable Random Vectors

We now consider stable random vectors in $\mathbb{R}^n$. We have a similar definition inspired by Definition 7.17.

Definition 7.18 A random vector X is stable if its characteristic function is as follows:

(i) $0 < \alpha < 1$:

$$\Phi_X(u) = \exp\left\{\int_{\mathbb{R}^n \setminus \{0\}} (e^{i\langle x,u\rangle} - 1)\nu(dx) + i\langle \gamma_0, u\rangle\right\};$$

(ii) $\alpha = 1$:

$$\Phi_X(u) = \exp\left\{\int_{\mathbb{R}^n \setminus \{0\}} (e^{i\langle x,u\rangle} - 1 - i\langle x, u\rangle I_{\{\|x\|<1\}}(x))\nu(dx) + i\langle \gamma_*, u\rangle\right\};$$

(iii) $1 < \alpha < 2$:

$$\Phi_X(u) = \exp\left\{\int_{\mathbb{R}^n \setminus \{0\}} (e^{i\langle x,u\rangle} - 1 - i\langle x, u >)\nu(dx) + i\langle \gamma_1, u\rangle\right\};$$

(iv) $\alpha = 2$:

$$\Phi_X(u) = \exp\left\{i\langle \gamma, u\rangle - \frac{1}{2}\sigma^2\langle u, u\rangle\right\},$$

where $\gamma_0, \gamma_*, \gamma_1$ are real vectors in $\mathbb{R}^n$ and $\nu(dx)$ is a Borel measure on $\mathbb{R}^n \setminus \{0\}$ (called jump measure).

A stable random vector in $\mathbb{R}^n$ is rotationally invariant or rotationally symmetric if its characteristic function has the following special form [7, Chapter 1]:

$$\Phi_X(u) = e^{-\sigma^\alpha \|u\|^\alpha}, \text{ if } \alpha \neq 2, \tag{7.27}$$

$$\Phi_X(u) = e^{-\frac{\sigma^2}{2}\|u\|^2}, \text{ if } \alpha = 2. \tag{7.28}$$

In this case, the jump measure is

$$\nu_\alpha(dx) = \frac{c(n, \alpha)}{\|x\|^{n+\alpha}} dx. \tag{7.29}$$

See also [244, pp. 114–115] for more details.

7.3.2 The α-Stable Lévy Motions in $\mathbb{R}^1$

We first discuss scalar symmetric and then nonsymmetric α-stable Lévy motions.

Symmetric α-Stable Lévy Motions

A symmetric α-stable scalar Lévy motion L_t^α, with $0 < \alpha < 2$, is a stochastic process with the following properties:

(i) $L_0^\alpha = 0$, a.s.
(ii) L_t^α has independent increments.
(iii) $L_t^\alpha - L_s^\alpha \sim S_\alpha((t-s)^{\frac{1}{\alpha}}, 0, 0)$.
(iv) L_t^α has stochastically continuous sample paths, that is, for every $s > 0$, $L_t^\alpha \to L_s^\alpha$ in probability, as $t \to s$.

From this definition, we see that $L_t^\alpha \sim S_\alpha(t^{\frac{1}{\alpha}}, 0, 0)$. By Theorem 7.12 (iii), we conclude that if $X \sim S_\alpha(1, 0, 0)$, then $t^{\frac{1}{\alpha}} X \sim S_\alpha(t^{\frac{1}{\alpha}}, 0, 0)$, for $t > 0$. Indeed, for every $c > 0$, L_{ct}^α and $c^{\frac{1}{\alpha}} L_t^\alpha$ have the same distribution. See [243, p. 113].

Using the facts that $\mathbb{P}(t^{\frac{1}{\alpha}} X \le x) = \mathbb{P}(X \le t^{-\frac{1}{\alpha}} x)$ and $\frac{d}{dx}\mathbb{P}(X \le t^{-\frac{1}{\alpha}} x) = t^{-\frac{1}{\alpha}} \frac{d}{d\tilde{x}}\mathbb{P}(X \le \tilde{x})|_{\tilde{x} = t^{-\frac{1}{\alpha}} x}$, we conclude that the probability density function for L_t^α is

$$t^{-\frac{1}{\alpha}} f_\alpha\left(t^{-\frac{1}{\alpha}} x\right), \tag{7.30}$$

where f_α is the probability density function for the standard symmetric α-stable random variable $X \sim S_\alpha(1, 0, 0)$, as in (7.21). The generalized time derivative $\frac{dL_t^\alpha}{dt}$ as a model for non-Gaussian white noise is discussed in [211, 172, 257].

A symmetric α-stable Lévy motion L_t^α, for $\alpha \in (0, 2)$, has the generating triplet $(0, 0, \nu_\alpha)$, where the jump measure $\nu_\alpha(du) = c_\alpha \frac{du}{|u|^{1+\alpha}}$, with

$$c_\alpha = \frac{\alpha}{2^{1-\alpha}\sqrt{\pi}} \frac{\Gamma\left(\frac{1+\alpha}{2}\right)}{\Gamma\left(1 - \frac{\alpha}{2}\right)}. \tag{7.31}$$

In this formula, Γ is the Gamma function as defined in Remark 2.5.

When $\alpha = 2$, this family reduces to the well-known Brownian motion B_t.

Generator of a Scalar Symmetric α-Stable Lévy Motion

The generator of a scalar symmetric α-stable Lévy motion L_t^α, with triplet $(0, 0, \nu_\alpha)$, is

$$A_\alpha \varphi = \int_{\mathbb{R}^1 \setminus \{0\}} [\varphi(x+y) - \varphi(x)]\ \nu_\alpha(dy), \tag{7.32}$$

where the right-hand side is understood as a Cauchy principal value. The domain of A is the collection of function φ such that this Cauchy principal value integral exists. Here

$$\nu_\alpha(\mathrm{dy}) = c_\alpha \frac{dy}{|y|^{1+\alpha}},$$

with c_α from (7.31). Note that the integrand $I_{\{|y|<1\}}\, y$ is an odd function in y and the corresponding integral in (7.10) is zero.

Simulation of Symmetric α-Stable Lévy Motion

By an algorithm in [140, p. 48], we can generate random numbers from a scalar standard symmetric α-stable random variable $S_\alpha(1, 0, 0)$, for $\alpha \in (0, 2)$:

First generate a uniform random variable V on $(-\frac{\pi}{2}, \frac{\pi}{2})$ and an exponential random variable W with parameter 1. Then a scalar standard symmetric α-stable random variable $X \sim S_\alpha(1, 0, 0)$ is produced by

$$X = \frac{\sin \alpha V}{(\cos V)^{\frac{1}{\alpha}}} \left\{ \frac{\cos(V - \alpha V)}{W} \right\}^{\frac{1-\alpha}{\alpha}}. \tag{7.33}$$

We call this algorithm stablernd.

In Matlab, UNIFRND(A,B) returns an array of random numbers chosen from the continuous uniform distribution on the interval from A to B (with $A < B$). UNIFRND(A,B,M,N) returns an $M \times N$ array of such random numbers. For example, UNIFRND(0, 1, 10, 1) produces a sequence of 10 random numbers chosen from the standard uniform distribution $U(0, 1)$. This generates V in (7.33).

Moreover, EXPRND(MU) returns an array of random numbers chosen from the exponential distribution with parameter MU. EXPRND(MU,M,N) returns an $M \times N$ array of such random numbers. For example, EXPRND(2, 10, 1) produces an array of 10 random numbers chosen from the exponential distribution with parameter 2. This generates W in (7.33). Thus, standard symmetric α-stable random numbers, $X \sim S_\alpha(1, 0, 0)$, can be generated.

Let us generate sample paths for a scalar symmetric α-stable Lévy motion L_t^α, $t \in [0, T]$ for some given $T > 0$. Partition the time interval $[0, T]$ into small subintervals: $t_0 = 0 < t_1 < t_2 < \cdots < t_k < \cdots < t_N = T$. To get an approximate sample path for L_t^α, we generate $L_{t_0}^\alpha, \ldots, L_{t_N}^\alpha$.

First we generate independent standard symmetric α-stable random numbers $Z_1, Z_2, \ldots, Z_N$ and compute the independent increments

$$L_{t_i}^\alpha - L_{t_{i-1}}^\alpha = (t_i - t_{i-1})^{\frac{1}{\alpha}} Z_i, \; i = 1, \ldots, N. \tag{7.34}$$

Then, to simulate the values $L_{t_1}^\alpha, \ldots, L_{t_N}^\alpha$, we use the following recursive algorithm:

$$L_{t_i}^\alpha = L_{t_{i-1}}^\alpha + \left(L_{t_i}^\alpha - L_{t_{i-1}}^\alpha\right) = L_{t_{i-1}}^\alpha + (t_i - t_{i-1})^{\frac{1}{\alpha}} Z_i, \; i = 1, \ldots, N. \tag{7.35}$$

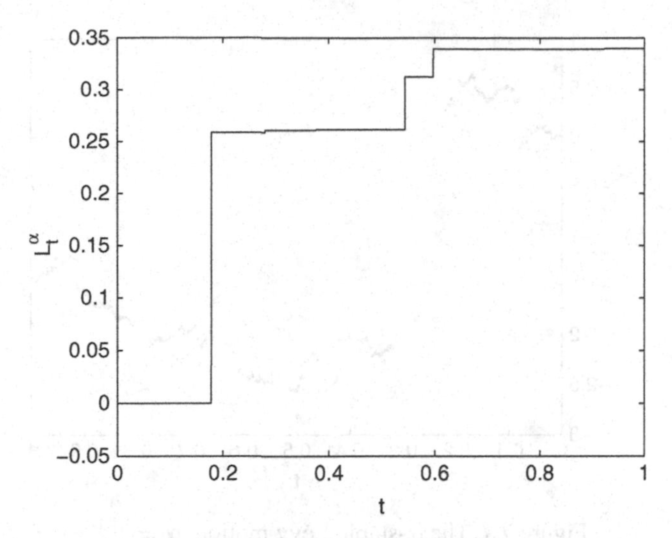

Figure 7.5 The α-stable Lévy motion: $\alpha = 0.25$.

A Matlab code can then be developed to plot sample paths for L_t^α, for $t \geq 0$.

Following Section 3.4 about how to simulate sample paths for a two-sided Brownian motion, we can compute sample paths for a two-sided symmetric α-stable Lévy motion defined in (7.36), L_t^α, for $t \in \mathbb{R}$.

See Figures 7.5–7.9 for a few sample paths for the symmetric α-stable Lévy motion L_t^α for various α values. For more details about numerical procedures to simulate α-stable Lévy motions, see [140].

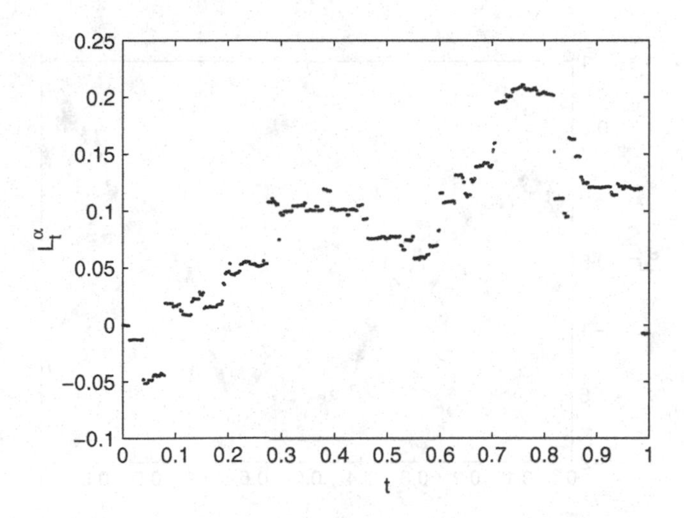

Figure 7.6 The α-stable Lévy motion: $\alpha = 0.75$.

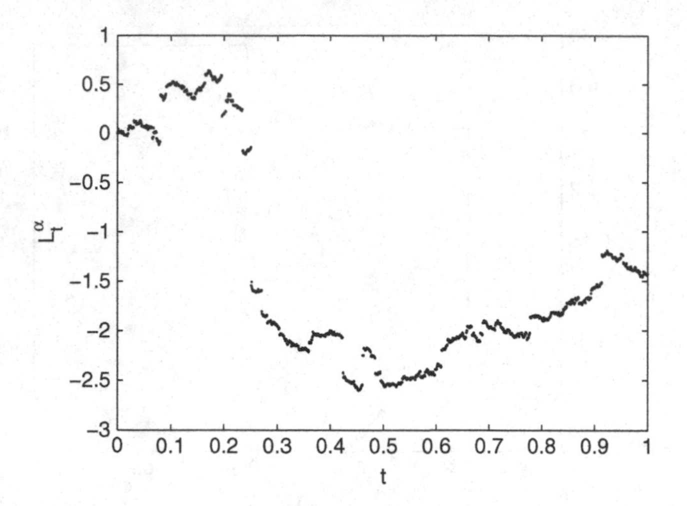

Figure 7.7 The α-stable Lévy motion: $\alpha = 1.4$.

Remark 7.19 From Section 7.2.4, we see that for a symmetric α-stable Lévy motion L_t^α with the generating triplet $(0, 0, \nu_\alpha)$, $-L_t^\alpha$ is also a symmetric α-stable Lévy motion with the same generating triplet.

Thus we can define a two-sided symmetric α-stable Lévy motion L_t^α, as in Problem 3.13:

$$L_t^\alpha \triangleq \begin{cases} \hat{L}_t^\alpha, & t \geq 0, \\ -\check{L}_{-t}^\alpha, & t < 0, \end{cases} \tag{7.36}$$

where $\hat{L}_t^\alpha$ and $\check{L}_t^\alpha$ are two independent symmetric α-stable Lévy motions for $t \geq 0$.

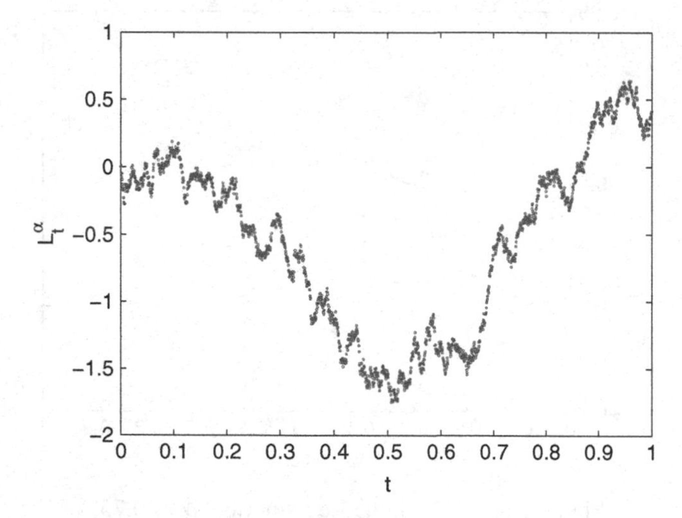

Figure 7.8 The α-stable Lévy motion: $\alpha = 1.9$.

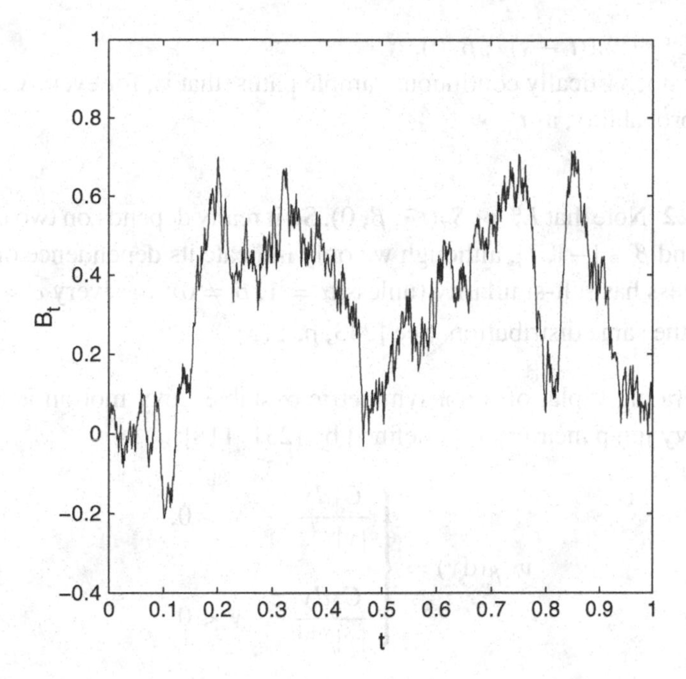

Figure 7.9 When $\alpha = 2$: Brownian motion with continuous path but nowhere differentiable.

Remark 7.20 We calculate the improper integral, for all $\alpha \in (0, 2)$:

$$\begin{aligned}
\nu_\alpha(\mathbb{R}^1) &= \int_{\mathbb{R}^1\setminus\{0\}} \nu_\alpha(dy) \\
&= \int_{\mathbb{R}^1\setminus\{0\}} \frac{1}{|y|^{1+\alpha}} dy \\
&= \lim_{\varepsilon_1 \to 0+} \int_{-\infty}^{-\varepsilon_1} \frac{1}{(-y)^{1+\alpha}} dy + \lim_{\varepsilon_2 \to 0+} \int_{\varepsilon_2}^{+\infty} \frac{1}{y^{1+\alpha}} dy = +\infty.
\end{aligned}$$

Thus, the jumps (in time) for a α-stable Lévy motion are countable but dense in every time interval.

Nonsymmetric α-Stable Lévy Motion

We now introduce nonsymmetric α-stable Lévy motions ([140, p. 30] and [243, p. 113]).

Definition 7.21 A nonsymmetric scalar α-stable Lévy motion L_t^α, with stability index $\alpha \in (0, 2)$ and skewness parameter $\beta \in [-1, 1]$, is a stochastic process with the following properties:

(i) $L_0^\alpha = 0$, a.s.
(ii) L_t^α has independent increments.

(iii) $L_t^\alpha - L_s^\alpha \sim S_\alpha((t-s)^{\frac{1}{\alpha}}, \beta, 0)$.
(iv) L_t^α has stochastically continuous sample paths, that is, for every $s > 0$, $L_t^\alpha \to L_s^\alpha$ in probability, as $t \to s$.

Remark 7.22 Note that $L_t^\alpha \sim S_\alpha(t^{\frac{1}{\alpha}}, \beta, 0)$. So it really depends on two parameters, $\alpha \in (0, 2)$ and $\beta \in [-1, 1]$, although we only indicate its dependence on α.

This process has self-similarity (unless $\alpha = 1, \beta \neq 0$): for every $c > 0$, L_{ct}^α and $c^{\frac{1}{\alpha}} L_t^\alpha$ have the same distribution. See [243, p. 113].

The generating triplet of a nonsymmetric α-stable Lévy motion is $(0, 0, \nu_{\alpha,\beta})$, with the Lévy jump measure $\nu_{\alpha,\beta}$ defined by [231, 118]:

$$\nu_{\alpha,\beta}(\mathrm{d}y) = \begin{cases} \dfrac{C_1 dy}{|y|^{\alpha+1}}, & y > 0, \\ \\ \dfrac{C_2 dy}{|y|^{\alpha+1}}, & y < 0, \end{cases} \tag{7.37}$$

with $\beta = (C_1 - C_2)/(C_1 + C_2)$, $C_1 = k_\alpha \frac{1+\beta}{2}$, and $C_2 = k_\alpha \frac{1-\beta}{2}$, where

$$k_\alpha = \begin{cases} \dfrac{\alpha(1-\alpha)}{\Gamma(2-\alpha)\cos(\frac{\pi\alpha}{2})}, & \alpha \neq 1, \\ \\ \dfrac{2}{\pi}, & \alpha = 1. \end{cases}$$

In this formula, Γ is the Gamma function as defined in Remark 2.5.

Generator of a Scalar Nonsymmetric α-Stable Lévy Motion

The generator of a scalar nonsymmetric α-stable Lévy motion, with triplet $(0, 0, \nu_{\alpha,\beta})$, is

$$A_{\alpha,\beta}\varphi = \int_{\mathbb{R}^1\setminus\{0\}} [\varphi(x+y) - \varphi(x) - I_{\{|y|<1\}}\, y\varphi'(x)]\ \nu_{\alpha,\beta}(dy) \tag{7.38}$$

for φ in the domain of definition of $A_{\alpha,\beta}$.

7.3.3 The α-Stable Lévy Motions in $\mathbb{R}^n$

In this subsection, we discuss rotationally symmetric α-stable Lévy motions L_t^α in $\mathbb{R}^n$.

Definition 7.23 For $\alpha \in (0, 2)$, an n-dimensional rotationally symmetric α-stable Lévy motion L_t^α is a Lévy motion with characteristic function

$$\mathbb{E}e^{i<u, L_t^\alpha>} = e^{-Ct\|u\|^\alpha}, \quad u \in \mathbb{R}^n, \tag{7.39}$$

where

$$C = \pi^{-1/2}\frac{\Gamma((1+\alpha)/2)\Gamma(n/2)}{\Gamma((n+\alpha)/2)}.$$

The value of C is 1 when the dimension $n = 1$.

We recall the following result ([56] and [244, Chapter 3]).

Theorem 7.24 (Properties of α-stable Lévy motions [7]) *A rotationally symmetric α-stable Lévy motion L_t^α in $\mathbb{R}^n$ has the generating triplet* $(0, 0, \nu_\alpha)$, *with the jump measure*

$$\nu_\alpha(du) = c(n, \alpha)\frac{du}{\|u\|^{n+\alpha}} \tag{7.40}$$

and the intensity constant

$$c(n, \alpha) = \frac{\alpha\Gamma((n+\alpha)/2)}{2^{1-\alpha}\pi^{n/2}\Gamma(1-\alpha/2)}, \tag{7.41}$$

where Γ is the Gamma function, defined in Remark 2.5.

When the spatial dimension n is clear from context, we often denote $c(n, \alpha)$ as c_α or just c. For example, in the case of $\alpha = 1$, $c = \frac{1}{\pi}$ (for $n = 1$) and $c = \frac{1}{2\pi}$ (for $n = 2$). See [56].

Remark 7.25 Thus, for an n-dimensional rotationally symmetric α-stable Lévy motion L_t^α, its diffusion matrix $Q = 0$ and the drift vector $\gamma = 0$. It is characterized by the jump measure ν_α, for $\alpha \in (0, 2)$.

Figures 7.10–7.12 show the intensity constant $c(n, \alpha)$ for $n = 1$, $n = 2$, and $n = 3$, respectively.

The generator for this rotationally symmetric α-stable Lévy motion L_t^α in $\mathbb{R}^n$ is ([7, Theorem 3.3.3] or [56])

$$A_\alpha\varphi(x) = \int_{\mathbb{R}^n\setminus\{0\}} [\varphi(x+y) - \varphi(x)]\ \nu_\alpha(dy), \tag{7.42}$$

where the right-hand side is understood as a Cauchy principal value. The domain of A is the collection of function φ such that this Cauchy principal value integral exists. This can also be shown directly, from the definition of generator $A_\alpha\varphi(x) = \frac{d}{dt}|_{t=0}\mathbb{E}\varphi(X_t)$ for $X_t = x + L_t^\alpha$ (a α-stable Lévy motion starting at x). See Problems 4.6 and 7.8.

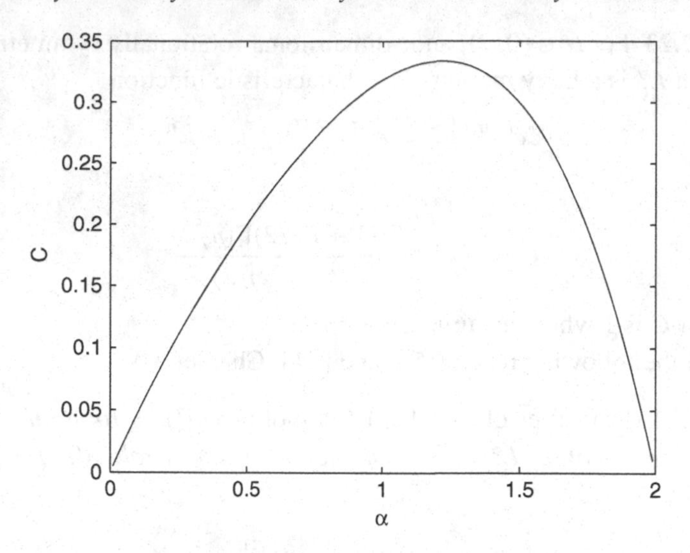

Figure 7.10 Intensity constant c v.s. $\alpha \in (0, 2)$: $n = 1$.

This generator A_α has a unique extension to a self-adjoint operator [3] in the domain of definition $W^{\alpha,2}(\mathbb{R}^n) \triangleq \{g \in L^2(\mathbb{R}^n) \colon \|k\|^\alpha \mathbb{F}(g) \in L^2(\mathbb{R}^n)\}$. Here the Fourier transform for g is defined by

$$\mathbb{F}(g)(k) = \frac{1}{(2\pi)^{\frac{n}{2}}} \int_{\mathbb{R}^n} e^{-i <k,x>} g(x)dx. \tag{7.43}$$

Sometimes $\mathbb{F}(g)(k)$ is also denoted by $\hat{g}(k)$. By [2, Theorem 7.39], $W^{\alpha,2}(\mathbb{R}^n) = W_0^{\alpha,2}(\mathbb{R}^n)$, that is, all functions in $W^{\alpha,2}(\mathbb{R}^n)$ have compact support in $\mathbb{R}^n$. Further

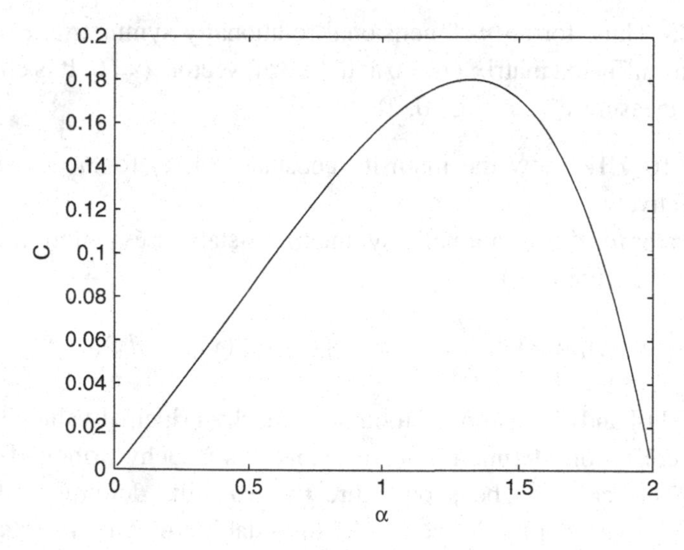

Figure 7.11 Intensity constant c v.s. $\alpha \in (0, 2)$: $n = 2$.

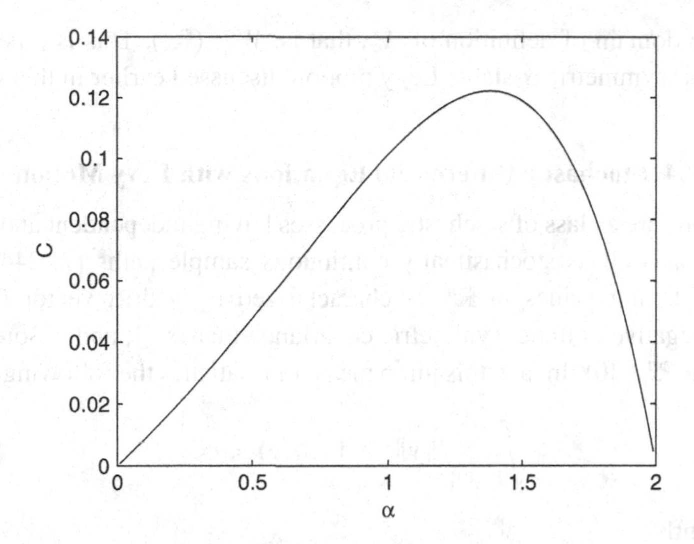

Figure 7.12 Intensity constant c v.s. $\alpha \in (0, 2)$: $n = 3$.

note that this integral operator is related to the fractional Laplacian operator [3]. Indeed, for $\alpha \in (0, 2)$, by Fourier inverse transform,

$$\begin{aligned} A_\alpha u &\triangleq \int_{\mathbb{R}^n\setminus\{0\}} [u(x+y) - u(x)]\ \nu_\alpha(dy) \\ &= \theta_{\alpha,n}\mathbb{F}^{-1}(\|k\|^\alpha \mathbb{F}(u)(k)) \\ &= \theta_{\alpha,n}\ (-\Delta)^{\frac{\alpha}{2}} u(x), \end{aligned} \tag{7.44}$$

where

$$\theta_{\alpha,n} \triangleq \int_{\mathbb{R}^n\setminus\{0\}} (\cos(e\cdot y) - 1)\ \nu_\alpha(dy) < 0, \tag{7.45}$$

with e being any unit vector in $\mathbb{R}^n$. Here we have used the notation for the fractional Laplacian operator:

$$\mathbb{F}^{-1}(\|k\|^\alpha \mathbb{F}(u)(k)) \triangleq (-\Delta)^{\frac{\alpha}{2}} u(x), \tag{7.46}$$

$$\mathbb{F}((-\Delta)^{\frac{\alpha}{2}} u(x)) = \|k\|^\alpha \mathbb{F}(u)(k). \tag{7.47}$$

Clearly this notation is inspired by the fact that

$$\mathbb{F}(-\Delta u(x)) = \|k\|^2 \mathbb{F}(u)(k).$$

Thus, the generator for the rotationally symmetric α-stable Lévy motion L_t^α in $\mathbb{R}^n$ is also written as

$$A_\alpha \varphi = \theta_{\alpha,n}\ (-\Delta)^{\frac{\alpha}{2}} \varphi, \quad \alpha \in (0, 2), \tag{7.48}$$

for φ in the domain of definition of A_α, that is, $W^{\alpha,2}(\mathbb{R}^n)$. This is especially true for the scalar symmetric α-stable Lévy motion discussed earlier in this section.

7.4 Stochastic Differential Equations with Lévy Motions

Lévy motions are a class of stochastic processes having independent and stationary increments as well as stochastically continuous sample paths [7, 244]. A Lévy motion L_t, taking values in $\mathbb{R}^n$, is characterized by a drift vector $b \in \mathbb{R}^n$; an $n \times n$ nonnegative-definite, symmetric covariance matrix Q; and a Borel measure ν defined on $\mathbb{R}^n \setminus \{0\}$. In fact, this jump measure ν satisfies the following condition:

$$\int_{\mathbb{R}^n\setminus\{0\}} (\|y\|^2 \wedge 1)\nu(dy) < \infty, \tag{7.49}$$

or equivalently,

$$\int_{\mathbb{R}^n\setminus\{0\}} \frac{\|y\|^2}{1+\|y\|^2}\nu(dy) < \infty. \tag{7.50}$$

Here $\|\cdot\|$ is the usual Euclidean norm in $\mathbb{R}^n$.

7.4.1 Stochastic Integration with Respect to Lévy Motions

By the Lévy-Itô decomposition, a Lévy motion with the generating triplet (b, Q, ν) has the following representation:

$$L_t = bt + Q^{\frac{1}{2}}B_t + \int_{\|y\|<1} y\tilde{N}(t, dy) + \int_{\|y\|\geq 1} yN(t, dy), \tag{7.51}$$

where $N(dt, dx)$ is the Poisson random measure, $\tilde{N}(dt, dx) = N(dt, dx) - \nu(dx)dt$ is the compensated Poisson random measure, $\nu(S) = \mathbb{E}N(1, S)$ is the jump measure, and B_t is an independent standard n-dimensional Brownian motion (i.e., Wiener process). The small jumps ($\|y\| < 1$) are controlled by $\tilde{N}(t, dy)$, whereas large jumps ($\|y\| \geq 1$) are controlled by $N(t, dy)$.

Recall that Q is nonnegative definite and symmetric. Thus Q has nonnegative eigenvalues $\lambda_1, \ldots, \lambda_n$ and an orthonormal basis $\{e_1, \ldots, e_n\}$ formed by the corresponding eigenvectors. The definition of $Q^{\frac{1}{2}}$ is via $Q^{\frac{1}{2}}x = \sum_{i=1}^n \lambda_i^{\frac{1}{2}} < x, e_i > e_i$. Moreover, $Q^{\frac{1}{2}}B_t$ has covariance matrix $Q^{\frac{1}{2}}(Q^{\frac{1}{2}})^{\mathrm{T}} = Q$.

Equation (7.51) can be formally rewritten in a differential form as

$$\mathrm{d}L_t = b\mathrm{d}t + Q^{\frac{1}{2}}\mathrm{d}B_t + \int_{\|y\|<1} y\tilde{N}(\mathrm{d}t, \mathrm{d}y) + \int_{\|y\|\geq 1} yN(\mathrm{d}t, \mathrm{d}y). \tag{7.52}$$

Convention. For notational simplicity, (7.52) is further equivalently rewritten as

$$dL_t = bdt + Q^{\frac{1}{2}}dB_t + y\tilde{N}(dt, dy) + yN(dt, dy). \tag{7.53}$$

That is, the specification on small jumps ($\|y\| < 1$) and large jumps ($\|y\| \geq 1$) is implicit in (7.53). Equations (7.52) and (7.53) are regarded as identical.

We now discuss stochastic integrals with respect to this Lévy motion. For simplicity, we just consider the one-dimensional case. In $\mathbb{R}^n$, the stochastic integral can be similarly defined, component by component. For a given $T > 0$, consider the stochastic integral with respect to this Lévy motion:

$$\begin{aligned}\int_0^T g(t,\omega)dL(t) = \int_0^T gdt &+ \int_0^T g(t,\omega)Q^{\frac{1}{2}}dB(t) \\ &+ \int_0^T \int_{\|y\|<1} g(t,\omega)y\tilde{N}(dt, dy) \\ &+ \int_0^T \int_{\|y\|\geq 1} g(t,\omega)yN(dt, dy). \end{aligned} \tag{7.54}$$

The first two integrals, the Lebesgue (or Riemann) integral and the stochastic integral with respect to Brownian motion, are familiar to us. The third and fourth integrals may be defined first for simple or step functions and then by extending to more general functions. For more details, see [7, Section 4.3].

The differential form for (7.54) is

$$\begin{aligned} g(t,\omega)dL(t) = gdt &+ gQ^{\frac{1}{2}}dB(t) \\ &+ \int_{\|y\|<1} g(t,\omega)y\tilde{N}(dt, dy) \\ &+ \int_{\|y\|\geq 1} g(t,\omega)yN(dt, dy), \end{aligned} \tag{7.55}$$

or equivalently, with our notational convention,

$$\begin{aligned} g(t,\omega)dL(t) = gdt &+ gQ^{\frac{1}{2}}dB(t) \\ &+ g(t,\omega)y\tilde{N}(dt, dy) + g(t,\omega)yN(dt, dy). \end{aligned} \tag{7.56}$$

For convenience, we could think of Q as being combined or absorbed into g, and we thus will not explicitly write Q in an SDE with the preceding L_t.

7.4.2 SDEs with Lévy Motions

We consider the following SDE in $\mathbb{R}^n$:

$$dX_t = f(X_{t-})dt + \sigma(X_{t-})dB_t + dL_t, \quad X_0 = x, \tag{7.57}$$

where B_t is a standard Brownian motion in $\mathbb{R}^n$, L_t is a Lévy motion in $\mathbb{R}^n$ with the generating triplet $(0, 0, \nu)$, and they are independent. The vector field (also called drift) f takes values in $\mathbb{R}^n$, and σ is an $n \times n$ matrix function. We could think that the matrix $Q^{\frac{1}{2}}$ (in the previous subsection) is combined with σ to give a new matrix, but we still denote it by σ. We could also think that b (in the previous subsection) is combined with f.

By the Lévy-Itô decomposition,

$$\mathrm{d}L_t = \int_{\|y\|<1} y\tilde{N}(\mathrm{d}t, \mathrm{d}y) + \int_{\|y\|\geq 1} yN(\mathrm{d}t, \mathrm{d}y), \tag{7.58}$$

where $N(dt, dx)$ is the Poisson random measure (quantifying the number of jumps of L_t), $\tilde{N}(dt, dx) = N(dt, dx) - \nu(dx)dt$ is the compensated Poisson random measure, and $\nu(S) = \mathbb{E}N(1, S)$ is the jump measure. Then, the SDE (7.57) becomes

$$\begin{aligned} dX_t &= f(X_{t-})dt + \sigma(X_{t-})dB_t \\ &\quad + \int_{\|y\|<1} y\tilde{N}(\mathrm{d}t, \mathrm{d}y) \\ &\quad + \int_{\|y\|\geq 1} yN(\mathrm{d}t, \mathrm{d}y), \quad X_0 = x. \end{aligned} \tag{7.59}$$

We make some appropriate assumptions on the vector field f, the noise intensity σ, and the jump measure ν, so that SDE (7.57) has a unique, adapted, cadlag solution [7].

In the following, $\|\cdot\|$ denotes the Euclidean norm in $\mathbb{R}^n$ and also a matrix norm $\|\sigma\|^2 \triangleq \sum_{i=1}^n \sum_{j=1}^n |\sigma_{ij}|^2$. We make the following assumptions.

Lipschitz Condition

There exists a positive constant K_1 such that, for all $x_1, x_2 \in \mathbb{R}^n$,

$$\|f(x_1) - f(x_2)\|^2 + \|\sigma(x_1) - \sigma(x_2)\|^2 \leq K_1\|x_1 - x_2\|^2. \tag{7.60}$$

Growth Condition

There exists a positive constant K_2 such that, for all $x \in \mathbb{R}^n$,

$$\|f(x)\|^2 + \|\sigma(x)\|^2 + \int_{\|y\|<1} \|y\|^2\nu(dy) \leq K_2(1 + \|x\|^2). \tag{7.61}$$

In view of the *Lipschitz condition*, the *growth condition* can actually be replaced by the following condition:

There exists a positive constant K_3 such that

$$\int_{\|y\|<1} \|y\|^2 \nu(dy) \leq K_3. \tag{7.62}$$

Theorem 7.26 (Existence and uniqueness) *Under the Lipschitz condition* (7.60) *and growth condition* (7.61)*, the SDE (7.57) has a unique, adapted, cadlag solution $X(t)$. Moreover, if $\mathbb{E}\|X_0\|^2 < \infty$, then $\mathbb{E}\|X(t)\|^2 < \infty$ for all $t > 0$, and there exists a positive quantity $C(t)$ such that*

$$\mathbb{E}\|X(t)\|^2 \leq C(t)(1 + \mathbb{E}\|X_0\|^2). \tag{7.63}$$

Proof First, rewrite the SDE (7.57) in the integral form

$$\begin{aligned} X_t = X_0 &+ \int_0^t f(X_{t-})dt + \int_0^t \sigma(X_{t-})dB_t \\ &+ \int_0^t \int_{\|y\|<1} y\tilde{N}(\mathrm{d}t, \mathrm{d}y) \\ &+ \int_0^t \int_{\|y\|\geq 1} yN(\mathrm{d}t, \mathrm{d}y). \end{aligned} \tag{7.64}$$

Then, conduct the Picard iteration and apply a fixed point argument. See [7, Section 6.2] for more details. □

Remark 7.27 When the Lipschitz condition and growth condition are only locally satisfied, that is, K_1 and K_2 depend on the domain where x_1, x_2, x live, then the SDE (7.57) has a unique, adapted, cadlag local solution $X(t)$, which is well defined up to a stopping time T_*. Under certain a priori estimates of the solution $X(t)$, it is still true that the solution is globally defined (i.e., $T_* = \infty$).

More discussions about the existence, uniqueness, and properties for solutions of SDEs with Lévy processes can be found in [7, 232]. For example, the solution process $X_t(x)$ of the SDE (7.93) can be shown to be a strong Markov process.

For the rest of this chapter, we assume that a Lipschitz condition and a growth condition are satisfied, so that an SDE system has a unique solution with every given initial condition.

Solution paths for SDEs with a symmetric α-stable Lévy motion can be simulated by, for example, Euler's method. Let us reconsider Example 4.34, but with α-stable Lévy motion.

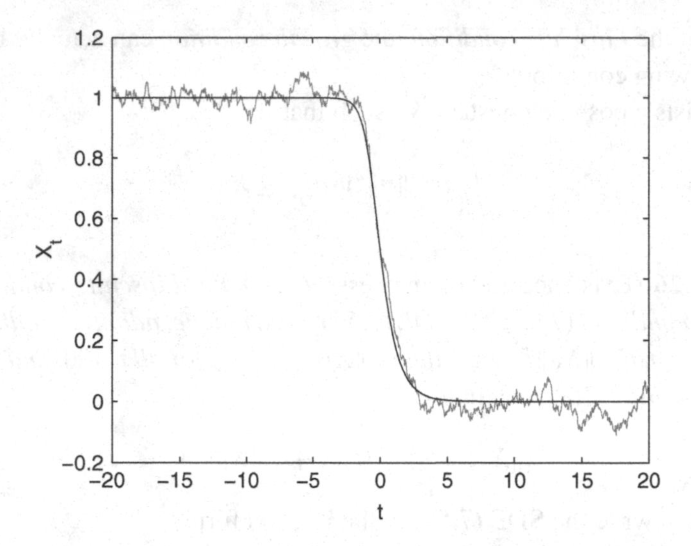

Figure 7.13 A solution path of $dX_t = (-X_t + X_t^3)dt + \varepsilon dB_t$ with "wigglings" when $\varepsilon = 0.05$, on top of the deterministic solution (i.e., when $\varepsilon = 0$).

Example 7.28 As in Example 4.34, we numerically compute solution paths for the following scalar SDE:

$$dX_t = (-X_{t-} + X_{t-}^3)dt + \varepsilon dB_t, \quad X_0 = 0.5.$$

A solution path is shown in Figure 7.13.

Now, replace the Brownian motion by a α-stable Lévy motion

$$dX_t = (-X_{t-} + X_{t-}^3)dt + \varepsilon dL_t^\alpha, \quad X_0 = 0.5.$$

Solution paths for $\alpha = 0.85, \alpha = 0.95$, and $\alpha = 1.5$ are shown in Figures 7.14–7.16, respectively.

7.4.3 Generators for SDEs with Lévy Motion

We consider a stochastic dynamical system described by the following SDE in $\mathbb{R}^n$:

$$dX_t = f(X_{t-})dt + \sigma(X_{t-})dB_t + dL_t, \quad X_0 = x, \tag{7.65}$$

where L_t is a Lévy motion, independent of the Brownian motion B_t, with the generating triplet $(0, 0, \nu)$. The vector field (also called drift) f takes values in $\mathbb{R}^n$ and σ is an $n \times n$ matrix function.

Recall that the generator for L_t is (see Sections 7.2 and 7.3)

$$A_0 g(x) = \int_{\mathbb{R}^n \setminus \{0\}} [g(x+y) - g(x) - I_{\{\|y\|<1\}}\, y \cdot \nabla g(x)]\, \nu(dy), \tag{7.66}$$

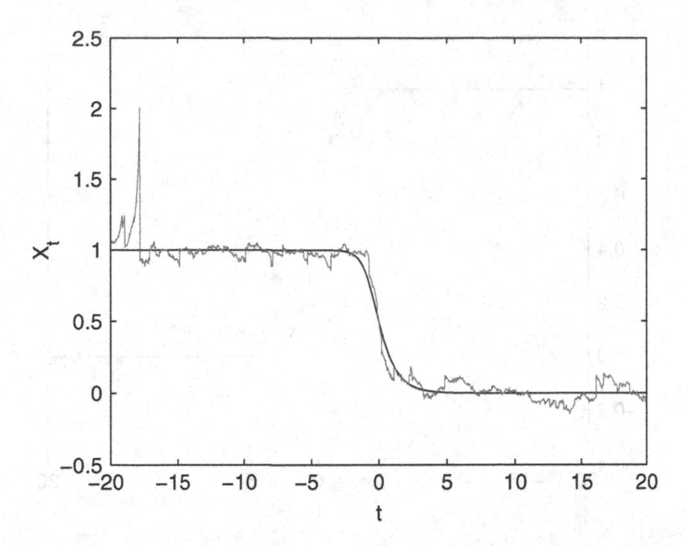

Figure 7.14 When $\alpha = 0.85$: a solution path of $dX_t = (-X_{t-} + X_{t-}^3)dt + \varepsilon dL_t^\alpha$ with "wigglings" when $\varepsilon = 0.05$, on top of the deterministic solution (i.e., when $\varepsilon = 0$).

where I_S is the indicator function of the set S, that is, it takes value 1 on this set and takes zero value otherwise.

The generator A of the solution process X_t is also defined by $A\varphi = \lim_{t\downarrow 0} \frac{P_t\varphi - \varphi}{t}$, where $P_t\varphi(x) = \mathbb{E}^x\varphi(X_t)$, for φ in the domain of definition for the operator A.

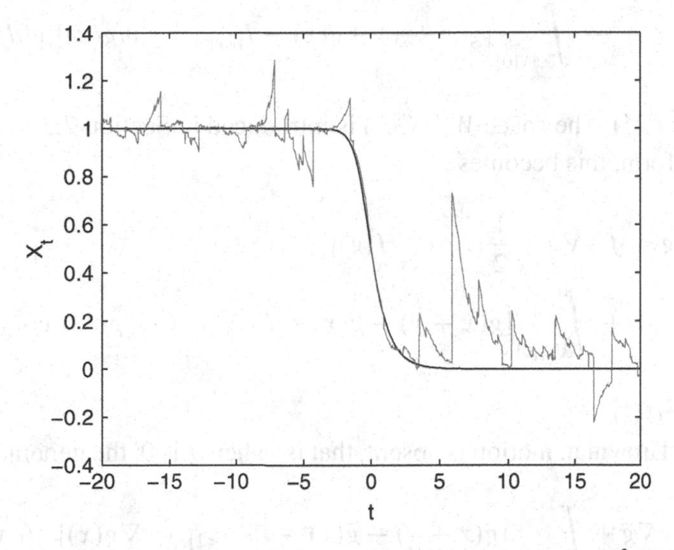

Figure 7.15 When $\alpha = 0.95$: a solution path of $dX_t = (-X_{t-} + X_{t-}^3)dt + \varepsilon dL_t^\alpha$ with "wigglings" when $\varepsilon = 0.05$, on top of the deterministic solution (i.e., when $\varepsilon = 0$).

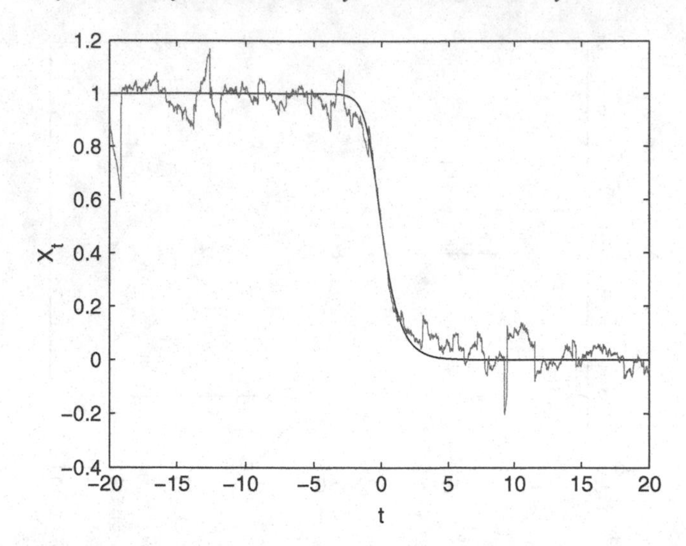

Figure 7.16 When $\alpha = 1.5$: a solution path of $dX_t = (-X_{t-} + X_{t-}^3)dt + \varepsilon dL_t^\alpha$ with "wigglings" when $\varepsilon = 0.05$, on top of the deterministic solution (i.e., when $\varepsilon = 0$).

The generator A for the SDE (7.65), or for the solution process X_t, is [7, p. 402]

$$Ag(x) \triangleq \sum_{i=1}^{n} f_i(x)\partial_i g(x) + \frac{1}{2}\sum_{i,j=1}^{n}(\sigma\sigma^{\mathrm{T}})_{ij}\partial_i\partial_j g(x)$$
$$+ \int_{\mathbb{R}^n\setminus\{0\}} [g(x+y) - g(x) - I_{\{\|y\|<1\}}\, y_i\partial_i g(x)]\, \nu(dy), \quad (7.67)$$

for $g \in W^{\alpha,2}(\mathbb{R}^n)$. The space $W^{\alpha,2}(\mathbb{R}^n)$ is introduced in Section 7.3.

In vector form, this becomes

$$Ag = f \cdot \nabla g + \frac{1}{2}\mathrm{Tr}[\sigma\sigma^{\mathrm{T}} H(g)]$$
$$+ \int_{\mathbb{R}^n\setminus\{0\}} [g(x+y) - g(x) - I_{\{\|y\|<1\}}\, y \cdot \nabla g(x)]\, \nu(dy) \quad (7.68)$$

for $g \in W^{\alpha,2}(\mathbb{R}^n)$.

When the Brownian motion is absent, that is, when σ is 0, the generator becomes

$$Ag = f \cdot \nabla g + \int_{\mathbb{R}^n\setminus\{0\}} [g(x+y) - g(x) - I_{\{\|y\|<1\}}\, y \cdot \nabla g(x)]\, \nu(dy), \quad (7.69)$$

for $g \in W^{\alpha,2}(\mathbb{R}^n)$.

7.5 Mean Exit Time

Let us consider an n-dimensional stochastic system with a non-Gaussian Lévy motion as well as a Brownian motion

$$dX_t = f(X_{t-})dt + \sigma(X_{t-})dB_t + dL_t, \quad X_0 = x, \tag{7.70}$$

where B_t is a Brownian motion in $\mathbb{R}^n$, σ is an $n \times n$ matrix function, and L_t is a Lévy motion in $\mathbb{R}^n$ with generating triplet $(0, 0, \nu)$. The generator for the solution process X_t is

$$Ah = f \cdot \nabla h + \frac{1}{2}\mathrm{Tr}[\sigma\sigma^{\mathrm{T}} H(h)] + \int_{\mathbb{R}^n \setminus \{0\}} [h(x+y) - h(x) \\ - I_{\{|y|<1\}}\, y \cdot \nabla h(x)]\, \nu(dy).$$

We make the following Hypothesis (H), as in [7, p. 396].

Hypothesis (H)

(i) Assume that the conditions in Section 7.4.2 for existence and uniqueness of solutions are satisfied.

(ii) Assume that $\mathrm{Tr}(\sigma\sigma^{\mathrm{T}} H(\cdot))$ is a uniformly elliptic operator, that is, that there exists a positive constant γ such that

$$\sum_{i,j=1}^{n} (\sigma\sigma^{\mathrm{T}})_{ij}\xi_i\xi_j \geq \gamma|\xi|^2, \quad \text{for } \xi \in \mathbb{R}^n. \tag{7.71}$$

(iii) Assume that there exists a real number $p \geq 2$ such that $\int_{\|y\|<1} \|y\|^p \nu(dy)$ is bounded (this is true for α-stable Lévy motion), and additionally, f, σ are continuous and bounded in $\mathbb{R}^n$.

Under **Hypothesis (H)**, by [7, Theorem 6.7.4, p. 402], the semigroup T_t associated with X_t is a Feller semigroup, that is,

(i) $T_t C_0(\mathbb{R}^n) \subset C_0(\mathbb{R}^n)$
(ii) $T_t g(x) \to g(x)$, as $t \to 0$, for $g \in C_0(\mathbb{R}^n)$ and $x \in \mathbb{R}^n$

Therefore, by [145, Lemma 19.21, p. 382], the following Dynkin formula holds: for a bounded stopping time τ,

$$\mathbb{E}^x g(X_\tau) = g(x) + \mathbb{E}^x\left[\int_0^\tau Ag(X_s)ds\right], \tag{7.72}$$

for g in the domain of definition for the generator A. This may be regarded as a fundamental theorem of stochastic calculus. Recall from Section 4.6 that $\mathbb{E}^x$ is the expectation with respect to the probability measure $\mathbb{P}^x$ induced by a solution process starting at x.

We study the first exit problem for the solution process X_t from bounded domains in $\mathbb{R}^n$.

Define the first exit time τ_x of a solution path (i.e., a "particle") starting at x from a bounded domain D as

$$\tau_x(\omega) \triangleq \inf\{t \geq 0, X_t(\omega, x) \notin D\}.$$

The mean exit time is then denoted by

$$u(x) = \mathbb{E}^x \tau_x(\omega),$$

for $x \in D$. We state the following result that characterizes the mean exit time as the solution of a deterministic (nonlocal) integral-partial differential equation, with an "external" Dirichlet boundary condition.

Theorem 7.29 (Mean exit time) *Under **Hypothesis (H)**, the mean exit time $u(x)$, for a solution path starting at $x \in D$, satisfies the nonlocal partial differential equation*

$$Au = -1, \tag{7.73}$$

with an external Dirichlet boundary condition

$$u|_{D^c} = 0, \tag{7.74}$$

where A is the generator

$$\begin{aligned} Au = {} & f \cdot \nabla u + \frac{1}{2} Tr[\sigma\sigma^{\mathrm{T}} H(u)] \\ & + \int_{\mathbb{R}^n \setminus \{0\}} [u(x+y) - u(x) - I_{\{\|y\|<1\}}\, y \cdot \nabla u(x)]\, \nu(dy) \end{aligned} \tag{7.75}$$

and D^c is the complement of the bounded domain D in $\mathbb{R}^n$.

Proof The first exit time τ_x is a stopping time. By Dynkin's formula 7.72, we have

$$\mathbb{E}^x u(X_{\tau_x}) = u(x) + \mathbb{E}^x \left[\int_0^{\tau_x} Au(X_s) ds \right]. \tag{7.76}$$

Note that $u(X_{\tau_x}) = 0$, because $\tau_x \in D^c$. If u satisfies the equation $Au = -1$, then $u(x) = \mathbb{E}^x \tau_x$. This completes the proof. □

Discussions on the existence and uniqueness of solutions to nonlocal systems similar to (7.73) and (7.74) may be found in [101].

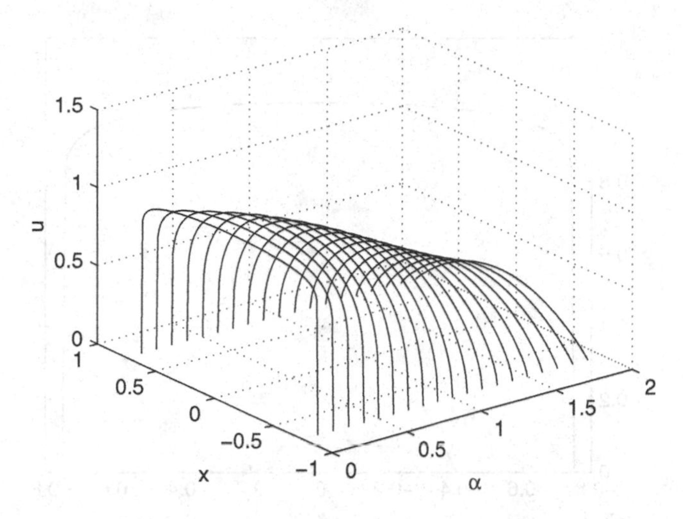

Figure 7.17 Mean exit time from the interval (−0.75, 0.75) under α-stable Lévy motion: $0 < \alpha < 2$.

7.5.1 Mean Exit Time for α-Stable Lévy Motion

Only in some simple cases, the mean exit time $u(x)$ can be found analytically. For example,

$$dX_t = dL_t^\alpha, \ X_0 = x,$$

where the scalar symmetric α-stable Lévy motion L_t^α has the generating triplet $(0, 0, \nu_\alpha)$. The jump measure $\nu_\alpha(du) = c_\alpha \frac{du}{|u|^{1+\alpha}}$, where c_α is in (7.41) with $n = 1$.

The mean exit time u from the interval $D = (-r, r)$ satisfies the nonlocal equation

$$\int_{\mathbb{R}^1 \setminus \{0\}} \frac{u(x+y) - u(x) - I_{\{|y|<1\}} \, yu'(x)}{|y|^{1+\alpha}} \, c_\alpha \, \mathrm{d}y = -1, \ x \in D, \tag{7.77}$$

and $u(x) = 0$ for $x \in D^c$. From [102], we obtain the exact solution for the mean exit time

$$u(x) = \frac{\sqrt{\pi}}{2^\alpha \Gamma\left(1 + \frac{\alpha}{2}\right) \Gamma\left(\frac{1}{2} + \frac{\alpha}{2}\right)} \, (r^2 - x^2)^{\frac{\alpha}{2}}, \tag{7.78}$$

where Γ is the Gamma function, which can be computed in Matlab. See Remark 2.5 for the definition of the Gamma function. Figures 7.17–7.21 show this mean exit time for various α values. We observe that the mean exit time, under the symmetric α-stable Lévy motion but with no drift, decreases as α increases. Figure 7.22 shows the mean exit time under Brownian motion (with no drift).

In particular, for $\alpha = 1$ (Cauchy distribution), the mean exit time is

$$u(x) = \sqrt{r^2 - x^2}. \tag{7.79}$$

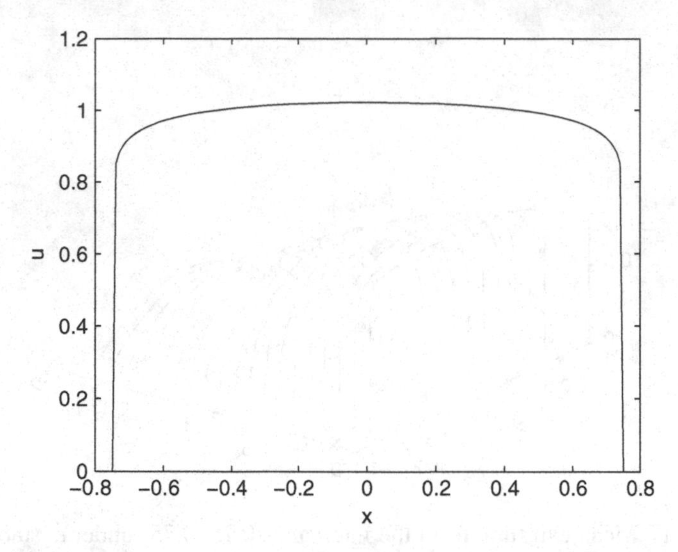

Figure 7.18 Mean exit time from the interval $(-0.75, 0.75)$ under α-stable Lévy motion: $\alpha = 0.1$.

7.5.2 Mean Exit Time for SDEs with α-Stable Lévy Motion

This subsection is adopted from our earlier work [97]. We consider a scalar SDE of the form (7.70),

$$dX_t = f(X_t)dt + \sqrt{d}\, dB_t + dL_t^{\alpha},\ X_0 = x, \tag{7.80}$$

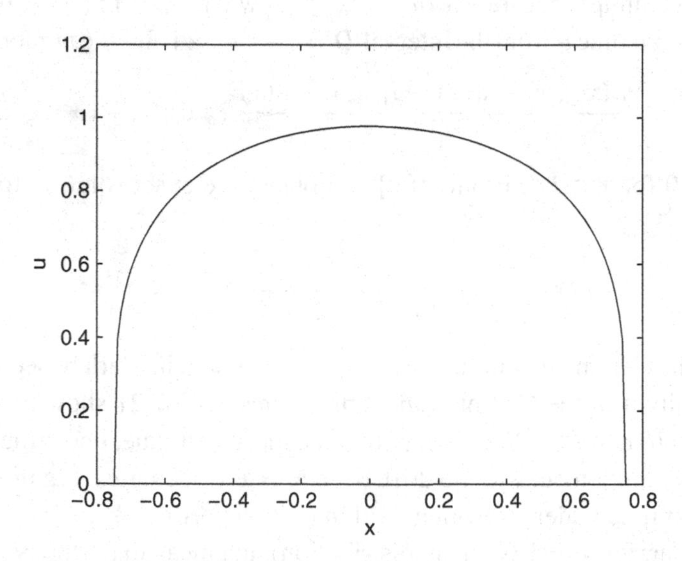

Figure 7.19 Mean exit time from the interval $(-0.75, 0.75)$ under α-stable Lévy motion: $\alpha = 0.5$.

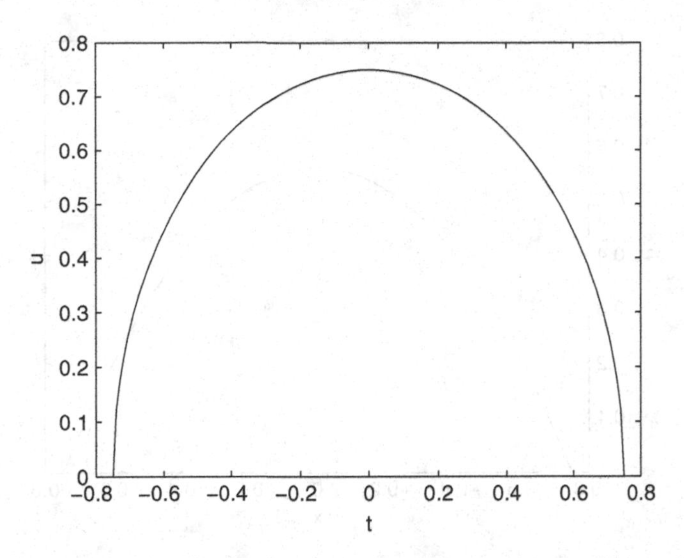

Figure 7.20 Mean exit time from the interval $(-0.75, 0.75)$ under pure α-stable Lévy motion: $\alpha = 1$.

with variance $d \geq 0$ and a α-stable symmetric Lévy motion L_t^α of generating triplet $(0, 0, \nu_\alpha)$. The jump measure $\nu_\alpha(du) = c_\alpha \frac{du}{|u|^{1+\alpha}}$, for $0 < \alpha < 2$. Here c_α is in (7.41) with $n = 1$.

We are interested in the mean exit time, $u(x)$, for an orbit (or solution path) starting at x, from a bounded interval D.

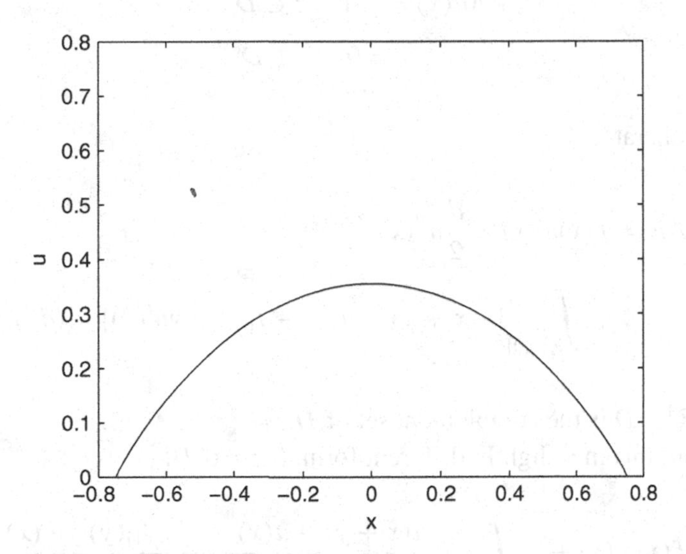

Figure 7.21 Mean exit time from the interval $(-0.75, 0.75)$ under pure α-stable Lévy motion: $\alpha = 1.8$.

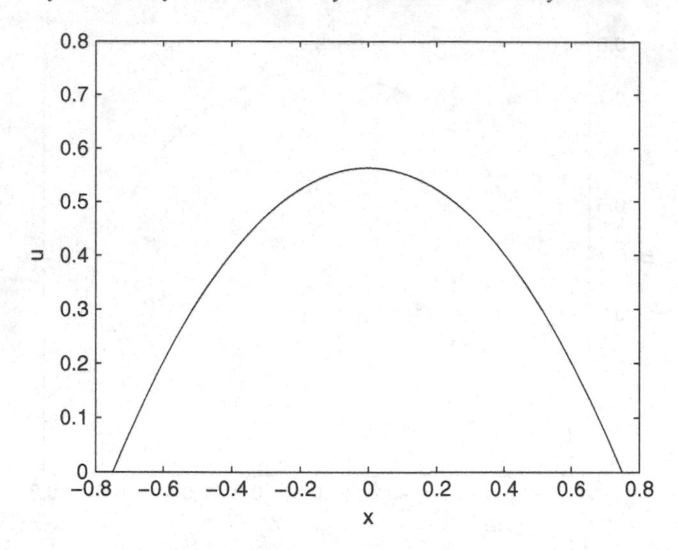

Figure 7.22 Mean exit time from the interval $(-0.75, 0.75)$ under Brownian motion: $\alpha = 2$.

The existing work on mean exit time is about an asymptotic estimate for $u(x)$ when the noise intensity is sufficiently small, that is, the noise term in (7.70) is $\varepsilon\, dL_t$ with $0 < \varepsilon \ll 1$. See, for example, [135, 136] and [302].

The mean exit time u satisfies the following differential-integral equation:

$$Au(x) = -1, \quad x \in D, \tag{7.81}$$

$$u = 0, \quad x \in D^c, \tag{7.82}$$

where the generator A is

$$Au = f(x)u'(x) + \frac{d}{2}u''(x) + \int_{\mathbb{R}^1\setminus\{0\}} [u(x+y) - u(x) - I_{\{|y|<1\}}\, yu'(x)]\ \nu_\alpha(dy), \tag{7.83}$$

and $D^c = \mathbb{R}^1 \setminus D$ is the complement set of D.

We can put this in a slightly different form: for $x \in D$,

$$\frac{d}{2}u''(x) + f(x)u'(x) + c_\alpha \int_{\mathbb{R}^1\setminus\{0\}} \frac{u(x+y) - u(x) - I_{\{|y|<1\}}(y)\ yu'(x)}{|y|^{1+\alpha}}\, \mathrm{d}y = -1, \tag{7.84}$$

and for $x \in D^c$, $u(x) = 0$. Here the integral is in the improper Riemann sense for $\alpha \in (0, 2)$.

Note that in this symmetric α-stable Lévy motion case, the final part in the integral operator is zero (in the sense of the Cauchy principal value):

$$\int_{\mathbb{R}^1\setminus\{0\}} I_{\{|y|<1\}}\, yu'(x)\, \nu_\alpha(dy) = 0. \tag{7.85}$$

Further note that this integral operator is related to the fractional Laplacian operator, as in (7.44) or [3, 109, 56, 44], for $\alpha \in (0, 2)$:

$$\int_{\mathbb{R}^1\setminus\{0\}} [u(x+y) - u(x)]\, \nu_\alpha(dy) = \theta_{\alpha,1}\, (-\Delta)^{\frac{\alpha}{2}} u(x), \tag{7.86}$$

where

$$\theta_{\alpha,1} \triangleq \int_{\mathbb{R}^1\setminus\{0\}} (\cos y - 1)\, \nu_\alpha(dy) < 0.$$

Let us look at an example [53, 54, 97].

Example 7.30 We consider an Ornstein-Uhlenbeck system

$$\mathrm{d}X_t = -X_t\, \mathrm{d}t + \sqrt{d}\, dB_t + \mathrm{d}L_t^\alpha, \;\; X(0) = x, \tag{7.87}$$

where the variance (or diffusion) $d > 0$ and the α-stable symmetric Lévy motion L_t^α has zero drift, zero diffusion, and jump measure $\varepsilon\nu_\alpha(du) = \varepsilon c_\alpha du/|u|^{1+\alpha}$ with coefficient $\varepsilon \geq 0$. The corresponding deterministic system $\dot{x} = -x$ has a global attractor $\{0\}$.

We simulate the mean exit time $u(x)$ for the solution orbit $X(t)$ starting at x and escape a bounded interval $D = (-0.75, 0.75)$ surrounding the attractor $\{0\}$.

We discuss Gaussian noise and non-Gaussian noise separately.

Gaussian Case: $\varepsilon = 0$

Let us first consider the Ornstein-Uhlenbeck system (7.87) with Gaussian noise (i.e., $\varepsilon = 0$). The equation (7.84) for the mean exit time $u(x)$ is now

$$\frac{d}{2}u''(x) - xu'(x) = -1 \tag{7.88}$$

for $D = (-0.75, 0.75)$, and $u(-0.75) = 0 = u(0.75)$. The numerical solution for $d = 0.1$ is shown in Figure 7.23. Moreover, for a smaller d value, such as $d = 0.01$, the peak value of the $u(x)$ is much bigger. The mean exit time is longer for smaller Gaussian noise.

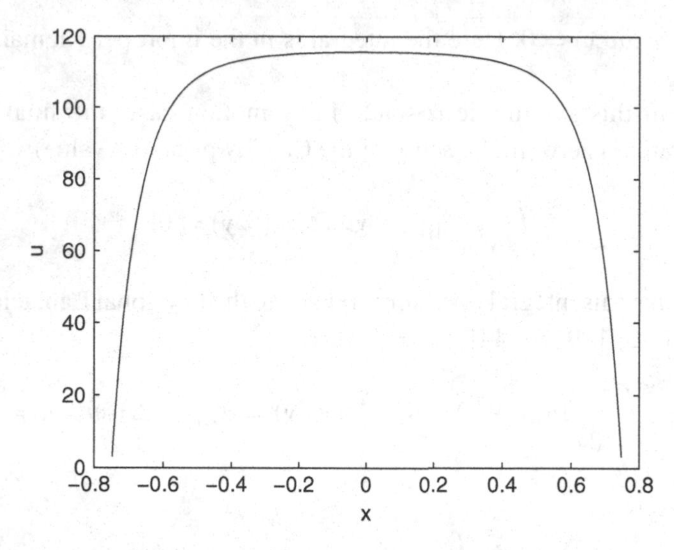

Figure 7.23 The solution $u(x)$ of Eq. (7.88) with $d = 0.1$.

Non-Gaussian Case: $\varepsilon \neq 0$

Now we look at the Ornstein-Uhlenbeck system (7.87) with non-Gaussian α-stable Lévy noise. The corresponding equation (7.84) for the mean exit time $u(x)$ is as follows:

$$\frac{d}{2}u''(x) - xu'(x) + \varepsilon c_\alpha \int_{\mathbb{R}^1\setminus\{0\}} \frac{u(x+y) - u(x) - I_{\{|y|<1\}}(y)\, yu'(x)}{|y|^{1+\alpha}}\, dy = -1 \tag{7.89}$$

for $x \in D = (-0.75, 0.75)$, and $u(x) = 0$ for $x \in D^c$.

We observe that the mean exit time u is much shorter (comparing with Gaussian $\varepsilon = 0$ case), even with a very small coefficient ε for the jump measure. See Figure 7.24 for the simulation of u for $\varepsilon = 0.1$ and compare with Figure 7.23.

7.6 Escape Probability and Transition Phenomena

In this section, we consider escape probability [234] in dynamical systems driven by Lévy motions.

Non-Gaussian random fluctuations are widely observed in various sysytems in physics, biology, seismology, electrical engineering, and finance [294, 156, 196]. Lévy motions are a large class of non-Gaussian stochastic processes whose sample paths have jumps in time. For a dynamical system driven by Lévy motion, almost all the orbits X_t have jumps in time. In fact, these orbits are cadlag (right continuous

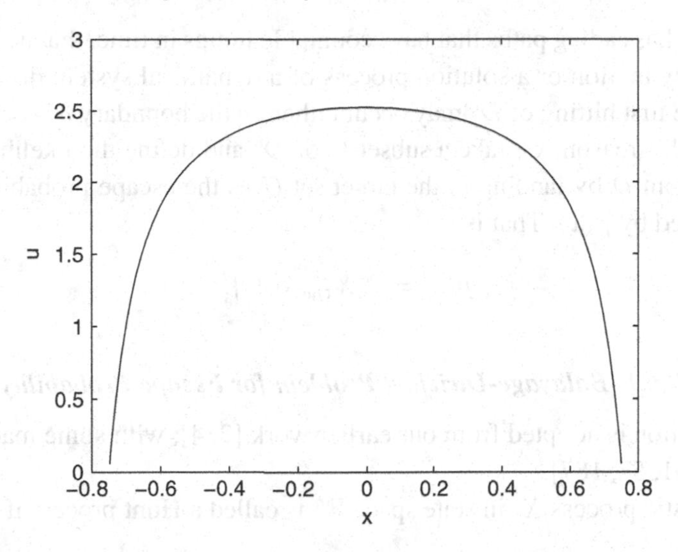

Figure 7.24 Mean exit time, i.e., the solution u of Eq. (7.89) with fixed $\alpha = 1$ and $\varepsilon = 0.1$: $d = 0.1$.

with left limit at each time instant), that is, each of these orbits has countable jumps in time. Owing to these jumps, an orbit could escape a domain without passing through its boundary. In this case, the *escape probability* is the likelihood that an orbit, starting inside an open domain D, exits this domain first by landing in a target domain U in D^c (the complement of domain D).

As we see, the escape probability is defined slightly differently for dynamical systems driven by Gaussian or non-Gaussian processes. Although the escape probability for the former has been investigated extensively, the characterization of the escape probability for the latter has not been well documented as a dynamical systems analysis tool for applied mathematics communities. See our recent works [53, 97] for numerical analysis of escape probability and mean exit time for dynamical systems driven by symmetric α-stable Lévy motions.

More precisely, let X_t be a stochastic process defined on a probability space $(\Omega, \mathcal{F}, \mathbb{P})$. Let D be a domain (by definition, it is open) in $\mathbb{R}^n$. Define the *first exit time*

$$\tau_{D^c}(\omega) \triangleq \inf\{t > 0\colon X_t \in D^c\}, \tag{7.90}$$

where D^c is the complement of D in $\mathbb{R}^n$. Namely, τ_{D^c} is the first time when X_t hits D^c. Note that the first exit time $\tau_{D^c}(\omega)$ also depends on the starting point of the stochastic process X_t: $X_0 = x \in D$. So it should be understood as $\tau_{D^c}(x, \omega)$. In the context of SDEs, we often denote the stochastic process X_t starting at x by $X_t(x)$.

When X_t has cadlag paths that have countable jumps in time, that is, X_t could be either a Lévy motion or a solution process of a dynamical system driven by Lévy motions, the first hitting of D^c may occur either on the boundary ∂D or somewhere in D^c. For this reason, we take a subset U of D^c and define the likelihood that X_t exits first from D by landing in the target set U as the escape probability from D to U, denoted by $p(x)$. That is,

$$p(x) = \mathbb{P}\{X_{\tau_{D^c}} \in U\}. \tag{7.91}$$

7.6.1 *Balayage-Dirichlet Problem for Escape Probability*

This subsection is adopted from our earlier work [234], with some materials from [34, 177, 261, 57, 187].

A stochastic process X_t in state space $\mathbb{R}^n$ is called a Hunt process if

(i) the path functions $t \to X_t$ are right continuous on $[0, \infty)$ and have left-hand limits on $[0, \zeta)$ almost surely, where $\zeta \triangleq \inf\{t\colon X_t = \infty\}$
(ii) X_t is a strong Markov process
(iii) X_t is quasi-left-continuous: if $\{\tau_n\}$ is an increasing sequence of $\mathcal{F}_t$-stopping times with limit τ, then almost surely $X_{\tau_n} \to X_\tau$ on $\{\tau < \infty\}$

Let D be an open subset of $\mathbb{R}^n$ and $X_t(x)$ be a Hunt process starting at $x \in D$. We denote a point on the boundary ∂D by z. Recall that D is called relatively compact if its closure $\bar{D}$ is compact (i.e., $\bar{D}$ is bounded and closed).

Definition 7.31 (Harmonic function with respect to a stochastic process) A non-negative function h, defined on $\mathbb{R}^n$, is said to be harmonic with respect to X_t in D if, for every compact set $K \subset D$,

$$\mathbb{E}[h(X_{\tau_{K^c}}(x))] = h(x), \quad x \in D. \tag{7.92}$$

Definition 7.32 (Balayage-Dirichlet problem with respect to a stochastic process) Let φ be a nonnegative function on D^c. We say a function h, defined on $\mathbb{R}^n$, solves the Balayage-Dirichlet problem for D with "boundary value" φ, denoted by (D, φ), if

(i) $h = \varphi$ on D^c
(ii) h is harmonic with respect to a Hunt process X_t in D
(iii) h further satisfies the following boundary condition:

$$\forall z \in \partial D,\ h(x) \to \varphi(z), \qquad \text{as } x \to z \text{ from inside } D$$

Definition 7.33 (Regular set with respect to a stochastic process) A boundary point z for a domain D is called regular with respect to a Hunt process X_t if the first exit time of $X_t(z)$ from the domain D is almost surely zero, that is,

$$\mathbb{P}\{\tau_{D^c} = 0\} = 1.$$

An open set D is said to be regular if each of its boundary points is regular.

Let I_D be the family of nonnegative functions g, bounded on $\mathbb{R}^n$ and lower semicontinuous in D, such that for every $x \in D$, there is a number $Ag(x)$ satisfying

$$\frac{\mathbb{E}[g(X_{\tau_\epsilon}(x))] - g(x)}{\mathbb{E}[\tau_\epsilon]} \to Ag(x), \text{ as } \epsilon \downarrow 0,$$

where $\tau_\epsilon \triangleq \inf\{t > 0\colon \|X_t(x) - x\| > \epsilon\}$. In the context of SDEs in this chapter, this linear operator A is just the (infinitesimal) generator [234], with I_D in the domain of A.

We quote the following result about the existence and uniqueness of the solution for the Balayage-Dirichlet problem [177].

Theorem 7.34 (Solution of the Balayage-Dirichlet problem) *Suppose that D is relatively compact and regular and φ is nonnegative and bounded on D^c. If φ is continuous at every point $z \in \partial D$, then $h(x) = \mathbb{E}[\varphi(X_{\tau_{D^c}}(x))]$ is the unique solution to the Balayage-Dirichlet problem (D, φ) and $Ah(x) = 0$ and $h \in I_D$.*

Consider the following stochastic differential equation in $\mathbb{R}^n$:

$$dX_t = f(X_{t-})dt + \sigma(X_{t-})dB_t + dL_t, \quad X_0 = x, \tag{7.93}$$

where f is a vector field (i.e., drift), σ is an $n \times n$ matrix function (i.e., diffusion), and L_t is a Lévy motion independent of the Brownian motion B_t in $\mathbb{R}^n$. The generating triplet for L_t is taken to be $(0, 0, \nu)$. Assume that the drift f, the diffusion σ, and the jump measure ν satisfy the conditions specified in Section 7.4. Then, the solution process X_t is a strong Markov process, and in fact, it is a Hunt process.

Because L_t has cadlag and quasi-left-continuous paths [244], $X_t(x)$ also has cadlag and quasi-left-continuous paths. Note that the solution process $X_t(x)$ of the SDE (7.93) is a strong Markov process (see Section 7.4). In fact, $X_t(x)$ is a Hunt process. Let D be a relatively compact and regular open domain (Figures 7.25 or 7.26). Theorem 7.34 implies that $\mathbb{E}[\varphi(X_{\tau_{D^c}}(x))]$ is the unique solution to the Balayage-Dirichlet problem (D, φ). Taking

$$\varphi(x) = \begin{cases} 1, & x \in U, \\ 0, & x \in D^c \setminus U, \end{cases}$$

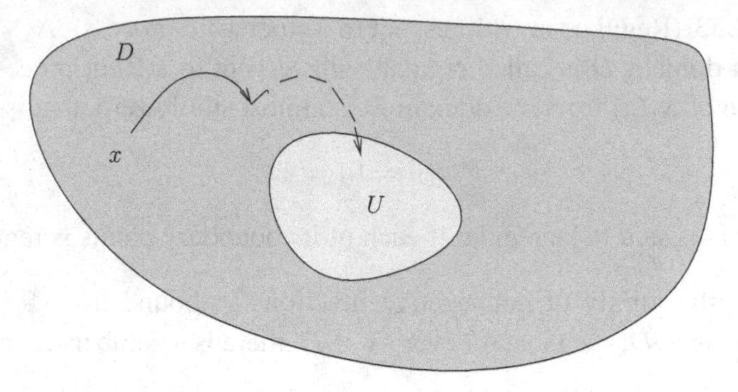

Figure 7.25 Escape probability for SDEs driven by Lévy motions: an open annular domain D, with its inner part U (which is in D^c) as a target domain.

we observe that

$$\begin{aligned}\mathbb{E}[\varphi(X_{\tau_{D^c}}(x))] &= \int_{\{\omega:\ X_{\tau_{D^c}}(x)\in U\}} \varphi(X_{\tau_{D^c}}(x))\mathrm{d}\mathbb{P}(\omega) \\ &\quad + \int_{\{\omega:\ X_{\tau_{D^c}}(x)\in D^c\setminus U\}} \varphi(X_{\tau_{D^c}}(x))\mathrm{d}\mathbb{P}(\omega) \\ &= \mathbb{P}\{\omega:\ X_{\tau_{D^c}}(x)\in U\} \\ &= p(x).\end{aligned}$$

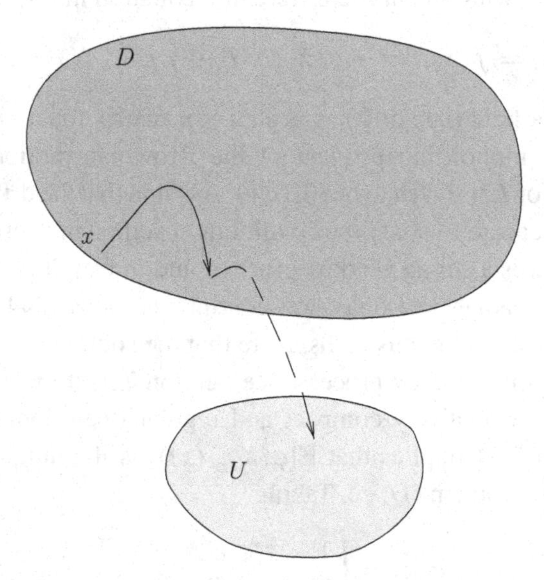

Figure 7.26 Escape probability for SDEs driven by Lévy motions: a general open domain D, with a target domain U in D^c.

This means that $\mathbb{E}[\varphi(X_{\tau_{D^c}}(x))]$ is the escape probability $p(x)$ that we are looking for. Note that $p \in I_D$, and thus by Theorem 7.34, we conclude that $Ap(x) = 0$. Thus, we obtain the following theorem.

Theorem 7.35 *Let D be a relatively compact and regular open domain, and let U be a set in D^c. Then the escape probability p, for the dynamical system driven by Lévy motion (7.93), from D to U, is the solution of the following Balayage-Dirichlet problem:*

$$Ap = 0, \tag{7.94}$$

$$p|_U = 1, \qquad p|_{D^c \setminus U} = 0, \tag{7.95}$$

where A is the generator for (7.93)

$$\begin{aligned} Ap = {} & f \cdot \nabla p + \frac{1}{2} Tr[\sigma \sigma^{\mathrm{T}} H(p)] \\ & + \int_{\mathbb{R}^n \setminus \{0\}} [p(x+y) - p(x) - I_{\{|y|<1\}}\, y \cdot \nabla p(x)]\, \nu(dy). \end{aligned} \tag{7.96}$$

Remark 7.36 Unlike the SDEs with Brownian motion in Section 5.5, a typical open domain D here could be a quite general open domain (Figure 7.26) as well as an annular domain (Figure 7.25). This is because the solution paths have jump discontinuities. It is also because, in Theorem 7.34, the function φ is only required to be continuous on the boundary ∂D (not on the domain D^c).

Results about the existence and uniqueness of solutions to nonlocal systems similar to (7.94) and (7.95) may be found in [101, 273].

7.6.2 Escape Probability for α-Stable Lévy Motion

We now consider an analytical example [234] for the escape probability under α-stable Lévy motion, with no drift.

Example 7.37 For a scalar symmetric α-stable Lévy motion L_t^α, take $D = (-r, r)$ and $U = [r, \infty)$. For each $x \in D$, the escape probability p of the solution path $X_t = x + L_t^\alpha$ for $dX_t = dL_t^\alpha$, $X_0 = x$, from D to U, satisfies the following differential-integral equation:

$$\begin{cases} -(-\Delta)^{\frac{\alpha}{2}} p(x) = 0, & x \in (-r, r), \\ p(x)|_{[r,\infty)} = 1, \\ p(x)|_{(-\infty,-r]} = 0. \end{cases}$$

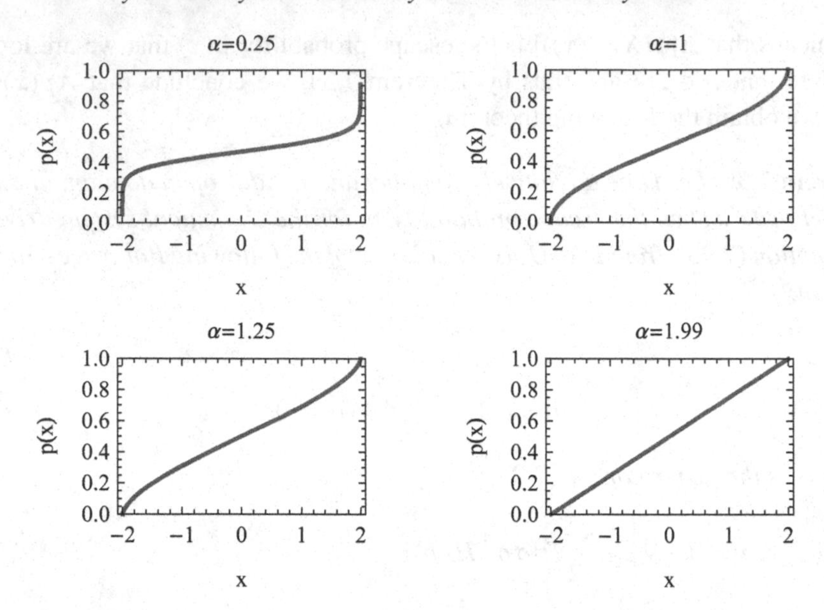

Figure 7.27 Escape probability in Example 7.37: the case of symmetric α-stable Lévy motion; $r = 2$.

By [137], the exact solution is

$$p(x) = \frac{\sin \frac{\pi\alpha}{2}}{\pi} \int_r^\infty \frac{(r^2 - x^2)^{\alpha/2}}{(y^2 - r^2)^{\alpha/2}} \frac{1}{(y - x)} \mathrm{d}y,$$

for $x \in (-r, r)$.

Obviously, $p(-r) = 0$. To justify $p(r) = 1$, we apply the substitution $y = (r^2 - xv)(x - v)^{-1}$ to obtain

$$\begin{aligned} p(r) &= \frac{\sin \frac{\pi\alpha}{2}}{\pi} \int_{-r}^{r} (r - v)^{\alpha-1} (r^2 - v^2)^{-\frac{\alpha}{2}} \mathrm{d}v \\ &= \frac{\sin \frac{\pi\alpha}{2}}{\pi} \int_0^1 (1 - v)^{\frac{\alpha}{2}-1} v^{1-\frac{\alpha}{2}-1} \mathrm{d}v \\ &= \frac{\sin \frac{\pi\alpha}{2}}{\pi} B\left(\frac{\alpha}{2}, 1 - \frac{\alpha}{2}\right) \\ &= \frac{\sin \frac{\pi\alpha}{2}}{\pi} \Gamma\left(\frac{\alpha}{2}\right) \Gamma\left(1 - \frac{\alpha}{2}\right) \\ &= 1, \end{aligned}$$

where the Beta and Gamma functions and their properties are used in the last two steps. The escape probability p is plotted in Figure 7.27 for $r = 2$, with various α values. They are all curves (see later for the case of Brownian motion).

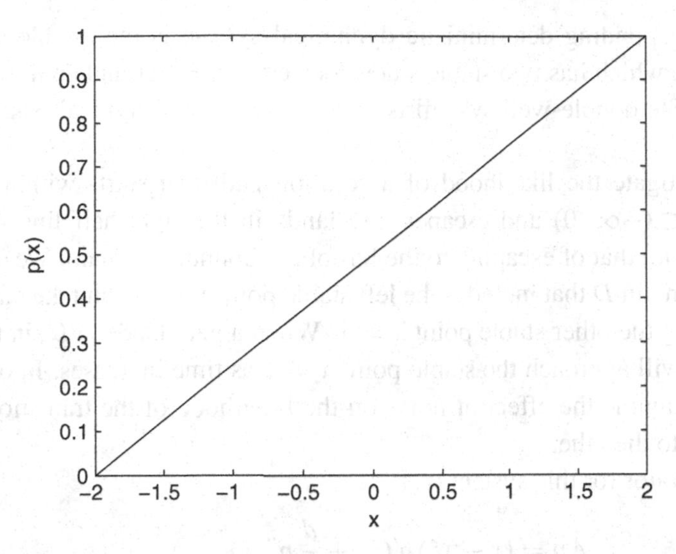

Figure 7.28 Escape probability in Example 7.37: the case of Brownian motion; $r = 2$.

We now compare with the case of Brownian motion. The escape probability $p(x)$ of solution path $X_t = x + B_t$ for $dX_t = dB_t$, $X_0 = x$, from $D = (-r, r)$ to $U = [r, \infty)$, satisfies the following differential equation:

$$\frac{1}{2}p''(x) = 0,$$

$$p(-r) = 0, \; p(r) = 1.$$

The general solution is $p(x) = ax + b$, and applying the boundary conditions, we get $p(x) = \frac{1}{2r}x + \frac{1}{2}$. See Figure 7.28 for the case of $r = 2$. It is a straight line, instead of a curve.

7.6.3 Escape Probability for SDEs with α-Stable Lévy Motion

This subsection is adopted from our earlier work [97]. Now we look at a numerical example for escape probability in a system described by a scalar SDE with an α-stable Lévy motion.

Example 7.38 Consider an SDE

$$dX_t = (X_{t-} - X_{t-}^3)dt + \sqrt{d}\, dB_t + dL_t^\alpha, \; X_0 = x, \tag{7.97}$$

where d is diffusion and L_t^α is a α-stable symmetric Lévy motion with generating triplet $(0, 0, \nu_\alpha)$. The jump measure $\nu_\alpha(du) = c_\alpha \frac{du}{|u|^{1+\alpha}}$, for $0 < \alpha < 2$. Here c_α is in (7.41) with $n = 1$.

The corresponding deterministic dynamical system is the double-well system $\dot{x} = x - x^3$, which has two stable states located at $x = \pm 1$ and an unstable steady state at 0. The double-well system is widely used in prototypical phase transition studies.

We investigate the likelihood of a solution path that starts within a bounded domain $D \subset (-\infty, 0)$ and escapes and lands in the right-half line $U = [0, \infty)$ compared with that of escaping to the left of the bounded domain. We consider the bounded domain D that includes the left stable point $x = -1$ and the target domain U containing the other stable point $x = 1$. When a path lands in U, in the absence of noise, it will approach the stable point $x = 1$ as time increases. In other words, we try to examine the effect of noise on the likelihood of the transition from one stable state to the other.

The generator for this system is

$$\begin{aligned} Ap &= (x - x^3)p'(x) + \frac{d}{2}p''(x) \\ &+ \int_{\mathbb{R}^1 \setminus \{0\}} [p(x+y) - p(x)]\ \nu_\alpha(dy), \end{aligned} \tag{7.98}$$

where the integral is understood as a Cauchy Principal Value (P. V.), and the jump measure $\nu_\alpha(dy) = c_\alpha \frac{dy}{|y|^{1+\alpha}}$, with c_α from (7.31), for $\alpha \in (0, 2)$. The escape probability p is the solution of the following boundary value problem:

$$Ap = 0, \quad x \in D, \tag{7.99}$$

$$p|_U = 1, \qquad p|_{D^c \setminus U} = 0. \tag{7.100}$$

Several numerical solutions for p are plotted in Figure 7.29.

Figure 7.29(a) shows the escape probability for $D = (-1.1, 0)$, $U = [0, \infty)$ and $\alpha = 0.5, 1, 1.5, 2$, when the stochastic effects are given by α-stable symmetric processes only ($d = 0$). The escape probability $P(x)$ deviates more from a straight line as α decreases, and it is smaller for smaller α when starting from $x > 0.5$. On the contrast, the probability is larger for smaller α when the starting point is close to the left boundary of the bounded domain. For the bigger domain $D = (-2, 0)$ shown in Figure 7.29(b), the likelihood of escape to the right is more than a half for most of the starting points $x > -1.5$ and $1 \leq \alpha \leq 2$; the probability stays close to a half for most of the starting points when $\alpha = 0.5$. Note that, in the case of $\alpha < 1$, the probability is discontinuous at the left boundary of the domain. As shown in Figures 7.29(c) and 7.29(d), the differences in escape probabilities among different values of α become smaller when an amount of Gaussian noise is added ($d = 0.1$), but otherwise probabilities have similar values and features compared to those of case $d = 0$.

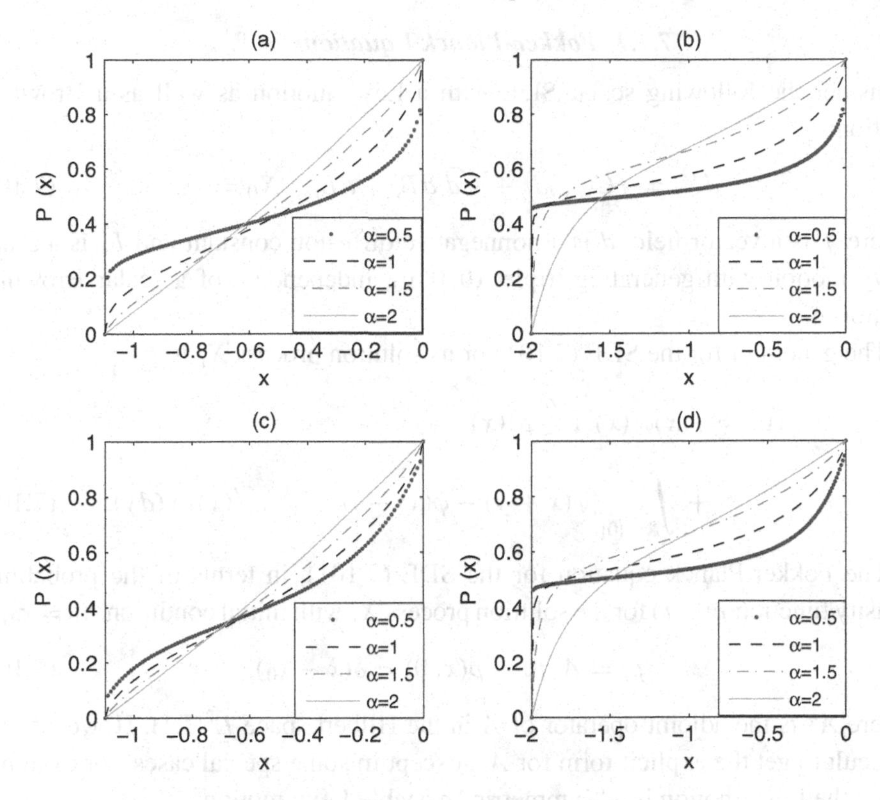

Figure 7.29 Escape probability for $P(x)$, Example 7.38, with target set $U = [0, \infty)$ and $\alpha = 0.5, 1, 1.5, 2$: (a) $D = (-1.1, 0)$, $d = 0$; (b) $D = (-2, 0)$, $d = 0$; (c) $D = (-1.1, 0)$, $d = 0.1$; and (d) $D = (-2, 0)$, $d = 0.1$.

7.7 Fokker-Planck Equations

The Fokker-Planck equations are one of the deterministic tools to quantify how randomness propagates or evolves in nonlinear dynamical systems. For SDEs with Gaussian processes such as Brownian motion, the Fokker-Planck equations are well established in Section 5.2. However, for SDEs with non-Gaussian processes, such as Lévy processes, explicit forms of the Fokker-Planck equations are not readily available, except in some special cases ([246, 199], and references therein). The difficulty is to obtain the expressions for the adjoint operators of the infinitesimal generators associated with these SDEs. Sun and Duan [268] recently developed a method to derive Fokker-Planck equations, in terms of an infinite series, in this context.

We consider Fokker-Planck equations associated with SDEs with symmetric α-stable Lévy motions, first in the scalar case and then in the vector case.

7.7.1 Fokker-Planck Equations in $\mathbb{R}^1$

Consider the following scalar SDE with a Lévy motion as well as a Brownian motion:

$$dX_t = f(X_{t-})dt + \sqrt{d}\, dB_t + dL_t, \quad X_0 = x_0, \tag{7.101}$$

where f is a vector field, d is a nonnegative diffusion constant and L_t is a scalar Lévy motion with generating triplet $(0, 0, \nu)$, independent of a scalar Brownian motion B_t.

The generator for the SDE (7.101) or its solution process X_t is

$$A\varphi = f(x)\varphi'(x) + \frac{d}{2}\varphi''(x) + \int_{\mathbb{R}^1\setminus\{0\}} [\varphi(x+y) - \varphi(x) - I_{\{|y|<1\}}\, y\varphi'(x)]\, \nu(dy). \tag{7.102}$$

The Fokker-Planck equation for the SDE (7.101), in terms of the probability density function $p(x,t)$ for the solution process X_t with initial condition $X_0 = x_0$, is

$$p_t = A^* p, \quad p(x,0) = \delta(x - x_0), \tag{7.103}$$

where A^* is the adjoint operator of A in the Hilbert space $L^2(\mathbb{R}^1)$. However, it is difficult to get the explicit form for A^*, except in some special cases, for example, when the Lévy motion is a "symmetric" α-stable Lévy motion.

Let us try to find the adjoint operator A^* in the Hilbert space $L^2(\mathbb{R}^1)$:

$$\int_{\mathbb{R}^1} A\varphi(x)\, v(x)dx = \int_{\mathbb{R}^1} \varphi(x)\, A^*v(x)dx, \tag{7.104}$$

for φ, v in the domains of definition for the operators A and A^*, respectively. The adjoint parts for the first two terms in A are easy to find via integration by parts: $-(f(x)\varphi(x))'$ and $\frac{d}{2}\varphi''(x)$. For the third term, denoted by $\tilde{A}\varphi$, we have

$$\begin{aligned}
&\int_{\mathbb{R}^1} \tilde{A}\varphi(x)\, v(x)dx \\
&= \int_{\mathbb{R}^1}\int_{\mathbb{R}^1\setminus\{0\}} [\varphi(x+y) - \varphi(x) - I_{\{|y|<1\}}\, y\varphi'(x)]\, \nu(dy)\, v(x)dx \\
&= \int_{\mathbb{R}^1\setminus\{0\}} \left\{ \int_{\mathbb{R}^1} [\varphi(x+y) - \varphi(x) - I_{\{|y|<1\}}\, y\varphi'(x)]\, v(x)dx \right\} \nu(dy) \\
&= \int_{\mathbb{R}^1\setminus\{0\}} \left\{ \int_{\mathbb{R}^1} \varphi(x)[v(x-y) - v(x) + I_{\{|y|<1\}}\, yv'(x)]\, dx \right\} \nu(dy) \\
&= -\int_{\mathbb{R}^1} \varphi(x) \left\{ \int_{\mathbb{R}^1\setminus\{0\}} [v(x) - v(x-y) - I_{\{|y|<1\}}\, yv'(x)]\, \nu(dy) \right\} dx.
\end{aligned} \tag{7.105}$$

Thus, the adjoint operator of the third term in A is

$$\tilde{A}^* v = -\int_{\mathbb{R}^1\setminus\{0\}} [v(x) - v(x-y) - I_{\{|y|<1\}}\, y v'(x)]\nu(dy), \tag{7.106}$$

for v in the domain of definition for A^*. Hence, the adjoint operator of the generator A is

$$A^*\varphi = -(f(x)\varphi(x))' + \frac{d}{2}\varphi''(x) - \int_{\mathbb{R}^1\setminus\{0\}} [\varphi(x) - \varphi(x-y) - I_{\{|y|<1\}}\, y\varphi'(x)]\, \nu(dy). \tag{7.107}$$

The Fokker-Planck equation for the SDE (7.101), in terms of the probability density function $p(x,t)$ for the solution process X_t with initial condition $X_0 = x_0$, is

$$p_t = A^* p, \quad p(x,0) = \delta(x - x_0), \tag{7.108}$$

that is,

$$p_t = -\partial_x(f(x)p(x)) + \frac{d}{2}\partial_{xx} p(x,t) - \int_{\mathbb{R}^1\setminus\{0\}} [p(x,t) - p(x-y,t) - I_{\{|y|<1\}}\, y\partial_x p(x,t)]\, \nu(dy). \tag{7.109}$$

Remark 7.39 When L_t is replaced by an α-stable Lévy motion L_t^α, the third term in the generator A is self-adjoint and the Fokker-Planck equation is easily written out, as shown in the following example.

Example 7.40 Consider the following scalar SDE:

$$dX_t = f(X_{t-})dt + \sqrt{d}\, dB_t + dL_t^\alpha, \quad X_0 = x_0, \tag{7.110}$$

where d is a nonnegative diffusion constant, B_t is a scalar Brownian motion, and L_t^α is a scalar symmetric α-stable Lévy motion with the generating triplet $(0, 0, \nu_\alpha)$. The jump measure is $\nu_\alpha(dy) = c_\alpha \frac{dy}{|y|^{1+\alpha}}$, with c_α from (7.31), for $\alpha \in (0,2)$. The processes B_t and L_t^α are taken to be independent.

The generator for this system is

$$Ap = f(x)p_x(x,t)) + \frac{d}{2}\partial_{xx} p(x,t) + \int_{\mathbb{R}^1\setminus\{0\}} [p(x+y,t) - p(x,t)]\, \nu_\alpha(dy).$$

Note that the integral part can be uniquely extended to a self-adjoint operator (see Section 7.3.3) in $W^{\alpha,2}(\mathbb{R}^1)$, and in fact, as seen by Fourier transform, it is equal to a nonlocal Laplacian operator $\theta_{\alpha,1}\,(-\Delta)^{\frac{\alpha}{2}}$, with

$$\theta_{\alpha,1} \triangleq \int_{\mathbb{R}^1\setminus\{0\}} (\cos y - 1)\,\nu_\alpha(dy) < 0.$$

Thus the adjoint operator of the generator A is

$$A^* p = -\partial_x(f(x)p(x,t)) + \frac{d}{2}\partial_{xx}p(x,t) + \int_{\mathbb{R}^1\setminus\{0\}} [p(x+y,t) - p(x,t)]\,\nu_\alpha(dy),$$

or

$$A^* p = -\partial_x(f(x)p(x,t)) + \frac{d}{2}\partial_{xx}p(x,t) + \theta_{\alpha,1}\,(-\Delta)^{\frac{\alpha}{2}} p. \tag{7.111}$$

The Fokker-Planck equation is then

$$p_t = -\partial_x(f(x)p(x,t)) + \frac{d}{2}\partial_{xx}p(x,t) + \int_{\mathbb{R}^1\setminus\{0\}} [p(x+y,t) - p(x,t)]\,\nu_\alpha(dy), \tag{7.112}$$

or

$$p_t = -\partial_x(f(x)p(x,t)) + \frac{d}{2}\partial_{xx}p(x,t) + \theta_{\alpha,1}\,(-\Delta)^{\frac{\alpha}{2}} p, \tag{7.113}$$

with initial condition $p(x,0) = \delta(x - x_0)$.

7.7.2 Fokker-Planck Equations in $\mathbb{R}^n$

More generally, we derive the Fokker-Planck equation for the following SDE in $\mathbb{R}^n$:

$$dX_t = f(X_{t-})dt + \sigma(X_{t-})dB_t + dL_t^\alpha, \quad X_0 = x_0, \tag{7.114}$$

where f is a vector field, σ is an $n \times n$ matrix, B_t is a Brownian motion in $\mathbb{R}^n$, and L_t^α is a symmetric α-stable Lévy motion in $\mathbb{R}^n$, with the generating triplet $(0, 0, \nu_\alpha)$. The jump measure

$$\nu_\alpha(\mathrm{d}y) = c(n,\alpha)\|y\|^{-(n+\alpha)}\,\mathrm{d}y,$$

with $c(n,\alpha) = \frac{\alpha\Gamma((n+\alpha)/2)}{2^{1-\alpha}\pi^{n/2}\Gamma(1-\alpha/2)}$. The processes B_t and L_t^α are taken to be independent.

The generator A for the SDE (7.114) is

$$Ag = f\cdot\nabla g + \frac{1}{2}\mathrm{Tr}[\sigma\sigma^{\mathrm{T}} H(g)] + \int_{\mathbb{R}^n\setminus\{0\}} [g(x+y) - g(x)]\nu_\alpha(dy). \tag{7.115}$$

The Fokker-Planck equation for the SDE (7.114) is then

$$p_t = -\nabla \cdot (fp) + \frac{1}{2}\mathrm{Tr}[H(\sigma\sigma^{\mathrm{T}} p)] + \int_{\mathbb{R}^n \setminus \{0\}} [p(x+y,t) - p(x,t)]\nu_\alpha(dy), \tag{7.116}$$

where $p(x,0) = \delta(x - x_0)$, and, as in Section 5.2, $H(\sigma\sigma^{\mathrm{T}} p)$ is interpreted as matrix multiplication of H and $\sigma\sigma^{\mathrm{T}} p$ (note that p is a scalar function). Here the integral in the right hand side is understood as a Cauchy principal value.

The Fokker-Planck equation (7.116) on a bounded domain D in $\mathbb{R}^n$ may be subject to the following absorbing boundary condition and an initial condition:

$$p(x,t) = 0 \quad \text{for } x \in D^c \ ; \ p(x,0) = p_0(x) \quad \text{for } x \in D. \tag{7.117}$$

This initial condition $p_0(x)$ needs to be non-negative and satisfies $\int_{\mathbb{R}^n} p_0(x) = 1$.

Results on the existence and uniqueness for nonlocal partial differential equations similar to the Fokker-Planck system (7.116) and (7.117) may be found in [100].

To obtain Fokker-Planck equations for SDEs with multiplicative Lévy noises, it is considerably more complicated to find out the adjoint operator for the generator A. See [268] for a way to achieve this goal.

7.8 Problems

7.1 Two definitions for a stable random variable

Show that two definitions for a stable random variable are equivalent, that is, Definition 7.9 implies Definition 7.17, and vice versa.

7.2 Lévy jump measure

Show that $\nu((-\varepsilon,\varepsilon)^c) < \infty$ for every Lévy jump measure ν. How about $\nu((-\varepsilon,\varepsilon))$? Here $(-\varepsilon,\varepsilon)^c$ is the complement of $(-\varepsilon,\varepsilon)$.

7.3 Stable random variables

Calculate the following integrals related to the jump measure $\nu_\alpha(dy) = c_\alpha \frac{dy}{|y|^{1+\alpha}}$ with $\alpha \in (0,2)$, of the scalar symmetric α-stable random variables:

(a) The improper integral

$$\int_{\mathbb{R}^1 \setminus \{0\}} \mathbf{I}_{\{|y|<1\}}(y)\, y\, \nu_\alpha(dy).$$

(b) The Cauchy principal value

$$\text{P. V.} \int_{\mathbb{R}^1 \setminus \{0\}} \mathbf{I}_{\{|y|<1\}}(y)\, y\, \nu_\alpha(dy).$$

7.4 Stable random variables again

Calculate the following integrals related to the jump measure $\nu_\alpha(dy) = c_\alpha \frac{dy}{|y|^{1+\alpha}}$ with $\alpha \in (0, 2)$ of the scalar symmetric α-stable random variables:

(a) The improper integral

$$\int_{\mathbb{R}^1\setminus\{0\}} \mathbf{I}_{\{|y|<1\}}(y)\, |y|\, \nu_\alpha(dy).$$

(b) The Cauchy principal value

$$\text{P. V.} \int_{\mathbb{R}^1\setminus\{0\}} \mathbf{I}_{\{|y|<1\}}(y)\, |y|\, \nu_\alpha(dy).$$

7.5 The α-Stable Jump Measure

For the jump measure $\nu_\alpha(dy) = c_\alpha \frac{dy}{|y|^{1+\alpha}}$, $\alpha \in (0, 2)$, of the scalar symmetric α-stable random variables, evaluate $\nu_\alpha((-\varepsilon, \varepsilon))$ and $\nu_\alpha((-\varepsilon, \varepsilon)^c)$. Here $(-\varepsilon, \varepsilon)^c$ is the complement of $(-\varepsilon, \varepsilon)$.

7.6

(a) Develop a random number generator with a symmetric α-stable distribution $S_\alpha(1, 0, 0)$ for $\alpha \in (0, 2)$. (For the normal distribution case $\alpha = 2$, Matlab has it: `randn`.)
(b) Conduct Matlab simulation of the scalar α-stable Lévy motion L_t^α for two-sided time and for various $\alpha \in (0, 2)$. Plot sample paths for various α values. What do you observe, that is, what is the impact of α on sample paths, if any? Why do some people say, "When $\alpha = 2$, L_t^α becomes the usual (Gaussian) Brownian motion B_t"?

7.7 Relation between Lévy motion L_t^α and Brownian motion B_t when $\alpha \to 2$

The characteristic functions for scalar Lévy motion L_t^α and Brownian motion B_t are $\Phi_\alpha(u, t) \triangleq e^{-t|u|^\alpha}$ and $\Phi_2(u, t) \triangleq e^{-t|u|^2}$, respectively. Thus at each $t > 0$, we see that $\Phi_\alpha \to \Phi_2$ for every $u \in \mathbb{R}^1$. Let $f_\alpha(x, t)$ and $f_2(x, t)$ be the probability density functions for L_t^α and B_t, respectively.

(a) At each $t > 0$, does $f_\alpha(x, t) \to f_2(x, t)$ as $\alpha \to 2$, almost everywhere, pointwise, or in $L^1(\mathbb{R}^1)$?
(b) At each $t > 0$, does $f_{\alpha_n}(x, t) \to f_2(x, t)$ for an increasing sequence $\alpha_n \to 2$ as $n \to \infty$, almost everywhere, pointwise, or in $L^1(\mathbb{R}^1)$?
(c) At each $t > 0$, does L_t^α converge to B_t in distribution (i.e., the corresponding probability measures converge weakly), as $\alpha \to 2$ or for an increasing sequence $\alpha_n \to 2$ as $n \to \infty$?

7.8 Find the generator A_α for a scalar symmetric α-stable Lévy motion L_t^α using the definition $A_\alpha\varphi(x) = \frac{d}{dt}|_{t=0}\mathbb{E}\varphi(X_t)$ for $X_t = x + L_t^\alpha$ (a scalar α-stable Lévy motion starting at x). See Problem 4.6.

7.9

(a) Matlab simulation of sample solution paths of $dX_t = (aX_{t-} - X_{t-}^3)dt + dL_t^\alpha$ with $X_0 = 0.1 \in \mathbb{R}$, where L_t^α has the generating triplet (drift, diffusion, jump measure)= $(0, 0, \nu_\alpha)$. Plot the sample solution paths when $a = 1$ for two-sided time $t \in [-1, 1]$. Calculate and plot the mean $\mathbb{E}(X_t)$ if it exists. What is the impact of α on solution paths and on the mean, if any?

(b) Solve the steady Fokker-Planck equation for possible stationary probability densities to detect possible bifurcation when a and α vary.

7.10 Calculate the mean exit time $u(x)$ for the scalar α-stable Lévy motion starting at x, $L_t^\alpha + x$, for x in the interval $(-3, 3)$. This leads to direct numerical simulation of the deterministic differential-integral equation satisfied by $u(x)$. Compare with the exact result for various α values.

7.11 Mean exit time under Lévy noise
Consider a scalar SDE $dX_t = (-X_{t-} + X_{t-}^3)dt + dL_t^\alpha$, where L_t^α is a scalar symmetric Lévy motion with the generating triplet $(0, 1, \nu_\alpha)$ and $\alpha \in (0, 2)$. Compute the mean exit time $u(x)$ for x in the interval $D = (-0.75, 0.75)$. What is the impact of α on the mean exit time, if any? Compare with the Gaussian (Brownian motion) case $dX_t = (-X_t + X_t^3)dt + dB_t$, which corresponds to the preceding case, but with the jump measure ν_α set to zero (no jumps).

7.12 Consider a two-dimensional system of SDEs

$$dX_t = (9X_{t-} - X_{t-}^3)dt + dL_t^1,$$
$$dY_t = -Y_{t-}dt + dL_t^2,$$

where (L_t^1, L_t^2) is a two-dimensional rotationally symmetric α-stable Lévy motion. For the corresponding deterministic dynamical system, $\mathcal{A} = \{(x, y)\colon x \in [-3, 3],\ y \in [-2, 2]\}$ is an attracting set, as shown in Figure 5.12. Devise a numerical scheme to compute the mean exit time $u(x, y)$ from the domain $D = \{(x, y)\colon x \in (-2, 2),\ y \in (-1, 1)\}$.

Note that if L_t^1, L_t^2 are scalar-independent α-stable Lévy motions, their (joint) two-dimensional jump measure could be a delicate issue; see [243, pp. 66–67].

7.13 Consider a stochastic system in $\mathbb{R}^n$ with multiplicative independent Gaussian and non-Gaussian noises

$$dX_t = f(X_{t-})dt + \sigma(X_{t-})dB_t + g(X_{t-})dL_t, \quad X_0 = x, \tag{7.118}$$

where f is a vector field, B_t is a Brownian motion in $\mathbb{R}^n$, σ and g are $n \times n$ matrix functions, and L_t is a Lévy motion in $\mathbb{R}^n$ with generating triplet $(0, 0, \nu)$. The processes B_t and L_t are independent. What is the generator for the solution process X_t? Discuss the formation and computation of nonlocal partial differential equations for mean exit time and escape probability.

Hints and Solutions

Problems of Chapter 1

1.1
(a) The equilibrium states are 0, which is unstable, and 1, which is stable.
(b) The equilibrium states are 0, which is stable, and 1, which is unstable.

1.2 The equilibrium state is $(0, 0)$. The energy $\frac{1}{2}kx^2 + \frac{1}{2}y^2$ is conserved, because $\frac{d}{dt}[\frac{1}{2}kx^2 + \frac{1}{2}y^2] = 0$. Thus $\frac{1}{2}kx^2 + \frac{1}{2}y^2 = c$, for an arbitrary nonnegative constant c. For various c values, we obtain solution orbits that are ellipses. The phase portrait is composed of a family of ellipses. When $k = 1$, these ellipses become circles, as we have seen in Example 1.3.

1.3 The matrix for this linear system is

$$A = \begin{pmatrix} -1 & 0 \\ 0 & 1 \end{pmatrix}.$$

The eigenvalues are -1 and 1. So the equilibrium state $(0, 0)$ is a saddle (one negative and one positive eigenvalue). See Figure H.1. The stable eigenspace is the x-axis (also the stable manifold W^s) and the unstable is the y-axis (also the unstable manifold W^u).

Dividing the two equations, we obtain

$$\frac{dy}{dx} = -\frac{y}{x}$$

or

$$\frac{dy}{y} + \frac{dx}{x} = 0.$$

Integrating this equation, we see that the solution curves $(x(t), y(t))$ satisfy the relation

$$xy = c$$

for an arbitrary constant of integration c. Thus the phase portrait is composed of two families of hyperbolic curves: $xy = c_1 \geq 0$ and $xy = c_2 \leq 0$.

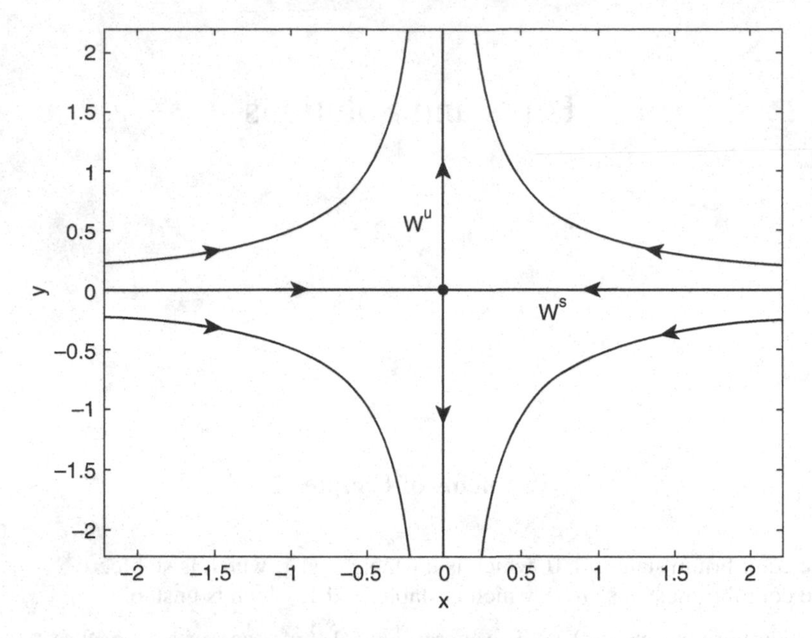

Figure H.1 Problem 1.3: phase portrait for $\dot{x} = -x,\ \dot{y} = y$.

Problems of Chapter 2

2.1 Note that for every Borel set $A \in \mathcal{B}(\mathbb{R}^1)$, the set $(f(X))^{-1}(A) = X^{-1}(f^{-1}(A))$ is an event in Ω. More precisely, because f is a Borel measurable function, for every Borel set $B \in \mathcal{B}(\mathbb{R}^1)$, we have $A = \{x \in \mathbb{R}^1 \colon f(x) \in B\}$, which is Borel measurable. Thus

$$\{\omega \in \Omega \colon f(X(\omega)) \in B\} = \{\omega \in \Omega \colon X(\omega) \in A\} \in \mathcal{F},$$

which implies that $f(X)$ is also a real-valued random variable.

2.2
(a) We need facts from calculus: $\int_{-\infty}^{\infty} e^{-z^2} dz = \sqrt{\pi}$ and $\int_{-\infty}^{\infty} e^{-z^2/2} dz = \sqrt{2\pi}$. First,

$$\begin{aligned}
\mathbb{E}(X) &= \int_{-\infty}^{\infty} x f(x) dx \\
&= \frac{1}{\sqrt{2\pi}\sigma} \int_{-\infty}^{\infty} x \exp \frac{-(x-\mu)^2}{2\sigma^2} dx \\
&= \frac{1}{\sqrt{2\pi}\sigma} \int_{-\infty}^{\infty} (\mu + \sigma z) \exp \frac{-z^2}{2} dz \\
&= \frac{\mu}{\sqrt{2\pi}} \int_{-\infty}^{\infty} \exp \frac{-z^2}{2} dz + \frac{1}{\sqrt{2\pi}} \int_{-\infty}^{\infty} z \exp \frac{-z^2}{2} dz \\
&= \frac{\mu}{\sqrt{2\pi}} \sqrt{2\pi} + 0 = \mu,
\end{aligned}$$

where $z = (x-\mu)/\sigma$ and the final integral is zero because the integrand is an odd function.

Then,

$$\begin{aligned}
\mathrm{Var}(X) = \mathbb{E}(X-\mu)^2 &= \int_{-\infty}^{\infty} (x-\mu)^2 f(x)dx \\
&= \frac{1}{\sqrt{2\pi}\sigma}\int_{-\infty}^{\infty} (x-\mu)^2 \exp\frac{-(x-\mu)^2}{2\sigma^2}dx \\
&= \frac{1}{\sqrt{2\pi}\sigma}\frac{1}{2}\int_{-\infty}^{\infty} (x-\mu) \exp\frac{-(x-\mu)^2}{2\sigma^2}d(x-\mu)^2 \\
&= -\frac{\sigma}{\sqrt{2\pi}}\int_{-\infty}^{\infty} (x-\mu) \exp\frac{-(x-\mu)^2}{2\sigma^2}d\frac{-(x-\mu)^2}{2\sigma^2} \\
&= -\frac{\sigma}{\sqrt{2\pi}}\int_{-\infty}^{\infty} (x-\mu) d\exp\frac{-(x-\mu)^2}{2\sigma^2} \\
&= \frac{\sigma}{\sqrt{2\pi}}\int_{-\infty}^{\infty} \exp\frac{-(x-\mu)^2}{2\sigma^2}dx \quad \text{(Integration by parts)} \\
&= \frac{\sigma^2}{\sqrt{2\pi}}\int_{-\infty}^{\infty} \exp\frac{-z^2}{2}dz \quad \text{(Change variable } z = (x-\mu)/\sigma) \\
&= \frac{\sigma^2}{\sqrt{2\pi}}\sqrt{2\pi} = \sigma^2.
\end{aligned}$$

(b) *Odd moments:*

$$\frac{1}{\sqrt{2\pi}\sigma}\int (x-\mu)^{2k+1} \exp\frac{-(x-\mu)^2}{2\sigma^2}\,\mathrm{d}x.$$

Making the substitution $\sigma z = x - \mu$, this integral becomes

$$\frac{\sigma^{2k+1}}{\sqrt{2\pi}}\int z^{2k+1} \exp\frac{-z^2}{2}\,\mathrm{d}z.$$

As the integrand is an odd function, it vanishes when integrated from $-\infty$ to ∞. Thus, the odd central moments are zero.

Even moments:

$$\frac{1}{\sqrt{2\pi}\sigma}\int (x-\mu)^{2k} \exp\frac{-(x-\mu)^2}{2\sigma^2}\,\mathrm{d}x.$$

Again making the substitution $\sigma z = x - \mu$, this integral becomes

$$\frac{\sigma^{2k}}{\sqrt{2\pi}}\int z^{2k} \exp\frac{-z^2}{2}\,\mathrm{d}z.$$

Integration by parts with $u = z^{2k-1}$, $v = -\exp\frac{-z^2}{2}$ to obtain

$$\frac{\sigma^{2k}}{\sqrt{2\pi}}\left[z^{2k-1}\exp\frac{-z^2}{2}\bigg|_{-\infty}^{\infty} + (2k-1)\int z^{2(k-1)}\exp\frac{-z^2}{2}\,\mathrm{d}x\right].$$

The first term vanishes. Integration by parts can be repeated on the second term, each time taking $v = -\exp\frac{-z^2}{2}$. It is found that the even central moments are $\mathbb{E}(X-\mu)^{2k} = 1\cdot 3\cdot 5\cdots(2k-1)\sigma^{2k}$ for $k = 1, 2, 3, \ldots$.

(c) First, $X+Y$ is not Gaussian in general. However, it is Gaussian when X and Y are independent or when they are components of a Gaussian vector in $\mathbb{R}^2$.
Second, XY is not Gaussian in general.

Moreover, XY may not be Gaussian even when X and Y are independent. In fact, if $Z = XY$ is a Gaussian random variable, then $\mathbb{E}(Z-\mu_1\mu_2)^{2k+1} = 0$, where $\mathbb{E}(X) = \mu_1$ and $\mathbb{E}(Y) = \mu_2$. But note that

$$\begin{aligned}
\mathbb{E}(Z) &= \mathbb{E}(X)\mathbb{E}(Y) = \mu_1\mu_2,\\
\mathbb{E}(Z^2) &= \mathbb{E}(X^2)\mathbb{E}(Y^2) = \left(\mu_1^2+\sigma_2^2\right)\left(\mu_1^2+\sigma_2^2\right),\\
\mathbb{E}(Z^3) &= \mathbb{E}(X^3)\mathbb{E}(Y^3) = \left(\mu_3 - 3\sigma_1^2\mu_1\right)\left(\mu_3 - 3\sigma_2^2\mu_2\right),\\
\mathbb{E}(Z-\mu_1\mu_2)^3 &= \mathbb{E}\left(Z^3 - 3\mu_1\mu_2 Z^2 + 3\mu_1^2\mu_2^2 Z - \mu_1^2\mu_2^2\right) \neq 0.
\end{aligned}$$

Therefore, $Z = XY$ is not a Gaussian random variable.

2.3

$$A = (a_{ij})_{3\times 3} = Q^{-1} = \begin{pmatrix} 1.5 & -0.5 & 0.5\\ -0.5 & 0.5 & -0.5\\ 0.5 & -0.5 & 1.5\end{pmatrix}.$$

The probability density function for $\boldsymbol{X} \sim \mathbb{N}(m, Q)$ is

$$p(x_1, x_2, x_3) = \frac{\sqrt{\det(A)}}{(2\pi)^{3/2}}\exp\left(-\frac{1}{2}\sum_{i=1}^{3}\sum_{j=1}^{3}(x_i - m_i)a_{ij}(x_j - m_j)\right),$$

where $m = (m_1, m_2, m_3) = (1, 2, 0)^{\mathrm{T}}$. The trace of Q is

$$\mathrm{Tr}(Q) = \sum_{i=1}^{3} q_{ii} = 1+4+1 = 6.$$

Moreover,

$$M_{\boldsymbol{X}}(u) = \mathbb{E}\left(e^{\langle u,\boldsymbol{X}\rangle}\right) = \mathbb{E}\left(e^{\sum_{j=1}^{3}u_j X_j}\right) = e^{\langle m,u\rangle + \frac{1}{2}\langle u,Qu\rangle}$$

and

$$\Phi_{\boldsymbol{X}}(u) = \mathbb{E}\left(e^{i\langle u,\boldsymbol{X}\rangle}\right) = \mathbb{E}\left(e^{\sum_{j=1}^{3}iu_j X_j}\right) = e^{i\langle m,u\rangle - \frac{1}{2}\langle u,Qu\rangle}.$$

So, for this specific problem, we have

$$M_X(u) = \exp\left\{(u_1 + 2u_2) + \frac{1}{2}\sum_{k=1}^{3}\sum_{j=1}^{3} u_k Q_{kj} u_j\right\},$$

and

$$\Phi_X(u) = \exp\left\{i(u_1 + 2u_2) - \frac{1}{2}\sum_{k=1}^{3}\sum_{j=1}^{3} u_k Q_{kj} u_j\right\},$$

where Q_{kj} is the (k, j)th element of the matrix Q.

2.4 Yes. Yes.

2.5 Let $Y \triangleq X^2$. Note that $F(x) = \mathbb{P}(X \le x)$ for each $x \in \mathbb{R}^1$. Thus,

$$\begin{aligned} F_Y(x) &= \mathbb{P}(Y \le x) = \mathbb{P}(X^2 \le x) \\ &= \begin{cases} \mathbb{P}(-\sqrt{x} \le X \le \sqrt{x}) = F(\sqrt{x}) - F(-\sqrt{x}), & x \ge 0, \\ \mathbb{P}(\emptyset) = 0, & x < 0. \end{cases} \end{aligned}$$

Let $Z \triangleq aX + b$. Then,

$$\begin{aligned} F_Z(x) &= \mathbb{P}(Z \le x) \\ &= \mathbb{P}(aX + b \le x) \\ &= \begin{cases} \mathbb{P}(X \le \frac{x-b}{a}) = F(\frac{x-b}{a}), & a > 0, \\ \mathbb{P}(X \ge \frac{x-b}{a}) = 1 - F(\frac{x-b}{a}), & a < 0, \\ \mathbb{P}(\Omega) = 1, & a = 0, b \le x, \\ \mathbb{P}(\emptyset) = 0, & a = 0, b > x. \end{cases} \end{aligned}$$

If $X \sim \mathbb{N}(0, 1)$, then

$$F(x) = \frac{1}{\sqrt{2\pi}} \int_{-\infty}^{x} \exp\left(-\frac{t^2}{2}\right) \mathrm{d}t.$$

Thus

$$F_Y(x) = \begin{cases} \frac{1}{\sqrt{2\pi}} \int_{-\sqrt{x}}^{\sqrt{x}} \exp\left(-\frac{t^2}{2}\right) \mathrm{d}t, & x \ge 0, \\ 0, & x < 0, \end{cases}$$

and

$$F_Z(x) = \begin{cases} \frac{1}{\sqrt{2\pi}} \int_{-\infty}^{\frac{x-b}{a}} \exp\left(-\frac{t^2}{2}\right) \mathrm{d}t, & a > 0, \\ \frac{1}{\sqrt{2\pi}} \int_{\frac{x-b}{a}}^{+\infty} \exp\left(-\frac{t^2}{2}\right) \mathrm{d}t, & a < 0, \\ 1, & a = 0, b \le x, \\ 0, & a = 0, b > x. \end{cases}$$

2.6 Let $\boldsymbol{x} \triangleq (x, y)^{\mathrm{T}}$. Recall the indicator function of a subset $A \subset \mathbb{R}^2$:

$$I_A(\boldsymbol{x}) = \begin{cases} 1, & \boldsymbol{x} \in A, \\ 0, & \boldsymbol{x} \notin A. \end{cases}$$

For every fixed $z \in \mathbb{R}^1$, we define

$$h(\boldsymbol{x}) = h(x, y) = I_{\{x+y \leq z\}}(\boldsymbol{x}).$$

Because $f(x, y)$ is the joint density function of X and Y, we have

$$\begin{aligned} F_Z(z) &= \mathbb{P}(Z \leq z) = \mathbb{P}(X + Y \leq z) = \mathbb{E}(h(X, Y)) \\ &= \int_{\mathbb{R}^2} I_{\{x+y\leq z\}}(\boldsymbol{x}) f(x, y) \mathrm{d}x\mathrm{d}y = \int_{\{x+y\leq z\}} f(x, y) \mathrm{d}x\mathrm{d}y \\ &= \int_{-\infty}^{\infty} \int_{-\infty}^{z-y} f(x, y) \mathrm{d}x\mathrm{d}y. \end{aligned}$$

Hence, the probability density function p_Z for Z is

$$p_Z(z) = \frac{dF_Z}{dz}(z) = \int_{-\infty}^{\infty} f(z - y, y) \mathrm{d}y.$$

Suppose that $X \sim \mathbb{N}(0, 4)$ and $Y \sim \mathbb{N}(0, 5)$ are independent. Then, their joint probability density function is

$$f(x, y) = p_X(x) p_Y(y),$$

where

$$p_X(x) = \frac{1}{2\sqrt{2\pi}} \exp\left(-\frac{x^2}{8}\right), \quad p_Y(y) = \frac{1}{\sqrt{10\pi}} \exp\left(-\frac{y^2}{10}\right).$$

Therefore,

$$p_Z(z) = \int_{-\infty}^{\infty} p_X(z - y) p_Y(y) \mathrm{d}y.$$

2.7 Let A be an open interval in $\mathbb{R}^1$. Then $A \in \mathcal{B}(\mathbb{R}^1)$. Because X is measurable from Ω to $\mathbb{R}^1$, we can define the probability distribution measure $\mathcal{L}_X$ induced by X on $(\mathbb{R}^1, \mathcal{B}(\mathbb{R}^1))$ by

$$\mathcal{L}_X(A) \triangleq \mathbb{P}(X \in A).$$

As $p(x)$ is the probability density function of X, we have

$$\mathcal{L}_X(A) = \mathbb{P}(X \in A) = \int_A p(x) \mathrm{d}x.$$

If $X \sim \mathbb{N}(0, 1)$, then

$$p(x) = \frac{1}{\sqrt{2\pi}} \exp\left(-\frac{x^2}{2}\right).$$

Thus,

$$\mathcal{L}_X(A) = \frac{1}{\sqrt{2\pi}} \int_A \exp\left(-\frac{x^2}{2}\right) \mathrm{d}x.$$

2.8 Let $Z \triangleq X + Y$. According to Problem 2.6,

$$\begin{aligned}\mathbb{P}(X + Y > 4) &= 1 - \mathbb{P}(Z \le 4) = 1 - F_Z(4) \\ &= 1 - \int_{-\infty}^{\infty} \int_{-\infty}^{4-y} p(x, y)\mathrm{d}x\mathrm{d}y \\ &= 1 - \int_0^2 \int_1^{4-x} \frac{1}{50}(x^2 + y^2)\mathrm{d}x\mathrm{d}y \\ &= 1 - \frac{7}{15} = \frac{8}{15}.\end{aligned}$$

2.10

(i) Suppose that X and Y are independent. Then

$$\begin{aligned}\mathrm{Cov}(X, Y) &= \mathbb{E}((X - \mathbb{E}(X))(Y - \mathbb{E}(Y))) \\ &= \mathbb{E}(XY - Y\mathbb{E}(X) - X\mathbb{E}(Y) + \mathbb{E}(X)\mathbb{E}(Y)) \\ &= \mathbb{E}(XY) - \mathbb{E}(X)\mathbb{E}(Y) - \mathbb{E}(Y)\mathbb{E}(X) + \mathbb{E}(X)\mathbb{E}(Y) \\ &= \mathbb{E}(XY) - \mathbb{E}(X)\mathbb{E}(Y) \\ &= \mathbb{E}(X)\mathbb{E}(Y) - \mathbb{E}(X)\mathbb{E}(Y) = 0.\end{aligned}$$

Therefore, X and Y are uncorrelated.

(ii) The uncorrelated X and Y may not be independent.

Consider a counterexample. Let $\Omega = (0, 1)$, $\mathcal{F} = \mathcal{B}((0, 1))$ and $\mathbb{P}$ be the Lebesgue measure on $(\Omega, \mathcal{B}((0, 1)))$. Define two random variables $X(\omega) \triangleq \sin(2\pi n\omega)$ and $Y(\omega) \triangleq \sin(4\pi n\omega)$. Then $\mathrm{Cov}(X, Y) = 0$, which shows that they are uncorrelated. However, X and Y are not independent.

(iii) In this case, X and Y may not be independent. This may be seen by a counterexample.

However, if additionally $\boldsymbol{X} \triangleq (X, Y)^{\mathrm{T}}$ is a two-dimensional Gaussian random vector, then X and Y are independent. This can be shown as follows. Let $X \sim \mathrm{N}(\mu_1, \sigma_1^2)$, $Y \sim \mathrm{N}(\mu_2, \sigma_2^2)$. Because X, Y are uncorrelated, the covariance matrix of $\boldsymbol{X}$ is

$$Q \triangleq (c_{jk})_{2\times 2} = \begin{pmatrix} \mathrm{Cov}(X, X) & \mathrm{Cov}(X, Y) \\ \mathrm{Cov}(Y, X) & \mathrm{Cov}(Y, Y) \end{pmatrix} = \begin{pmatrix} \mathrm{Var}(X) & 0 \\ 0 & \mathrm{Var}(Y) \end{pmatrix} = \begin{pmatrix} \sigma_1^2 & 0 \\ 0 & \sigma_2^2 \end{pmatrix}.$$

As $\boldsymbol{X}$ is Gaussian, the characteristic function of $\boldsymbol{X}$ is

$$\begin{aligned}\mathbb{E}(e^{iu\cdot X}) &= \exp\left(-\frac{1}{2}\sum_{j,k=1}^{2} u_j c_{jk} u_k + i\sum_{j}^{2} u_j m_j\right)\\ &= \exp\left(-\frac{1}{2}\sum_{j=1}^{2} u_j^2 c_{jj} + i\sum_{j}^{2} u_j m_j\right)\\ &= \exp\left(-\frac{1}{2}u_1^2\sigma_1^2 + iu_1m_1\right)\exp\left(-\frac{1}{2}u_2^2\sigma_2^2 + iu_2m_2\right)\\ &= \mathbb{E}(e^{iu_1X})\mathbb{E}(e^{iu_2Y}),\end{aligned}$$

where $u = (u_1, u_2)^{\mathrm{T}}$, $m_1 = \mathbb{E}(X)$, and $m_2 = \mathbb{E}(Y)$. This further implies that

$$p_X(x, y) = p_X(x)p_Y(y),$$

where p_X is the joint probability density function of X, Y and p_X and p_Y are the probability density functions of X and Y, respectively. Hence, X and Y are independent.

2.11 Suppose that $X, Y \in L^2(\Omega)$. Then, by the Cauchy-Schwarz inequality,

$$\begin{aligned}|\mathbb{E}(XY)| \le \mathbb{E}|XY| &= \int_\Omega |X(\omega)Y(\omega)|\mathrm{d}\mathbb{P}(\omega)\\ &\le \left(\int_\Omega |X|^2\mathrm{d}\mathbb{P}\right)^{1/2}\left(\int_\Omega |Y|^2\mathrm{d}\mathbb{P}\right)^{1/2} = \left(\mathbb{E}(X^2)\right)^{1/2}\left(\mathbb{E}(Y^2)\right)^{1/2}\\ &= \left(\mathrm{Var}(X) - (m(X))^2\right)^{1/2}\left(\mathrm{Var}(Y) - (m(Y))^2\right)^{1/2},\end{aligned}$$

where $m(X) = \mathbb{E}(X)$ and $m(Y) = \mathbb{E}(Y)$.

If $X \sim \mathbb{N}(\mu_1, \sigma_1^2)$ and $Y \sim \mathbb{N}(\mu_2, \sigma_2^2)$, then

$$\begin{aligned}|\mathbb{E}(XY)| &\le \left(\mathrm{Var}(X) - (m(X))^2\right)^{1/2}\left(\mathrm{Var}(Y) - (m(Y))^2\right)^{1/2}\\ &= \left(\sigma_1^2 - \mu_1^2\right)^{1/2}\left(\sigma_2^2 - \mu_2^2\right)^{1/2}.\end{aligned}$$

2.12 By direct calculations and the Cauchy-Schwarz inequality, we have

$$\begin{aligned}\mathbb{E}(X_n + Y_n - X - Y)^2 &= \mathbb{E}(X_n - X)^2 + \mathbb{E}(Y_n - Y)^2 + 2\mathbb{E}[(X_n - X)(Y_n - Y)]\\ &\le \mathbb{E}(X_n - X)^2 + \mathbb{E}(Y_n - Y)^2\\ &\quad + 2\sqrt{\mathbb{E}(X_n - X)^2}\sqrt{\mathbb{E}(Y_n - Y)^2}\\ &\to 0, \quad \text{as } n \to \infty.\end{aligned}$$

2.13 Without loss of the generality, we assume that $X = 0$ and $Y = 0$. Let us show that X_nY_n converges to 0 in probability as $n \to \infty$. In fact, if $|X_nY_n| > \delta$, then $|X_n| > \sqrt{\delta}$ or $|Y_n| > \sqrt{\delta}$ for every positive δ. Thus, we have the relation between events

$$\{\omega\colon |X_nY_n| > \delta\} \subset \{\omega\colon |X_n| > \sqrt{\delta}\} \cup \{\omega\colon |Y_n| > \sqrt{\delta}\}.$$

Therefore,

$$\mathbb{P}\{\omega\colon |X_nY_n| > \delta\} \le \mathbb{P}\{\omega\colon |X_n| > \sqrt{\delta}\} + \mathbb{P}\{\omega\colon |Y_n| > \sqrt{\delta}\}.$$

Letting $n \to \infty$, we conclude that

$$\mathbb{P}\{\omega\colon |X_nY_n| > \delta\} \to 0$$

for every positive δ.

2.14 The probability distribution function is

$$F_Y(y) = \mathbb{P}(Y \le y) = \begin{cases} 0, & y < a, \\ \frac{y-a}{b-a}, & a \le x \le b, \\ 1, & y > b. \end{cases}$$

The probability density function is thus

$$f_Y(y) = \frac{d}{dy} F_Y(y) = \begin{cases} 0, & y < a, \\ \frac{1}{b-a}, & a \le x \le b, \\ 0, & y > b. \end{cases}$$

2.15

$Y = \mu + \sigma .*$ `randn`.

2.16

Use the inverse transform algorithm.
$X = -2\ln(1 - U)$, where U has the standard uniform distribution.

Problems of Chapter 3

3.1

(a) They are independent and uncorrelated. In fact,

$$\begin{aligned} &\mathbb{E}[(B(t_2) - B(t_1))(B(t_4) - B(t_3))] \\ &= \mathbb{E}[B(t_2)B(t_4) - B(t_2)B(t_3) - B(t_1)B(t_4) + B(t_1)B(t_3)] \\ &= t_2 - t_2 - t_1 + t_1 (\mathbb{E}(B(t)B(s)) = \min(t, s)) \\ &= 0. \end{aligned}$$

So they are uncorrelated, and because they have a joint Gaussian distribution, they are also independent.

(b) They are independent and uncorrelated.
Recall that $\sigma(f(B(t_2) - B(t_1)))$ and $\sigma(f(B(t_4) - B(t_3)))$ are the smallest σ-field containing all sets of the form $C_1 = \{\omega \in \Omega\colon B(t_2) - B(t_1) \in A_1\}$, $C_2 = \{\omega \in \Omega\colon B(t_4) - B(t_3) \in A_2\}$, respectively. By (a), C_1, C_2 are independent. So the two σ-fields are independent, that is, $f(B(t_2) - B(t_1))$ and $f(B(t_4) - B(t_3))$ are independent, and therefore they are uncorrelated.

3.2
(a)

$$
\begin{aligned}
\Phi_t(u) &= \mathbb{E}(e^{iuB_t}) \\
&= \int_{-\infty}^{\infty} e^{iux} \frac{1}{\sqrt{2\pi t}} e^{-\frac{x^2}{2t}} dx \\
&= \int_{-\infty}^{\infty} \frac{1}{\sqrt{2\pi t}} e^{-\frac{1}{2t}(x-itu)^2 - \frac{tu^2}{2}} dx \\
&= e^{-\frac{tu^2}{2}}.
\end{aligned}
$$

(b)

$$
\begin{aligned}
M_t(u) &= \mathbb{E}(e^{uB_t}) \\
&= \int_{-\infty}^{\infty} e^{ux} \frac{1}{\sqrt{2\pi t}} e^{-\frac{x^2}{2t}} dx \\
&= \int_{-\infty}^{\infty} \frac{1}{\sqrt{2\pi t}} e^{-\frac{1}{2t}(x-tu)^2 + \frac{tu^2}{2}} dx \\
&= e^{\frac{tu^2}{2}}.
\end{aligned}
$$

3.3
(a)

$$
\begin{aligned}
\Phi_t(u) &= \mathbb{E}(e^{i\langle u, B_t\rangle}) \\
&= \int_{\mathbb{R}^n} e^{i\langle u,x\rangle} f(x_1, \ldots, x_n) dx \\
&= \int_{\mathbb{R}^n} \exp(i \sum_{k=1}^{n} u_k x_k) \frac{1}{(2\pi t)^{n/2}} \exp\left(-\frac{1}{2t} \sum_{k=1}^{n} x_k^2\right) dx \\
&= \exp\left(-\frac{t\sum_{k=1}^{n} u_k^2}{2}\right) \\
&= \exp\left(-\frac{t\|u\|^2}{2}\right).
\end{aligned}
$$

(b) $M_t(u) = \mathbb{E}(e^{\langle u, B_t\rangle}) = \exp(\frac{t\|u\|^2}{2})$.

3.4 Note that $B_t \sim \mathcal{N}(0, t)$. Thus,

$$
\mathbb{E}e^{2B_t} = \int_{\mathbb{R}} e^{2x} \frac{1}{\sqrt{2\pi t}} e^{-\frac{x^2}{2t}} dx = e^{2t},
$$

and

$$
\begin{aligned}
\mathbb{E}(e^{2B_t^2}) &= \int_{-\infty}^{\infty} e^{2x^2} \frac{e^{-\frac{x^2}{2t}}}{\sqrt{2\pi t}} dx \\
&= \frac{1}{\sqrt{2\pi t}} \int_{-\infty}^{\infty} e^{\frac{4t-1}{2t}x^2} dx.
\end{aligned}
$$

When $t \geq \frac{1}{4}$:

$$\mathbb{E}(e^{2B_t^2}) = \infty.$$

When $t < \frac{1}{4}$:

$$\begin{aligned}\mathbb{E}(e^{2B_t^2}) &= \sqrt{\frac{2t}{1-4t}} \frac{1}{\sqrt{2\pi t}} \int_{-\infty}^{\infty} e^{\frac{4t-1}{2t}x^2} d\left(\sqrt{\frac{1-4t}{2t}}x\right) \\ &= \frac{1}{\sqrt{(1-4t)\pi}} \int_{-\infty}^{\infty} e^{-x^2} dx \\ &= \frac{1}{\sqrt{1-4t}},\end{aligned}$$

where we have used the fact that $\int_{-\infty}^{\infty} e^{-x^2} dx = \sqrt{\pi}$.

3.5 $\sqrt{\frac{2t}{\pi}}$.

3.6 See Problem 2.2.

3.7
(a) Yes. They are independent and uncorrelated.
(b) $B_t^1 + B_t^2 \sim \mathcal{N}(0, 2t)$, and it is Gaussian.
The joint probability density function of B_t^1, B_t^2 is

$$f(x, y) = \frac{1}{2\pi t} e^{-\frac{x^2+y^2}{2t}},$$

and the probability density function of $Z = B_t^1 B_t^2$ is

$$f(z) = \int_{-\infty}^{\infty} f(x, z/x) \frac{1}{|x|} dx.$$

Therefore, $B_t^1 B_t^2$ is not Gaussian.

3.8
(a) $\mathbb{E}[(B_t - B_s)B_s] = \mathbb{E}[B_t B_s - B_s^2] = s - s = 0$. Because $\mathbb{E}[(B_t - B_s)B_s] = 0 = \mathbb{E}(B_t - B_s)\mathbb{E}B_s$, and both $B_t - B_s$ and B_s are Gaussian, they are thus independent and also uncorrelated.
(b) Weakly stationary, by definition.

3.9 By the Fubini theorem, $\mathbb{E}\int_0^s B_u du = \int_0^s \mathbb{E}B_u du = 0$, and similarly, $\mathbb{E}\int_0^t B_v dv = 0$. Thus, $\mathrm{Cov}(\int_0^s B_u du, \int_0^t B_v dv) = \mathbb{E}[\int_0^s B_u du \int_0^t B_v dv] = \mathbb{E}\int_0^s\int_0^t B_u B_v du dv = \int_0^s\int_0^t \times \mathbb{E}(B_u B_v) du dv$. Using the fact that $\mathbb{E}(B_u B_v) = u \wedge v$, we hence complete the proof.

3.10
(a) By the Chebyshev inequality, for every $\delta > 0$,

$$\mathbb{P}\left(\left|\frac{B_t}{t}\right| > \delta\right) \leq \frac{\mathbb{E}(B_t^2/t^2)}{\delta^2} = \frac{1}{t\delta^2} \to 0,$$

as $t \to \infty$.

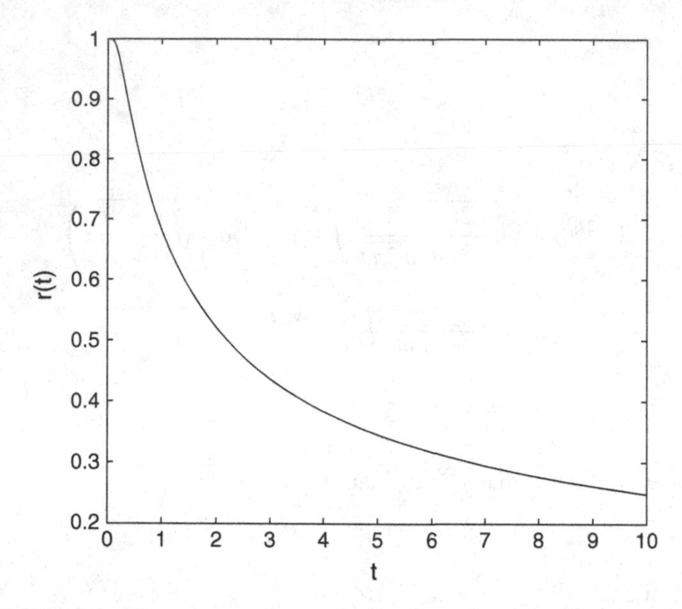

Figure H.2 Problem 3.11: plot of $r(t)$, the likelihood that one-dimensional Brownian motion B_t will stay within $(-1, 1)$ for every time $t \in [0, 10]$.

(b) We need to show that the distribution function $F_t(x)$ for $\frac{B_t}{t}$ converges to the distribution $F(x)$ of 0 at all continuous point x, as $t \to \infty$. Recall that F is actually the well-known Heaviside function:

$$F(x) \triangleq \mathbb{P}(0 \leq x) = \begin{cases} 1, & x \geq 0, \\ 0, & x < 0. \end{cases}$$

Note that

$$\begin{aligned} F_t(x) &\triangleq \mathbb{P}\left(\frac{B_t}{t} \leq x\right) \\ &= \mathbb{P}(B_t \leq tx) \\ &= \int_{-\infty}^{tx} \frac{1}{\sqrt{2\pi t}} e^{-\frac{y^2}{2t}} dy \\ &\quad \text{(change variables: } z = y/\sqrt{t}\text{)} \\ &= \int_{-\infty}^{\sqrt{t}x} \frac{1}{\sqrt{2\pi}} e^{-\frac{z^2}{2}} dz, \end{aligned}$$

which converges to $F(x)$ as $t \to \infty$, except for $x = 0$, which is a discontinuous point for both F_t and F.

(c) $r > \frac{1}{2}$.

3.11 This is a computational problem. Let $p_t(x)$ be the probability density function for B_t. For $k = 1, 2, 3$,

$$r_k(t) \triangleq \mathbb{P}(B_t \in D_k) = \int_{D_k} p_t(x)dx.$$

Calculate $r_i(t)$ via Matlab.

For example, in the one-dimensional case, $r(t)$ is plotted in Figure H.2. Note that $r(0) = 1$ because $B_0 = 0 \in D_1 = (-1, 1)$, a.s.

3.13 Yes, as B^2_{-t} is also a Brownian motion.

3.14 For each sample path, compute the part for $t > 0$ and the part for $t < 0$ separately and patch them together.

3.15 Numerically sum up this series, using as many terms as needed, so that the convergence in mean square is reached with a prescribed toterance.

Problems of Chapter 4

4.1 See Example 4.3.

4.2 Consider a partition of $[0, t]$: $t_0 = 0 < t_1 < \cdots < t_i < t_{i+1} < \cdots < t_n = t$.
As explained in Section 4.1 of [213], this symbolic notation indicates the following fact:

$$\sum_i a(t_i)(\Delta B_i)^2 \to \int_0^t a(s)ds$$

in $L^2(\Omega)$, as $\delta \triangleq \max_i\{t_{i+1} - t_i\} \to 0$. Here $\Delta B_i \triangleq B(t_{i+1}) - B(t_i)$.

4.3 Apply Itô isometry to obtain $\mathbb{E}\int_0^T f^2(t, \omega)dt = 0$. Discuss what we can imply about f, under suitable conditions on f.

4.4 $\frac{1}{2}B_T^2 + \frac{1}{2}T$.
In fact,

$$\begin{aligned}
&\sum_{i=0}^{n-1} B(t_{i+1}^n)\big(B(t_{i+1}^n) - B(t_i^n)\big) \\
&= \sum_{i=0}^{n-1} \big(B(t_{i+1}^n) - B(t_i^n) + B(t_i^n)\big)\big(B(t_{i+1}^n) - B(t_i^n)\big) \\
&= \sum_{i=0}^{n-1} \big(B(t_{i+1}^n) - B(t_i^n)\big)^2 + \sum_{i=0}^{n-1} B(t_i^n)\big(B(t_{i+1}^n) - B(t_i^n)\big).
\end{aligned}$$

The first term on the right-hand side converges to T in mean square (see [213, p. 20]). The second term corresponds to the Itô integral and thus it converges to $\frac{1}{2}B_T^2 - \frac{1}{2}T$ (see 4.13). Therefore, the final answer is $T + (\frac{1}{2}B_T^2 - \frac{1}{2}T) = \frac{1}{2}B_T^2 + \frac{1}{2}T$.

4.5 The mean square limit does not exist and thus the integral does not exist.

4.6 This solution is provided by Hansen Ha.
In this solution, all integrals are on $(-\infty, \infty)$. First, for $X_t = x + B_t$,

$$\begin{aligned}
\mathbb{E}f(X_t) &= \frac{1}{\sqrt{2\pi t}}\int f(y)\, e^{-\frac{(y-x)^2}{2t}}dy \\
&= \frac{1}{\sqrt{2\pi}}\int f(x + z\sqrt{t})\, e^{-\frac{z^2}{2}}dz,
\end{aligned}$$

where we have changed variables via $z = \frac{y-x}{\sqrt{t}}$.

Then, by Taylor expansion,

$$f(x+z\sqrt{t}) = f(x) + z\sqrt{t}f'(x) + \frac{1}{2}z^2 t f''(x+\theta z\sqrt{t})$$

for some $\theta \in (0,1)$. Thus,

$$\begin{aligned}\frac{\mathbb{E}f(X_t)-f(x)}{t} &= \frac{1}{\sqrt{2\pi}}\int \frac{z\sqrt{t}f'(x)+\frac{1}{2}z^2 t f''(x+\theta z\sqrt{t})}{t}e^{-\frac{z^2}{2}}dz \\ &= \frac{1}{\sqrt{2\pi}}\frac{f'(x)}{t}\int z\,e^{-\frac{z^2}{2}}dz + \frac{1}{2}\frac{1}{\sqrt{2\pi}}\int z^2 f''(x+\theta z\sqrt{t})\,e^{-\frac{z^2}{2}}dz \\ &= 0 + \frac{1}{2}\frac{1}{\sqrt{2\pi}}\int z^2 f''(x+\theta z\sqrt{t})\,e^{-\frac{z^2}{2}}dz.\end{aligned}$$

Finally, we get

$$\begin{aligned}Af(x) &= \frac{d}{dt}\Big|_{t=0}\mathbb{E}f(X_t) \\ &= \lim_{t\downarrow 0}\frac{\mathbb{E}f(X_t)-f(x)}{t} \\ &= \frac{1}{2}f''(x)\end{aligned}$$

when f'' exists and is bounded.

4.7 By the Itô formula, $dB_t^2 = dt + 2B_t dB_t$. Integrate both sides from 0 to T.

4.8 The mean $\mathbb{E}(X_t) = e^{-t}\mathbb{E}X_0 = 0$, the variance $\mathrm{Var}(X_t) = 2$, and the covariance $\mathrm{Cov}(X_t, X_s) = e^{-|t-s|}$.

4.9 There are a few methods to solve this problem. For example, define $X_t = e^{\int_0^t f(s)dB_s}$ with $X_0 = 1$. Then, by the Itô formula, we obtain an SDE satisfied by X_t. Solve this SDE, and we get $\mathbb{E}X_t$.

4.12 The Itô SDE is $dX_t = (\sin(X_t) - X_t^4 + \frac{1}{2}\cdot 5X_t^2 \cdot 10X_t)dt + 5X_t^2 dB_t$.

4.16 Note the facts $A^2 = -kI$, $A^3 = -kA$, $A^4 = k^2 I$, and so on, with I the 2×2 identity matrix. Also note the Taylor series expansions for $\sin(\sqrt{k}t)$ and $\cos(\sqrt{k}t)$. Then calculate $e^{At} = \sum_{n=0}^{\infty}\frac{A^n}{n!}t^n$

4.17 For this linear system with a deterministic initial condition, the solution is Gaussian. Thus "X and Y independent" and "X and Y uncorrelated" are equivalent. The exact solution is

$$X(t) = x_0\cos t + y_0\sin t + \int_0^t \sin(t-s)dB_s,$$

$$Y(t) = -x_0\sin t + y_0\cos t + \int_0^t \cos(t-s)dB_s.$$

The covariance matrix is

$$Q = \begin{pmatrix} \frac{1}{2}t - \frac{1}{4}\sin(2t) & \frac{1}{4}\sin^2 t \\ \frac{1}{4}\sin^2 t & \frac{1}{2}t + \frac{1}{4}\sin(2t) \end{pmatrix}.$$

The inverse of Q is

$$Q^{-1} = \frac{1}{\det(Q)} \begin{pmatrix} \frac{1}{2}t + \frac{1}{4}\sin(2t) & -\frac{1}{4}\sin^2 t \\ -\frac{1}{4}\sin^2 t & \frac{1}{2}t - \frac{1}{4}\sin(2t) \end{pmatrix},$$

where $\det(Q) = \frac{1}{4}(t^2 - \cos^2 t)$. The joint probability density function for (X, Y) is

$$p(t, x, y) = \frac{1}{2\pi\sqrt{\det(Q)}} e^{-\frac{1}{2}\bar{X}^T Q^{-1} \bar{X}},$$

where

$$\bar{X} = \begin{pmatrix} x - x_0 \cos t - y_0 \sin t \\ y + x_0 \sin t - y_0 \cos t \end{pmatrix}.$$

When $t \gg 1$,

$$Q^{-1} \simeq \begin{pmatrix} \frac{2}{t} + O(\frac{1}{t^2}) & O(\frac{1}{t^2}) \\ O(\frac{1}{t^2}) & \frac{2}{t} + O(\frac{1}{t^2}) \end{pmatrix},$$

that is,

$$Q^{-1} \simeq \begin{pmatrix} \frac{2}{t} & 0 \\ 0 & \frac{2}{t} \end{pmatrix}.$$

When Q^{-1} is diagonal (i.e., $\mathrm{Cov}(X, Y) = 0$), p can then be expressed as the product of probability density functions for X and Y. Thus, asymptotically, the position X and velocity Y are independent.

4.19 For this linear stochastic system with a deterministic initial condition, the solution is Gaussian. Thus "X and Y independent" and "X and Y uncorrelated" are equivalent. The exact solution is

$$X(t) = x_0 \cos t + y_0 \sin t + \int_0^t a \cos(t - s) dB_s^1 + \int_0^t b \sin(t - s) dB_s^2,$$

$$Y(t) = -x_0 \sin t + y_0 \cos t - \int_0^t a \sin(t - s) dB_s^1 + \int_0^t b \cos(t - s) dB_s^2,$$

and the covariance is

$$\mathrm{Cov}(X, Y) = \frac{1}{2}(\sin t)^2(b^2 - a^2).$$

When $a = b$, the position $X(t)$ and the velocity $Y(t)$ are uncorrelated (and thus independent). To make the position and velocity independent, we need to force the two equations with the noise of the same intensity.

Problems of Chapter 5

5.1 For this linear SDE, we take the mean $\mathbb{E}$ on both sides to get $\frac{d}{dt}\mathbb{E}X_t = -b\mathbb{E}X_t$. Thus $\mathbb{E}X_t = e^{-bt}\mathbb{E}X_0 = 0$. The noise does not affect the mean dynamics of a linear system. Then, applying Itô's formula to X_t^2 and taking $\mathbb{E}$, $\mathbb{E}X_t^2 = \text{Var}(X_t) = \sigma^2 e^{-2bt} + \frac{a^2}{2b}[1 - e^{-2bt}]$. Finally, applying Itô formula to $\sin(X_t)$ and take $\mathbb{E}$. In this case, we could only estimate $\mathbb{E}\sin(X_t)$.

5.3 ***Hint:*** Try to get $dX_t^2 = \ldots$ and then $d\mathbb{E}X_t^2 = \ldots$. Because this is a nonlinear SDE, we cannot get a clean-cut (or "closed") ordinary differential equation for $\mathbb{E}X_t^2$. However, we can get a differential inequality for $\mathbb{E}X_t^2$: $\frac{d}{dt}\mathbb{E}X_t^2 \leq \ldots$. By integrating both sides with respect to t over $[0, T]$, we get an integral inequality. Finally, apply the Gronwall inequality. Answer: $\mathbb{E}X_t^2 \leq e^{2t}\mathbb{E}X_0^2 + \frac{1}{2}\varepsilon^2(e^{2t} - 1)$.

5.4 The generator $A = \frac{1}{2}\varepsilon^2\frac{d^2}{dx^2} + (x - x^3)\frac{d}{dx}$ and its adjoint operator is $A^*g = \frac{1}{2}\varepsilon^2\frac{d^2g}{dx^2} - \frac{d}{dx}((x - x^3)g)$. The Fokker-Planck equation is $p_t = A^*p$ and Kolmogorov's backward equation is $u_t = Au$. The mean exit time u satisfies $Au(x) = -1, \quad u(-1) = u(1) = 0$.

5.6

(a) The generator $A = \frac{1}{2}a^2\frac{d^2}{dx^2} - bx\frac{d}{dx}$ and the adjoint operator is $A^*g = \frac{1}{2}a^2\frac{d^2g}{dx^2} + \frac{d}{dx}(bxg)$. The Fokker-Planck equation is $p_t = A^*p$, that is, $p_t = \frac{1}{2}a^2\frac{d^2p}{dx^2} + b\frac{d}{dx}(xp)$. Note that $p(x, t) = p(x, t; x_0, t_0)$ is the probability density function for the solution X_t with the initial condition $X(t_0) = x_0$.

(b) $p_t = \frac{1}{2}p_{xx}$, $p(x, 0) = \delta(x - x^*)$. Check a textbook on partial differential equations. This equation can be solved by Fourier transforms: $p(x, t) = \frac{1}{\sqrt{2\pi t}}e^{-\frac{(x-x^*)^2}{2t}}$. Recall the definition of Fourier transform: $\hat{g}(k) = \frac{1}{\sqrt{2\pi}}\int_{\mathbb{R}^1} e^{-ikx}g(x)dx$. Taking the Fourier transform on both sides of the equation and the initial condition, we obtain

$$\frac{d}{dt}\hat{p}(k, t) = \frac{1}{2}(-ik)^2\hat{p}(k, t), \quad \hat{p}(k, 0) = \frac{1}{\sqrt{2\pi}}e^{-ikx^*}.$$

Thus,

$$\hat{p}(k, t) = \frac{1}{\sqrt{2\pi}}e^{-ikx^*}e^{-\frac{1}{2}k^2t}.$$

By the inverse Fourier transform,

$$\begin{aligned} p(x, t) &= \frac{1}{\sqrt{2\pi}}\int_{\mathbb{R}^1}\frac{1}{\sqrt{2\pi}}e^{ikx}e^{-ikx^*}e^{-\frac{1}{2}k^2t}dk \\ &= \frac{1}{2\pi}\int_{\mathbb{R}^1} e^{ik(x-x^*)}e^{-\frac{1}{2}k^2t}dk \\ &= \frac{1}{2\pi}\int_{\mathbb{R}^1} e^{-\frac{1}{2}t[k^2 - \frac{2i(x-x^*)}{t}k + (i(x-x^*)/t)^2 - (i(x-x^*)/t)^2]}dk \\ &= \frac{1}{2\pi}\int_{\mathbb{R}^1} e^{-\frac{1}{2}t[k - i(x-x^*)/t]^2}e^{\frac{1}{2}t\cdot\frac{-(x-x^*)^2}{t^2}}dk \end{aligned}$$

$$= \frac{1}{2\pi}\int_{\mathbb{R}^1} \frac{1}{t} e^{-\frac{1}{2t}[kt-i(x-x^*)]^2} e^{-\frac{(x-x^*)^2}{2t}} d(kt)$$

$$= \frac{1}{\sqrt{2\pi t}} e^{-\frac{(x-x^*)^2}{2t}} \int_{\mathbb{R}^1} \frac{1}{\sqrt{2\pi t}} e^{-\frac{1}{2t}[kt-i(x-x^*)]^2} d(kt)$$

$$= \frac{1}{\sqrt{2\pi t}} e^{-\frac{(x-x^*)^2}{2t}}.$$

This is just the probability density function for $X_t = B_t + x^* \sim \mathcal{N}(x^*, t)$, which is the solution of $dX_t = dB_t,\ X_0 = x^*$.

5.7 See [178, Theorem 5.18] or [94, Theorem 2, Chapter5].

5.8 The generator $A = \frac{1}{2}a^2\frac{d^2}{dx^2} - bx\frac{d}{dx}$. To get mean exit time, numerically solve $Au(x) = -1, u(-1) = u(1) = 0$. For escape probability to exit through the left end point, solve $Ap(x) = 0,\ p(-1) = 1,\ p(1) = 0$. For escape probability to exit through the right end point, solve $Ap(x) = 0,\ p(-1) = 0,\ p(1) = 1$.

5.10 Solve the Fokker-Planck equation on a large domain, say, $(-100, 100) \times (-100, 100)$, with zero boundary condition for p and initial condition $p(x, y, 0)$ being the probability density function for the uniform distribution on $\bar{D}$.

Problems of Chapter 6

6.1 Eigenvalues $\lambda_s = -1; \lambda_c = 0, \lambda_u = 3$, with the corresponding eigenvectors $V_s = (1, 0, 0)^T,\ V_c = (0, 1, 0)^T$, and $V_u = (0, 2, 3)^T$. The solution is $x(t) = x_0e^{-t},\ z(t) = z_0e^{3t},\ y(t) = y_0 + \frac{2}{3}z_0(e^{3t} - 1)$. The flow is $\varphi(t, x_0, y_0, z_0) = (x_0e^{-t}, y_0 + \frac{2}{3}z_0(e^{3t} - 1), z_0e^{3t})^T$. For a linear system, stable, center, and unstable manifolds are identical to the corresponding stable, center, and unstable eigenspaces, respectively. Answer: $W^s(0, 0, 0)$ is the x-axis, $W^c(0, 0, 0)$ is the y-axis, and $W^u(0, 0, 0) = \{k(0, 2, 3)^T : k \in \mathbb{R}\}$.

6.2 The solution mapping or flow $\varphi(t, x_0, y_0, z_0) = (x_0 \cos t + y_0 \sin t, -x_0 \sin t + y_0 \cos t, z_0e^{-t})^T$. For a linear system, stable, center, and unstable manifolds are identical to the corresponding stable, center, and unstable eigenspaces, respectively. Invariant manifolds are as follows: $W^s(0, 0, 0)$ is the z-axis, $W^c(0, 0, 0)$ is the xy-plane, and $W^u(0, 0)$ is empty.

6.3 Find the exact solution and then check the definition of exponential dichotomy.

6.4 Find the exact solution and then check the definition of pseudo-exponential dichotomy.

6.5 The solution mapping, or flow, is $\varphi(t, x_0, y_0) = (x_0e^{-t} + \frac{1}{3}y_0^2(e^{2t} - e^{-t}), y_0e^t)^T$. Recall the fact that for $(x_0, y_0)^T \in W^s(0, 0),\ \varphi(t, x_0, y_0) \to (0, 0)^T$ as $t \to +\infty$, whereas for $(x_0, y_0)^T \in W^u(0, 0), \varphi(t, x_0, y_0) \to (0, 0)^T$ as $t \to -\infty$. Thus, $W^s(0, 0) = \{(x_0, y_0)^T : y_0 = 0\}$, $W^c(0, 0)$ is empty, and $W^u(0, 0) = \{(x_0, y_0)^T : x_0 - \frac{1}{3}y_0^2 = 0\}$.
See Figure H.3.

6.6
(a) Equilibrium point $(0, 0)^T$. The linearized system $\dot{x} = -x,\ \dot{y} = 0$. The stable eigenspace E^s is the x-axis, and the center eigenspace E^c is the y-axis. Nonhyperbolic.
(b) The center manifold $W^c(0, 0) = \{(x, y)^T : x = h(y) \approx -y^2 - 4y^4\}$. The dynamical system restricted on the center manifold is (approximately) $\dot{y} = -2y^3 - 8y^5$. This

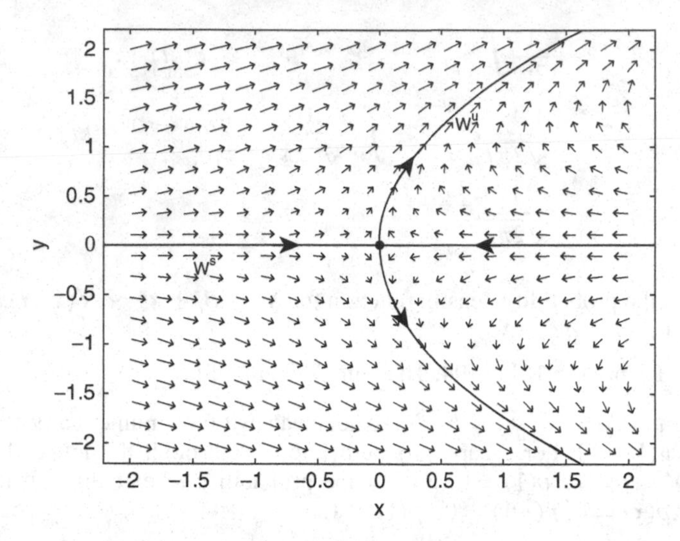

Figure H.3 Problem 6.5: phase portrait for $\dot{x} = -x + y^2$, $\dot{y} = y$.

system has only one equilibrium point 0, which is asymptotically stable. Thus the equilibrium point $(0, 0)^{\mathrm{T}}$ is locally asymptotically stable for the original dynamical system.

6.7 Differentiating both sides.

6.8 Let us define

$$F_t = e^{-B_t + \frac{t}{2}}$$

as an integrating factor and set $Y_t = F_t X_t$. Then

$$\dot{Y}_t = e^{B_t - \frac{t}{2}} Y_t^2, \; Y_0 = x.$$

This is a random differential equation (i.e., a differential equation with random coefficients), and its solution is

$$Y_t = \frac{x}{1 - x \int_0^t e^{B_s - \frac{s}{2}} ds}.$$

Thus

$$X_t = \frac{x e^{B_t - \frac{t}{2}}}{1 - x \int_0^t e^{B_s - \frac{s}{2}} ds}.$$

Let $\varphi(t, \omega, x) = \frac{x e^{B_t - \frac{t}{2}}}{1 - x \int_0^t e^{B_s - \frac{s}{2}} ds}$. Then, $\varphi(0, \omega, x) = x$ and $\varphi(t + s, \omega, x) \neq \varphi(t, \omega, \varphi(s, \omega, x))$. But we will see in this chapter that $\varphi(t + s, \omega, x) = \varphi(t, \theta_s \omega, \varphi(s, \omega, x))$, where θ_t is the Wiener shift.

6.9

(a) We have

$$dX_t = X_t^\gamma dt + X_t dB_t$$

with $X_0 = x$.

Let us define

$$F_t = e^{-B_t + \frac{t}{2}}$$

as an integrating factor and set $Y_t = F_t X_t$. Then

$$\begin{aligned} dY_t &= d(F_t X_t) = F_t dX_t + X_t dF_t + dF_t dX_t \\ &= F_t \left(X_t^\gamma dt + X_t dB_t\right) + X_t F_t (-B_t + dt) + (-X_t F_t dt) \\ &= F_t X_t^\gamma dt, \end{aligned}$$

which is

$$d(F_t X_t) = F_t X_t^\gamma dt.$$

So,

$$\frac{d(F_t X_t)}{(F_t X_t)^\gamma} = F_t^{\,1-\gamma} dt.$$

Integrating both sides, we obtain

$$\frac{(F_t X_t)^{1-\gamma} - x^{1-\gamma}}{1-\gamma} = \int_0^t F_u^{\,1-\gamma} du,$$

which can be rewritten as

$$(F_t X_t)^{1-\gamma} = x^{1-\gamma} + (1-\gamma) \int_0^t F_u^{\,1-\gamma} du.$$

Thus,

$$X_t = \left[F_t^{\gamma-1} \left(x^{1-\gamma} + (1-\gamma) \int_0^t F_u^{\,1-\gamma} du \right) \right]^{\frac{1}{1-\gamma}},$$

which is

$$X_t = e^{B_t - \frac{t}{2}} \left[x^{1-\gamma} + (1-\gamma) \int_0^t e^{(1-\gamma)(-B_u + \frac{u}{2})} du \right]^{\frac{1}{1-\gamma}}$$

for $0 < \gamma < 1$.

(b) The solution mapping is

$$\varphi(t, \omega, x) = e^{B_t - \frac{t}{2}} \left[x^{1-\gamma} + (1-\gamma) \int_0^t e^{(1-\gamma)(-B_u + \frac{u}{2})} du \right]^{\frac{1}{1-\gamma}}.$$

Let us check the cocycle property. First,

$$\varphi(0, \omega, x) = x.$$

Then,

$$\varphi(t+s, \omega, x) = e^{B_{t+s} - \frac{t+s}{2}} \left[x^{1-\gamma} + (1-\gamma) \int_0^{t+s} e^{(1-\gamma)(-B_u + \frac{u}{2})} du \right]^{\frac{1}{1-\gamma}},$$

and moreover,

$$\begin{aligned}
&\varphi(t, \theta_s\omega, \varphi(s, \omega, x)) \\
&= e^{B_t(\theta_s\omega)-\frac{t}{2}} \left[\left(e^{B_s-\frac{s}{2}} \left[x^{1-\gamma} + (1-\gamma) \int_0^s e^{(1-\gamma)(-B_u+\frac{u}{2})} du \right]^{\frac{1}{1-\gamma}} \right)^{1-\gamma} \right. \\
&\quad \left. + (1-\gamma) \int_0^t e^{(1-\gamma)(-B_u(\theta_s\omega)+\frac{u}{2})} du \right]^{\frac{1}{1-\gamma}},
\end{aligned}$$

where θ_t is the Wiener shift implicitly defined by

$$B_t(\theta_s\omega) = B_{t+s}(\omega) - B_s(\omega).$$

So, we have

$$\begin{aligned}
&\varphi(t, \theta_s\omega, \varphi(s, \omega, x)) \\
&= e^{B_{t+s}-B_s-\frac{t}{2}} \left[\left(e^{B_s-\frac{s}{2}} \left[x^{1-\gamma} + (1-\gamma) \int_0^s e^{(1-\gamma)(-B_u+\frac{u}{2})} du \right]^{\frac{1}{1-\gamma}} \right)^{1-\gamma} \right. \\
&\quad \left. + (1-\gamma) \int_0^t e^{(1-\gamma)(-(B_{u+s}-B_s)+\frac{u}{2})} du \right]^{\frac{1}{1-\gamma}} \\
&= e^{B_{t+s}-B_s-\frac{t}{2}} \left[e^{(1-\gamma)(B_s-\frac{s}{2})} \left[x^{1-\gamma} + (1-\gamma) \int_0^s e^{(1-\gamma)(-B_u+\frac{u}{2})} du \right] \right. \\
&\quad \left. + (1-\gamma) \int_0^t e^{(1-\gamma)(-(B_{u+s}-B_s)+\frac{u}{2})} du \right]^{\frac{1}{1-\gamma}} \\
&= \left[e^{(1-\gamma)(B_{t+s}-\frac{t+s}{2})} \left[x^{1-\gamma} + (1-\gamma) \int_0^s e^{(1-\gamma)(-B_u+\frac{u}{2})} du \right] \right. \\
&\quad \left. + (1-\gamma) e^{(1-\gamma)(B_{t+s}-\frac{t+s}{2})} \int_0^t e^{(1-\gamma)(-B_{u+s}+\frac{u+s}{2})} du \right]^{\frac{1}{1-\gamma}} \\
&= \left[e^{(1-\gamma)(B_{t+s}-\frac{t+s}{2})} x^{1-\gamma} + (1-\gamma) e^{(1-\gamma)(B_{t+s}-\frac{t+s}{2})} \right. \\
&\quad \left. \cdot \left(\int_0^s e^{(1-\gamma)(-B_u+\frac{u}{2})} du + \int_s^{t+s} e^{(1-\gamma)(-B_u+\frac{u}{2})} du \right) \right]^{\frac{1}{1-\gamma}} \\
&= \left[e^{(1-\gamma)(B_{t+s}-\frac{t+s}{2})} x^{1-\gamma} + (1-\gamma) e^{(1-\gamma)(B_{t+s}-\frac{t+s}{2})} \int_0^{t+s} e^{(1-\gamma)(-B_u+\frac{u}{2})} du \right]^{\frac{1}{1-\gamma}}
\end{aligned}$$

and

$$\begin{aligned}
\varphi(t, \theta_s\omega, \varphi(s, \omega, x)) &= e^{B_{t+s}-\frac{t+s}{2}} \left[x^{1-\gamma} + (1-\gamma) \int_0^{t+s} e^{(1-\gamma)(-B_u+\frac{u}{2})} du \right]^{\frac{1}{1-\gamma}} \\
&= \varphi(t+s, \omega, x).
\end{aligned}$$

(c) Now consider the stationary orbit. Let us calculate

$$Y(\omega) = \left[(1-\gamma)\int_{-\infty}^{0} e^{(1-\gamma)(-B_u+\frac{u}{2})}du\right]^{\frac{1}{1-\gamma}}.$$

We have

$$\begin{aligned}
\varphi(t,\omega,Y(\omega)) &= e^{B_t-\frac{t}{2}}\left[Y^{1-\gamma}(\omega) + (1-\gamma)\int_0^t e^{(1-\gamma)(-B_u+\frac{u}{2})}du\right]^{\frac{1}{1-\gamma}} \\
&= e^{B_t-\frac{t}{2}}\left[(1-\gamma)\int_{-\infty}^{0} e^{(1-\gamma)(-B_u+\frac{u}{2})}du + (1-\gamma)\int_0^t e^{(1-\gamma)(-B_u+\frac{u}{2})}du\right]^{\frac{1}{1-\gamma}} \\
&= e^{B_t-\frac{t}{2}}\left[(1-\gamma)\int_{-\infty}^{t} e^{(1-\gamma)(-B_u+\frac{u}{2})}du\right]^{\frac{1}{1-\gamma}} \\
&= e^{B_t-\frac{t}{2}}\left[(1-\gamma)\int_{-\infty}^{t} e^{(1-\gamma)(-B_u+\frac{u}{2})}du\right]^{\frac{1}{1-\gamma}},
\end{aligned}$$

whereas

$$\begin{aligned}
Y(\theta_t\omega) &= \left[(1-\gamma)\int_{-\infty}^{0} e^{(1-\gamma)(-B_u(\theta_t\omega)+\frac{u}{2})}du\right]^{\frac{1}{1-\gamma}} \\
&= \left[(1-\gamma)\int_{-\infty}^{0} e^{(1-\gamma)(-(B_{u+t}-B_t)+\frac{u}{2})}du\right]^{\frac{1}{1-\gamma}} \\
&= \left[(1-\gamma)e^{(1-\gamma)(B_t-\frac{t}{2})}\int_{-\infty}^{0} e^{(1-\gamma)(-B_{u+t}+\frac{u+t}{2})}du\right]^{\frac{1}{1-\gamma}} \\
&= e^{B_t-\frac{t}{2}}\left[(1-\gamma)\int_{-\infty}^{t} e^{(1-\gamma)(-B_u+\frac{u}{2})}du\right]^{\frac{1}{1-\gamma}} \\
&= \varphi(t,\omega,Y(\omega)).
\end{aligned}$$

6.10 The solution mapping is $\varphi(t,\omega,x,y) = (xe^{-t} + 2\int_0^t e^{-(t-s)}dB_s^1,\ ye^{-3t} + 4\int_0^t e^{-3(t-s)}dB_s^2)^{\mathrm{T}}$. Yes, it defines a random dynamical system (cocycle).

6.11 Define $Y(t) = \frac{1}{X(t)}$. So $f(x) = \frac{1}{x}$, $f'(x) = -\frac{1}{x^2}$, $f''(x) = \frac{2}{x^3}$. Then, by Itô's formula,

$$\begin{aligned}
dY(t) &= -\frac{1}{X^2(t)}dX(t) + \frac{1}{X^3(t)}\beta^2X^2(t)dt \\
&= -\frac{1}{X^2(t)}(rX(t)(K-X(t))dt + \beta X(t)dB(t)) + \frac{1}{X(t)}\beta^2dt \\
&= (r-(rK-\beta^2)Y(t))dt - \beta Y(t)dB(t), \text{ with } Y(0) = \frac{1}{x}.
\end{aligned}$$

This is a linear SDE. By Example 4.22, the solution is

$$Y(t) = \exp\left\{-\left(rK - \frac{1}{2}\beta^2\right)t - \beta B(t)\right\}\left(\frac{1}{x} + \int_0^t \frac{r}{\exp\{-(rK - \frac{1}{2}\beta^2)s - \beta B(s)\}}ds\right).$$

Thus, the solution mapping is $\varphi(t, \omega, x) = \frac{1}{Y(t)} = \frac{x\exp\{\alpha t + \sigma B_t\}}{1 + x\int_0^t \exp\{\alpha s + \sigma B_s\}ds}$. Yes, it is a cocycle. A stationary orbit is $Y(\omega) = (\int_{-\infty}^0 \exp\{\alpha s + \sigma B_s\}ds)^{-1}$.

6.12 In Itô form, the SDE becomes

$$dX_t = \left(\alpha X_t - X_t^3 + \frac{1}{2}\sigma^2 X_t\right)dt + \sigma X_t dB_t,\ X_0 = x.$$

First we use the method of integrating factors to solve this equation. Let $F_t = e^{\frac{1}{2}\sigma^2 t - \sigma B_t}$ be the integrating factor. Then $d(F_t X_t) = F_t(\alpha X_t - X_t^3 + \frac{1}{2}\sigma^2 X_t)dt$. Set $Y_t = F_t X_t$ with $Y_0 = x$. We note that $\dot{Y}_t = (\alpha + \frac{1}{2}\sigma^2)Y_t - F_t^{-2}Y_t^3$.

This random differential equation can be solved by introducing a new state variable $U_t = Y_t^{-2}$. The solution is $X_t = \frac{e^{\alpha t + \sigma B_t}}{(x^{-2} + 2\int_0^t e^{2\alpha s + 2\sigma B_s}ds)^{\frac{1}{2}}}$. Let $\varphi(t, \omega, x) = \frac{e^{\alpha t + \sigma B_t}}{(x^{-2} + 2\int_0^t e^{2\alpha s + 2\sigma B_s}ds)^{\frac{1}{2}}}$. In fact, we better write $\varphi(t, \omega, x) = \frac{x^2 e^{\alpha t + \sigma B_t}}{(1 + 2x^2\int_0^t e^{2\alpha s + 2\sigma B_s}ds)^{\frac{1}{2}}}$.

Clearly $\varphi(0, \omega, x) = x$. We now verify the cocycle property $\varphi(t + r, \omega, x) = \frac{e^{\alpha(t+r) + \sigma B_{t+r}}}{(x^{-2} + 2\int_0^{t+r} e^{2\alpha s + 2\sigma B_s}ds)^{\frac{1}{2}}}$; $\varphi(r, \omega, x) = \frac{e^{\alpha r + \sigma B_r}}{(x^{-2} + 2\int_0^r e^{2\alpha s + 2\sigma B_s}ds)^{\frac{1}{2}}}$; and $\varphi(t, \theta_r\omega, \varphi(r, \omega, x)) = \frac{e^{\alpha t + \sigma(B_{t+r} - B_r)}}{(\varphi(r,\omega,x)^{-2} + 2\int_0^t e^{2\alpha s + 2\sigma(B_{s+r} - B_r)}ds)^{\frac{1}{2}}}$. Therefore $\varphi(t + r, \omega, x) = \varphi(t, \theta_r\omega, \varphi(r, \omega, x))$.

Moreover, $Y(\omega) = (2\int_{-\infty}^0 e^{2\alpha s + 2\sigma B_s}ds)^{-\frac{1}{2}}$ is a stationary orbit.

6.13 In Itô form, the SDE becomes

$$dX_t = \left(\alpha - X_t^2 + \frac{1}{2}\sigma^2 X_t\right)dt + \sigma X_t dB_t,\ X_0 = x,$$

where α and σ are real constants. Introduce the integrating factor $F_t = e^{\frac{1}{2}\sigma^2 t - \sigma B_t}$ and define $Y_t = F_t X_t$. We get a random differential equation

$$\dot{Y}_t = F_t \cdot \left(\alpha - F_t^{-2}Y_t^2 + \frac{1}{2}\sigma^2 F_t Y_t\right),\ Y_0 = x.$$

But this differential equation is difficult to solve. It should define a random dynamical system. (This is related to how noise may affect a saddle-node bifurcation.)

6.17 $r - \frac{1}{2}\alpha^2$.

6.20

$$\Phi(t, \omega) = \begin{pmatrix} e^{2t} & 8e^{2t}\int_0^t e^{-\frac{5}{2}s - B_s^2}dB_s^1 \\ 0 & e^{-\frac{1}{2}t - B_t^2} \end{pmatrix}.$$

6.22

$$\Phi(t, \omega) = \begin{pmatrix} e^{2t} & 0 \\ 0 & e^{-\frac{1}{2}t - B_t^2} \end{pmatrix}.$$

6.24
(a)

$$dX_t = 9X_t dt + 2Y_t dB_t^1, \ X_0 = x$$
$$dY_t = -5Y_t dt, \ Y_0 = y.$$
$$\Phi(t,\omega) = \begin{pmatrix} e^{9t} & 2\int_0^t e^{9t-14s} dB_s^1 \\ 0 & e^{-5t} \end{pmatrix}.$$

We take the matrix norm for Φ to be the maximal absolute value of eigenvalues. Then,

$$\mathbb{E} \sup_{0\le t\le 1} \ln^+ \|\Phi(t,\omega)\| = 9 \in L^1(\Omega)$$
$$\mathbb{E} \sup_{0\le t\le 1} \ln^+ \|\Phi(-t,\omega)\| = 5 \in L^1(\Omega).$$

Hence, the multiplicative ergodic theorem applies. The Lyapunov exponents are

$$\lambda_1 = 9, \ \lambda_2 = -5.$$

and the corresponding Oseledets spaces are

$$E_1(\omega) = \left\{ k \begin{pmatrix} 1 \\ 0 \end{pmatrix} : k \in \mathbb{R}^1 \right\}$$
$$E_2(\omega) = \left\{ k \begin{pmatrix} -2\int_0^\infty e^{-14s} dB_s^1 \\ 1 \end{pmatrix} : k \in \mathbb{R}^1 \right\}.$$

Moreover,

$$\mathbb{R}^2 = E_1(\omega) \oplus E_2(\omega).$$

(b) Yes.

6.26 The solution formula for a nonhomogeneous first-order differential equation $\dot{z} = az + f(t)$ is $z(t) = e^{at} z_0 + \int_0^t e^{a\,(t-s)} f(s) ds$.

Starting the stochastic system on $W^s(0,0)$, does the random orbit $(X_t^\varepsilon, Y_t^\varepsilon)$ run away (i.e., deviate) from it? To quantify this "deviation," calculate $\mathbb{E}(X_t^\varepsilon - X_t^0)$ and $\mathbb{E}(Y_t^\varepsilon - Y_t^0)$.

6.27 Refer to [80, 269].

Problems of Chapter 7

7.1 This solution is provided by Huijie Qiao.

Step 1: (i) $\Leftrightarrow$ (i')
Recall the fact: for $0 < \alpha < 1$,

$$\int_0^\infty (e^{ir} - 1) r^{-1-\alpha} dr = \Gamma(-\alpha) e^{-i\pi\alpha/2},$$

and its complex conjugate is

$$\int_0^\infty (e^{-ir} - 1)r^{-1-\alpha}dr = \Gamma(-\alpha)e^{i\pi\alpha/2}.$$

Note that

$$\int_{\mathbb{R}^1\backslash\{0\}} (e^{ixu} - 1)\nu(dx) = \int_0^\infty (e^{ixu} - 1)\frac{c_1}{x^{1+\alpha}}dx + \int_0^\infty (e^{-ixu} - 1)\frac{c_2}{x^{1+\alpha}}dx.$$

Considering the case for $u > 0$ and $u < 0$ separately, eventually we obtain

$$\int_{\mathbb{R}^1\backslash\{0\}} (e^{ixu} - 1)\nu(dx) = -\sigma|u|^\alpha[1 - i\beta\text{sgn}(u)\tan\frac{\pi\alpha}{2}],$$

where $\sigma = -\Gamma(-\alpha)(c_1 + c_2)\cos\frac{\pi\alpha}{2}$, $\beta = \frac{c_1 - c_2}{c_1 + c_2}$, and $\gamma = \gamma_0$.

Step 2: (ii) ⇔ (ii′)

$$\int_{\mathbb{R}^1\backslash\{0\}} (e^{ixu} - 1 - ixuI_{\{|x|<1\}}(x))\nu(dx)$$

$$= \int_0^\infty (e^{ixu} - 1 - iuxI_{(0,1)}(x))\frac{c_1}{x^2}dx + \int_{-\infty}^0 (e^{ixu} - 1 - iuxI_{(-1,0)}(x))\frac{c_2}{x^2}dx$$

$$= -\sigma|u|[1 + i\beta\frac{2}{\pi}\text{sgn}(u)\log|u|] + icu(c_1 - c_2),$$

where $c = \int_1^\infty r^{-2}\sin r dr + \int_0^1 r^{-2}(\sin r - r)dr$, $\sigma = \frac{\pi}{2}(c_1 + c_2)$, $\beta = \frac{c_1 - c_2}{c_1 + c_2}$ and $\gamma = \gamma_* + c(c_1 - c_2)$.

Step 3: (iii) ⇔ (iii')
Recall the fact: for $1 < \alpha < 2$,

$$\int_0^\infty (e^{ir} - 1 - ir)r^{-1-\alpha}dr = \Gamma(-\alpha)e^{-i\pi\alpha/2},$$

$$\int_{\mathbb{R}^1\backslash\{0\}} (e^{ixu} - 1 - ixu)\nu(dx)$$

$$= \int_0^\infty (e^{ixu} - 1 - iux)\frac{c_1}{x^{1+\alpha}}dx + \int_0^\infty (e^{-ixu} - 1 + iux)\frac{c_2}{x^{1+\alpha}}dx$$

$$= -\sigma|u|^\alpha[1 - i\beta\text{sgn}(u)\tan\frac{\pi\alpha}{2}],$$

where $\sigma = -\Gamma(-\alpha)(c_1 + c_2)\cos\frac{\pi\alpha}{2}$, $\beta = \frac{c_1 - c_2}{c_1 + c_2}$, and $\gamma = \gamma_1$.

Step 4: (vi) ⇔ (vi')
They are already the same.

7.2 For $\varepsilon \geq 1$, this is true by the definition of the Lévy jump measure.
For $\varepsilon < 1$, we have

$$\begin{aligned}
\nu((-\varepsilon,\varepsilon)^c) &= \int_{|u|\geq\varepsilon} \nu(du) = \int_{\mathbb{R}^1\setminus\{0\}} I_{\{|u|\geq\varepsilon\}}\nu(du) \\
&= \int_{|u|\leq 1} I_{\{|u|\geq\varepsilon\}}\nu(du) + \int_{|u|>1} I_{\{|u|\geq\varepsilon\}}\nu(du) \\
&\leq \int_{|u|\leq 1} \frac{|u|^2}{\varepsilon^2} I_{\{|u|\geq\varepsilon\}}\nu(du) + \int_{|u|>1} 1\ \nu(du) \\
&\leq \frac{1}{\varepsilon^2}\int_{|u|\leq 1} |u|^2\nu(du) + \int_{|u|>1} 1\ \nu(du) < \infty.
\end{aligned}$$

Note that $\nu((-\varepsilon,\varepsilon))$ may or may not be finite, depending on whether $\nu(\mathbb{R}^1\setminus\{0\})$ is finite.

7.3

(a) For any $\varepsilon_1 > 0$ and $\varepsilon_2 > 0$,

$$\begin{aligned}
\int_{-1}^{1} \frac{y}{|y|^{1+\alpha}}dy &= \lim_{\varepsilon_1\to 0+}\int_{-1}^{-\varepsilon_1} \frac{y}{(-y)^{1+\alpha}}dy + \lim_{\varepsilon_2\to 0+}\int_{\varepsilon_2}^{1} \frac{y}{y^{1+\alpha}}dy \\
&= \frac{1}{1-\alpha}\left[\lim_{\varepsilon_1\to 0+}(\varepsilon_1^{1-\alpha}-1) + \lim_{\varepsilon_2\to 0+}(1-\varepsilon_2^{1-\alpha})\right].
\end{aligned}$$

So the improper integral is 0 for $\alpha \in (0,1)$, but it does not exist (or diverges) for $\alpha \in [1,2)$.

(b) For any $\varepsilon > 0$,

$$\begin{aligned}
\text{P. V.}\int_{-1}^{1} \frac{y}{|y|^{1+\alpha}}dy &= \lim_{\varepsilon\to 0+}\left[\int_{-1}^{-\varepsilon} \frac{y}{(-y)^{1+\alpha}}dy + \int_{\varepsilon}^{1} \frac{y}{y^{1+\alpha}}dy\right] \\
&= \lim_{\varepsilon\to 0+} 0 = 0.
\end{aligned}$$

Hence, the Cauchy principal value is 0 for all $\alpha \in (0,2)$.

7.4

(a) The improper integral value is $c_\alpha\frac{2}{1-\alpha}$ for $\alpha \in (0,1)$, while it does not exist for $\alpha \in [1,2)$.

(b) The Cauchy principal value is $c_\alpha\frac{2}{1-\alpha}$ for $\alpha \in (0,1)$, while it does not exist for $\alpha \in [1,2)$.

7.5 This is an exercise about improper integrals in undergraduate calculus.

7.6 Refer to [140].

7.7

(a) The characteristic function is the inverse Fourier transform of the probability density function f_α. That is, f_α is the Fourier transform of Φ_α; then apply Lebesgue's dominated convergence theorem. Look at the proof of the dominated convergence theorem, and you will get an idea about convergence in $L^1(\mathbb{R}^1)$.

(b) A special case of (a).

(c) Examine the convergence of the probability distribution functions $F_\alpha(x) = \int_{-\infty}^{x} f_\alpha(z,t)dz$ as $\alpha \to 2$.

7.9
(a) Refer to [140].
(b) Refer to [54].

7.10
See Section 7.5.1. Numerically solve the deterministic integral equation (7.77), compute the exact result (7.78), and plot $u(x)$ for $\alpha = 0.1, 0.3, 0.5, \ldots, 1.99$.

7.11 Refer to [97].

7.12 Need to generalize [97] to two-dimensional case.

7.13 See [7, p. 402] for the generator A in the case of multiplicative noise.

Further Readings

We provide some additional references for each chapter so that more advanced readers may find further information.

Further Readings for Chapter 1

Biological, chemical, and physical systems are often subject to random influences [203, 125, 98, 134, 190]. Noise may not be negligible [28, 279] or could even be beneficial [119].

Zwanzig [312, Section 1.1] and Ebeling and Sokolov [86, Section 7.2] have discussed Brownian motion from a physical point of view.

Advanced references on stochastic differential equations include [104, 128, 93, 87].

Other books on dynamical systems include [123, 12].

Various phenomena of oscillations under random forces are also modeled by this type of stochastic differential equation [182, 311, 297, 298].

Other SDE models for complex phenomena under random influences include climate evolution [10, 134, 131, 51], stochastic parameterization of geophysical processes [191, 181, 221, 291, 82, 75, 52], and many more [217, 218, 99, 98, 114, 115, 112, 28].

Further Readings for Chapter 3

Other references about complex systems under random influences include [188, 203, 279, 280, 293, 301].

For more information about mathematical aspects of Brownian motion, see [207, 160, 202, 147, 121, 271, 200, 240, 126].

Gaussian white noise is widely discussed in physical and engineering literature; see, for example, [293, 219, 159, 279, 280, 125].

Further Readings for Chapter 6

For discussions on the cocycle property for SDEs, see [87, 22, 32, 128, 163, 170, 173, 176].

Investigations of SDEs from a dynamical systems point of view include slow and inertial manifolds [248, 249, 96], invariant manifolds [285, 38, 47, 74, 269, 270], stochastic bifurcation [263, 69, 245, 55], random attractors [264, 69, 250], slow-fast stochastic motions [92, 143, 26, 144, 262, 284], stochastic flows and ergodic theory [149, 304, 184, 106], monotone random dynamical systems [62], topological aspects of random dynamics [65], special random dynamical features [309, 174, 175], and many more [67, 81, 295].

References for elementary differential equations and dynamical systems include [123, 12, 266]. For advanced topics in dynamical systems, we refer to [110, 290, 226, 59].

Further Readings for Chapter 7

In addition to Applebaum's book *Lévy Processes and Stochastic Calculus* [7], many research works on SDEs with Lévy noise have been emerging in physical, geophysical, and biophysical literature, including research articles in *Physical Review E*, *Physical Review Letters*, *Physica A*, *Physica D*, *Nature*, and *Science*.

References

1. R. Abraham, J. E. Marsden, and T. Ratiu, *Manifolds, Tensor Analysis, and Applications*. 2nd ed. Springer, New York, 1988.
2. R. A. Adams and J. J. F. Fourier, *Sobolev Spaces*. 2nd ed. Academic Press, New York, 2003.
3. S. Albeverrio, B. Rüdiger, and J. L. Wu, Invariant measures and symmetry property of Lévy type operators. *Potential Anal.*, 13 (2000), 147–168.
4. E. Allen, *Modeling with Itô Stochastic Differential Equations*. Springer, New York, 2007.
5. L. J. S. Allen, *An Introduction to Stochastic Processes with Applications to Biology*. 2nd ed. CRC Press, Taylor & Francis Group. Boston, 2011.
6. T. M. Apostol, *Mathematical Analysis*. 2nd ed. Addison-Wesley, Boston, 1974.
7. D. Applebaum, *Lévy Processes and Stochastic Calculus*. 2nd ed. Cambridge University Press, Cambridge, 2009.
8. L. Arnold, *Stochastic Differential Equations: Theory and Applications*. Dover, Mineola, New York, 2012. Originally published: John Wiley & Sons, New York, 1974.
9. L. Arnold, *Random Dynamical Systems*. Springer, New York, 1998. Corrected 2nd printing, 2003.
10. L. Arnold, Hasselmann's program revisited: The analysis of stochasticity in deterministic climate models. In J.-S. von Storch and P. Imkeller, editors, *Stochastic Climate Models*, pages 141–158. Birkhäuser, Boston, 2001.
11. L. Arnold and M. Scheutzow, Perfect cocycles through stochastic differential equations. *Prob. Theory Relat. Fields* 101 (1995), 65–88.
12. V. I. Arnold, *Ordinary Differential Equations*. MIT Press, Cambridge, 1978.
13. R. B. Ash, *Probability and Measure Theory*. 2nd ed. Academic Press, New York, 2000.
14. L. Bachelier, Théorie de la spéculation, *Annales Scientifiques de lÉcole Normale Supérieure* 3 (1900), 21–86.
15. L. Barreira and Ya. B. Pesin, *Nonuniform Hyperbolicity: Dynamics of Systems with Nonzero Lyapunov Exponents*. Cambridge University Press, New York, 2007.
16. L. Barreira and C. Valls, Nonuniform exponential dichotomies and Lyapunov regularity. *J. Dyn. Diff. Eqns.* 19 (2007), 215–241.
17. L. Barreira and Ya. B. Pesin, *Lyapunov Exponents and Smooth Ergodic Theory*. Amer. Math. Soc., Providence, 2002.
18. R. F. Bass, *Diffusions and Elliptic Operators*. Springer, New York, 1997.

19. R. F. Bass, K. Burdzy, and Z.-Q. Chen, Stochastic differential equations driven by stable processes for which pathwise uniqueness fails. *Stochastic Process. Appl.* 111 (2004), 1–15.
20. P. W. Bates and C. K. R. T. Jones, Invariant manifolds for semilinear partial differential equations. *Dyn. Rep.* 2 (1989), 138.
21. P. W. Bates, K. Lu, and C. Zeng, Existence and persistence of invariant manifolds for semiflows in Banach space. *Mem. Amer. Math. Soc.* 645 (1998), 129.
22. P. Baxendale, Wiener processes on manifolds of maps. *Proc. R. Soc. Edinburgh*, 87A (1980), 127–152.
23. D. A. Beard and H. Qian, *Chemical biophysics: Quantitative Analysis of Cellular Systems*. Cambridge Univ. Press, New York, 2008.
24. H. C. Berg, *Random Walks in Biology.* Expanded ed. Princeton University Press, Princeton, NJ, 1993.
25. M. S. Berger, *Nonlinearity and Functional Analysis*. Academic Press, New York, 1977.
26. N. Berglund and B. Gentz, *Noise-Induced Phenomena in Slow-Fast Dynamical Systems: A Sample-Paths Approach*. Springer-Verlag, New York, 2005.
27. J. Bertoin, *Lévy Processes*. Cambridge University Press, Cambridge, 1998.
28. W Bialek, *Biophysics: Searching for Principles*. http://www.princeton.edu/~wbialek/PHY562.html.
29. K. Bichteler and S. J. Lin, On the stochastic Fubini theorem. *Stochastics Stochastics Rep.* 54 (1995), 271–279.
30. P. Billingsley, *Weak Convergence of Probability Measures*. 2nd ed. John Wiley & Sons, New York, 1999.
31. L. Billings, I. B. Schwartz, D. S. Morgan, E. M. Bollt, R. Meucci, and E. Allaria, Stochastic bifurcation in a driven laser system: Experiment and theory. *Phys. Rev. E.* 70 (2004), 026220.
32. J.-M. Bismut, Flots stochastiques et formule de Ito-Stratonovitch généralisée. *C. R. Acad. Sci., Paris, Ser. I*, 290 (1980), 483–486.
33. R. M. Blumenthal, R. K. Getoor, and D. B. Ray, On the distribution of first hits for the symmetric stable processes. *Trans. Am. Math. Soc.* 99 (1961), 540–554.
34. R. M. Blumenthal and R. K. Getoor, *Markov Processes and Potential Theory*. Academic Press, New York, 1968.
35. A. Bobrowski, *Functional Analysis for Probability and Stochastic Processes*. Cambridge University Press, Cambridge, 2005.
36. W. M. Boothby, *An Introduction to Differentiable Manifolds and Riemannian Geometry*. 2nd ed. Elsevier, New York, 2003.
37. T. Bose and S. Trimper, Stochastic model for tumor growth with immunization. *Phys. Rev. E* 79 (2009), 051903.
38. P. Boxler, A stochastic version of center manifold theory. *Prob. Theory Related Fields* 83 (1989), 509–545.
39. P. Boxler, How to construct stochastic center manifolds on the level of vector fields. *Lecture Notes Math.* 1486 (1991), 141–158.
40. J. Brannan, J. Duan, and V. Ervin, Escape probability, mean residence time and geophysical fluid particle dynamics, *Phys. D* 133 (1999), 23–33.
41. J. Brannan, J. Duan, and V. Ervin, Escape probability and mean residence time in random flows with unsteady drift. *Math. Problems Eng.* 7 (2001), 55–65. doi:10.1155/S1024123X01001521
42. E. Brown and P. Holmes, Modelling a simple choice task: Stochastic dynamics of mutually inhibitory neural groups. *Stochastics Dyn.* 1 (2001), 159–191.

43. Z. Brzezniak and T. Zastawniak, *Basic Stochastic Processes*. Springer, New York, 1999.
44. L. Caffarelli and L. Silvestre, An extension problem related to the fractional Laplacian. *Commun. Partial Differential Eqs.* 32 (2007), 1245–1260.
45. R. E. Caflisch, Stochastic modelling of a dilute fluid-particle suspension, IMA Vol. Math. Appl. 9 (1987), 1–12.
46. M. Capinski and E. Kopp, *Measure, Integral and Probability*. Springer, New York, 1999.
47. T. Caraballo, J. Duan, K. Lu, and B. Schmalfuss, Invariant manifolds for random and stochastic partial differential equations. *Adv. Nonlinear Stud.* 10 (2009), 23–52.
48. A. Carverhill, Flows of stochastic dynamical systems: Ergodic theory. *Stochastics* 14 (1985), 273–317.
49. J. Carr, *Applications of Centre Manifold Theory*. Springer, New York, 1981.
50. P. Castiglione, M. Falcioni, A. Lesne, and A. Vulpiani. *Chaos and Coarse-Graining in Statistical Mechanics*. Cambridge University Press, Cambridge, 2008.
51. M. Chekroun, E. Simonnet, and M. Ghil, Stochastic climate dynamics: Random attractors and time-dependent invariant measures, *Phy. D.* 240 (2011), 1685–1700.
52. B. Chen and J. Duan, Stochastic quantification of missing mechanisms in dynamical systems. *Interdisciplinary Math. Sci.* 8 (2010), 67–76.
53. H. Chen, J. Duan, X. Li, and C. Zhang, A computational analysis for mean exit time under non-Gaussian Lévy noises. *Appl. Math. Comput.* 218 (2011), 1845–1856.
54. H. Chen, J. Duan, and C. Zhang, Elementary bifurcations for a simple dynamical system under non-Gaussian Lévy noises. *Acta Math. Sci.* 32B (2012), 1391–1398.
55. X. Chen, J. Duan, and X. Fu, A sufficient condition for bifurcation in random dynamical systems. *Proc. Am. Math. Soc.* 138 (2010), 965–973.
56. Z.-Q. Chen, P. Kim, and R. Song, Heat kernel estimates for Dirichlet fractional Laplacian. *J. Eur. Math. Soc.* 12 (2010), 1307–1329.
57. Z.-Q. Chen, On notions of harmonicity, *Proc. Am. Math. Soc.* 137 (2009), 3497–3510.
58. Z.-Q. Chen and Z. Zhao, Diffusion processes and second order elliptic operators with singular coefficients for lower order terms. *Math. Ann.* 302 (1995), 323–357.
59. C. Chicone, *Ordinary Differential Equations with Applications.* 2nd ed. Springer, New York, 2006.
60. C. Chicone and Yu Latushkin, *Evolution Semigroups in Dynamicals Systems and Differential Equations*. American Mathematical Society. Providence, RI, 1999.
61. P. L. Chow, *Stochastic Partial Differential Equations*. Chapman & Hall/CRC, New York, 2007.
62. I. Chueshov, *Monotone Random Systems Theory and Applications*. Springer-Verlag, Berlin, 2002.
63. E. Coayla-Teran, S.-E. A., and P. R. Ruffino, Hartman-Grobman theorems along hyperbolic stationary trajectories. *Discrete Continuous Dyn. Syst.* 17 (2007), 287–292.
64. E. A. Coddington and N. Levinson, *Theory of Ordinary Differential Equations.* McGraw-Hill, New York, 1955.
65. N. D. Cong, *Topological Dynamics of Random Dynamical Systems*. Clarendon Press, Oxford, 1997.
66. R. Cont and P. Tankov, *Financial Modelling with Jump Processes*. Chapman & Hall/CRC, London, 2004.
67. H. Crauel and M. Gundlach (eds.), *Stochastic Dynamics*. Springer, New York, 1999.
68. H. Crauel and F. Flandoli, Attractors for random dynamical systems. *Prob. Theory Relat. Fields* 100 (1994), 1095–1113.

69. H. Crauel and F. Flandoli, Additive noise destroys a pitchfork bifurcation, *J. Dyn. Diff. Eqn.* 10 (1998), 259–274.
70. G. Da Prato and J. Zabczyk, *Ergodicity for Infinite Dimensional Systems*. Cambridge University Press, Cambridge, 1996.
71. A. Debussche, M. Högele, and P. Imkeller, *The Dynamics of Nonlinear Reaction-Diffusion Equations with Small Lévy Noise. Lecture Notes in Mathematics*, Vol. 2085. Springer, New York, 2013.
72. M. Denker, J. Duan, and M. McCourt, Pseudorandom numbers for conformal measures. *Dyn. Syst.* 24 (2009), 439–457.
73. P. D. Ditlevsen, Observation of α-stable noise induced millennial climate changes from an ice record. *Geophys. Res. Lett.* 26 (1999), 1441–1444.
74. A. Du and J. Duan, Invariant manifold reduction for stochastic dynamical systems. *Dyn. Syst. Appl.* 16 (2007), 681–696.
75. J. Duan, Stochastic modeling of unresolved scales in complex systems. *Frontiers Math. China* 4 (2009), 425–436.
76. J. Duan, Predictability in nonlinear dynamical systems with model uncertainty. In T. N. Palmer and P. Williams, editors, *Stochastic Physics and Climate Modeling*, pages 105–132. Cambridge University Press, Cambridge, 2009.
77. J. Duan, Predictability in spatially extended systems with model uncertainty I & II. *Eng. Simul.* 31 (2009), 17–32.
78. J. Duan, X. Kan, and B. Schmalfuss, Canonical sample spaces for stochastic dynamical systems. *Interdisciplinary Math. Sci.* 9 (2010), 53–70.
79. J. Duan, K. Lu, and B. Schmalfuß, Invariant manifolds for stochastic partial differential equations. *Ann. Prob.* 31 (2003), 2109–2135.
80. J. Duan, K. Lu, and B. Schmalfuss, Smooth stable and unstable manifolds for stochastic evolutionary equations, *J. Dyn. Diff. Eqns.* 16 (2004), 949–972.
81. J. Duan, S. Luo, and C. Wang (Eds.), *Recent Development in Stochastic Dynamics and Stochastic Analysis*. World Scientific, New Jersey, 2009.
82. J. Duan and B. Nadiga, Stochastic parameterization of large eddy simulation of geophysical flows. *Proc. Am. Math. Soc.* 135 (2007), 1187–1196.
83. R. M. Dudley, *Real Analysis and Probability*. Cambridge University Press, Cambridge, 2002.
84. R. Durrett, *Probability: Theory and Examples*. 2nd ed. Duxbury Press, Boston, 1996.
85. R. Durrett, *Stochastic Calculus: A Practical Introduction.* CRC Press, Boston, 1996.
86. W. Ebeling and I. M. Sokolov, *Statistical Thermodynamics and Stochastic Theory of Nonlinear Systems Far from Equilibrium.* World Scientific, New Jersey, 2005.
87. K. D. Elworthy, *Stochastic Differential Equations on Manifolds*, Cambridge University Press, Cambridge, 1982.
88. L. C. Evans, *An Introduction to Stochastic Differential Equations*. American Mathematical Society, Providence, RI, 2014.
89. L. C. Evans, *Partial Differential Equations*, 2nd ed. American Mathematical Society, Providence, RI, 2010.
90. S. Fang, P. Imkeller, and T. Zhang, Global flows for stochastic differential equations without global Lipschitz conditions. *Ann. Prob.* 35 (2007), 180–205.
91. E. Fedrizzi and F. Flandoli, Pathwise uniqueness and continuous dependence for SDEs with nonregular drift. *Stochastics*, 83 (2011), 241–257.
92. M. I. Freidlin and A. D. Wentzell. *Random Perturbations of Dynamical Systems*. 2nd ed. Springer-Verlag, New York, 1998.
93. A. Friedman, *Stochastic Differential Equations and Applications*. 2 vols. Academic Press, New York, 1975–1976.

94. A. Friedman, *Partial Differential Equations of Parabolic Type*. Dover, Mineola, New York, 2008
95. T. Fujiwara and H. Kunita, Canonical SDE's based on semimartingales with spatial parameters. Part 1: Stochastic flows of diffeomorphisms, *Kyushu J. Math.* 53 (1999), 265–300.
96. H. Fu, X. Liu, and J. Duan, Slow manifolds for multi-time-scale stochastic evolutionary systems. *Comm. Math. Sci.* 11 (2013), 141–162.
97. T. Gao, X. Li, J. Duan, and R. Song, Mean exit time and escape probability for dynamical systems driven by non-Gaussian noise. *SIAM J. Sci. Comput.* 36 (2014), A887–A906.
98. J. Garcia-Ojalvo and J. M. Sancho, *Noise in Spatially Extended Systems*. Springer, New York, 1999.
99. C. Gardiner, *Stochastic Methods*. 4th ed. Springer, New York, 2009. Previously published as *Handbook of Stochastic Methods*, 3rd ed., Springer, New York, 2004.
100. M. G. Garroni and J. L. Menaldi, *Green Functions for Second Order Parabolic Integro-differential Problems*. Longman House, England, 1992.
101. M. G. Garroni and J. L. Menaldi, *Second Order Elliptic Integro-differential Problems*. Chapman & Hall/CRC, Boca Raton, 2002.
102. R. K. Getoor, First passage times for symmetric stable processes in space. *Trans. Am. Math. Soc.* 101 (1961), 75–90.
103. I. I. Gikhman and A. V. Skorokhod, *Introduction to the Theory of Random Processes*. Dover, New York, 1969.
104. I. I. Gihman and A. V. Skorohod, *Stochastic Differential Equations*. New York, Springer, 1972.
105. D. Gilbarg and N. S. Trudinger, *Elliptic Partial Differential Equations of Second Order*. 2nd ed. Springer, New York, 1983.
106. F. Gong and J. Zhang, Flows associated to adapted vector fields on the Wiener space. *J. Funct. Anal.* 253 (2007), 647–674.
107. J. Grasman and O. A. van Herwaarden, *Asymptotic Methods for the Fokker-Planck Equation and the Exit Problem in Applications*. Springer, New York, 1999.
108. A. Griffa, A. D. Kirwan, A. J. Mariano, T. M. Ozgokmen, and T. Rossby. *Lagrangian Analysis and Prediction of Coastal and Ocean Dynamics*. Cambridge University Press, Cambridge, 2007.
109. Q. Y. Guan and Z.-M. Ma, Boundary problems for fractional Laplacians. *Stochastics and Dynamics* 5 (2005), 385–424.
110. J. Guckenheimer and P. Holmes. *Nonlinear Oscillations, Dynamical Systems and Bifurcations of Vector Fields*. Springer-Verlag, New York, 1983.
111. R. Haberman, *Applied Partial Differential Equations*. 4th ed. Prentice Hall, Mahwah, NJ, 2003.
112. P. S. Hagan, C. R. Doering, and D. C. Levermore, Mean exit times for particles driven by weakly colored noise. *SIAM J. Appl. Math.* 49 (1989), 1480–1513.
113. M. Hairer, A. M. Stuart, and J. Voss, Sampling conditioned hypoelliptic diffusions. *Ann. Appl. Prob.* 21 (2011), 669–698.
114. P. Hanggi, A. Alvarez-Chillida, and M. Morillo Buzon (Eds.), New horizons in stochastic complexity (Special Issue). *Phys. A* 351 (2005).
115. P. Hanggi and F. Marchesoni (Eds.), Stochastic systems: From randomness to complexity (Special Issue). *Phys. A* 325 (2003).
116. D. C. Hanselman and B. L. Littlefield, *Mastering MATLAB 7*. Prentice Hall, Mahwah, NJ, 2004.

117. K. Hasselmann, Stochastic climate models: Part I. Theory. *Tellus* 28 (1976), 473–485.
118. C. Hein, P. Imkeller, and I. Pavlyukevich, Limit theorems for p-variations of solutions of SDEs driven by additive stable Lévy noise and model selection for paleo-climatic data. *Interdisciplinary Math. Sci.* 8 (2009), 137–150.
119. J. Hemminger (Chair), *Directing Matter and Energy: Five Challenges for Science and the Imagination*. A Report from the Basic Energy Sciences Advisory Committee, U.S. Dept of Energy, Washington, DC, 2007.
120. S. Herrmann, P. Imkeller, I. Pavlyukevich, and D. Peithmann, *Stochastic Resonance – A Mathematical Approach in the Small Noise Limit, Mathematical Surveys and Monographs*, Vol. 194, American Mathematical Society, Providence, RI, 2014.
121. T. Hida, *Brownian Motion*. Springer, New York, 1974.
122. D. J. Higham, An algorithmic introduction to numerical simulation of stochastic differential equations. *SIAM Rev.* 43 (2001), 525–546.
123. M. Hirsch, S. Smale, and R. L. Devaney, *Differential Equations, Dynamical Systems, and an Introduction to Chaos*. 2nd ed. Academic Press, New York, 2003.
124. D. Hochberg, Mirror symmetry breaking and restoration: The role of noise and chiral bias. *Phys. Rev. Lett.* 102 (2009), 248101.
125. W. Horsthemke and R. Lefever, *Noise-Induced Transitions: Theory and Applications in Physics, Chemistry, and Biology*. Springer-Verlag, Berlin, 2007.
126. E. P. Hsu and G. Qin, Volume growth and escape rate of Brownian motion on a complete Riemannian manifold. *Ann. Prob.* 38 (2010), 1570–1582.
127. Z. Huang, *Foundation of Stochastic Analysis* (in Chinese). 2nd ed. Science Press, Beijing, 2001.
128. N. Ikeda and S. Watanabe, *Stochastic Differential Equations and Diffusion Processes*. 2nd ed. North-Holland, New York, 1989.
129. P. Imkeller and B. Schmalfuss, The conjugacy of stochastic and random differential equations and the existence of global attractors. *J. Dyn. Differential Eqns.* 13 (2001), 215–249.
130. P. Imkeller and I. Pavlyukevich, Model reduction and stochastic resonance. *Stochastics Dyn.* 2 (2002), 463–506.
131. P. Imkeller and Jin-Song von Storch, *Stochastic Climate Models. Progress in Probability*, Vol. 49. Birkhöuser, Berlin, 2001.
132. P. Imkeller and C. Lederer, On the cohomology of flows of stochastic and random differential equations. *Prob. Theory Related Fields* 120 (2001), 209–235.
133. P. Imkeller and C. Lederer, The cohomology of stochastic and random differential equations and local linearization of stochastic flows. *Stochastics Dyn.* 2 (2002), 131–159.
134. P. Imkeller and A. Monahan (Eds.), *Conceptual stochastic climate models*. (Special Issue). *Stochastics Dyn.* 2 (2002).
135. P. Imkeller and I. Pavlyukevich, First exit time of SDEs driven by stable Lévy processes. *Stochastic Proc. Appl.* 116 (2006), 611–642.
136. P. Imkeller, I. Pavlyukevich, and T. Wetzel, First exit times for Lévy-driven diffusions with exponentially light jumps. *Ann. Prob.* 37 (2009), 530–564.
137. N. Jacob, *Pseudo Differential Operators and Markov Processes, Volume III, Markov Processes and Application*. Imperial College Press, London, 2005.
138. J. Jacod and P. Protter, *Probability Essentials*. 2nd ed. Springer, New York, 2004.
139. J. Jacod and A. N. Shiryaev, *Limit Theorems for Stochastic Processes*. Springer, New York, 1987.
140. A. Janicki and A. Weron, *Simulation and Chaotic Behavior of α-Stable Stochastic Processes*. Marcel Dekker, New York, 1994.

141. J. Jost, *Partial Differential Equations*. 2nd ed. Springer, New York, 2007.
142. B. Jourdain, S. Méléard, and W. A. Woyczynski. Lévy flights in evolutionary ecology. *J. Math. Biol.* 65 (2012), 677–707.
143. W. Just, H. Kantz, C. Rodenbeck, and M. Helm, Stochastic modelling: Replacing fast degrees of freedom by noise. *J. Phys. A* 34 (2001), 3199–3213.
144. Y. Kabanov and S. Pergamenshchikov, *Two-Scale Stochastic Systems: Asymptotic Analysis and Control*. Springer-Verlag, New York, 2003.
145. O. Kallenberg, *Foundations of Modern Probability*. 2nd ed. Applied Probability Trust, Sheffield, UK, 2002.
146. G. Kager and M. Scheutzow, Generation of one-sided random dynamical systems by stochastic differential equations. *Electr. J. Prob.* 2 (1997), 1–17.
147. I. Karatzas and S. E. Shreve, *Brownian Motion and Stochastic Calculus*. 2nd ed. Springer, New York, 1991.
148. Y. Kifer and P.-D. Liu. Random dynamics. In B. Hasselblatt and A. Katok editors, *Handbook of Dynamical Systems*, Vol. 1B, pp. 379–499, Elsevier, New York, 2006.
149. Y. Kifer, *Ergodic Theory of Random Transformations*. Birkhäuser, Boston, 1986.
150. M. Kittisopikul and G. M. Suel, Biological role of noise encoded in a genetic network motif. *Proc. Natl. Acad. Sci. USA* 107 (2010), 13300–13305.
151. F. C. Klebaner, *Introduction to Stochastic Calculus with Applications*. 2nd ed. Imperial College Press, London, 2005.
152. P. E. Kloeden and E. Platen, *Numerical Solution of Stochastic Differential Equations*, Springer, New York, 1992.
153. E. Knobloch and K. A. Wiesenfeld, Bifurcations in fluctuating systems: The center-manifold approach. *J. Stat. Phys.* 33 (1983), 611–637.
154. S. Kogan, *Electronic Noise and Fluctuations in Solids*. Reissue ed. Cambridge University Press, Cambridge 2008.
155. P. Komjath and V. Totik, *Problems and Theorems in Classical Set Theory*. Springer, New York, 2006.
156. T. Koren, A. V. Chechkin, and J. Klafter, On the first passage time and leapover properties of Lévy motions. *Phys. A* 379 (2007), 10–22.
157. P. Kotelenez, *Stochastic Ordinary and Stochastic Partial Differential Equations: Transition from Microscopic to Macroscopic Equations*. Springer, New York, 2008.
158. E. Kreyszig, *Introductory Functional Analysis with Applications*. John Wiley. New York, 1989.
159. V. Krishnan, *Nonlinear Filtering and Smoothing: An Introduction to Martingales, Stochastic Integrals and Estimation*. John Wiley, New York, 1984.
160. N. V. Krylov, *Introduction to the Theory of Random Processes*. American Mathematical Society, Providence, RI, 2002.
161. N.V. Krylov and M. Rockner, Strong solutions to stochastic equations with singular time dependent drift. *Prob. Theory Related Fields* 131 (2005), 154–196.
162. H. Kunita, Itô's stochastic calculus: Its surprising power for applications. *Stochastic Process. Appl.* 120 (2010), 622–652.
163. H. Kunita, *Stochastic Flows and Stochastic Differential Equations*. Cambridge University Press, New York, 1990.
164. H. Kunita, Stochastic differential equations based on Lévy processes and stochastic flows of diffeomorphisms. In M. M. Rao, editor, *Real and Stochastic Analysis*, pages 305–373, Birkhäuser, Boston, 2004.
165. H.-H. Kuo, *Introduction to Stochastic Integration*. Springer, New York, 2006.
166. T. G. Kurtz, E. Pardoux, and P. Protter, Stratonovich stochastic differential equations driven by general semimartingales. *Ann. Inst. H. Poincar'e Prob. Stat.* 23 (1995), 351–377.

167. R. Kuske and J. B. Keller, Rate of convergence to a stable law. *SIAM J. Appl. Math.* 61 (2000–2001), 1308–1323.
168. A. Lasota and M. C. Mackey. *Chaos, Fratals and Noise–Stochastic Aspects of Dynamics*. 2nd ed. Springer, New York, 1994.
169. G. F. Lawler, *Introduction to Stochastic Processes*. 2nd edn. Chapman & Hall/CRC, Boca Raton, Florida, 2006.
170. Y. Le Jan and S. Watanabe, Stochastic flows of diffeomorphisms. In K. Itö editor, *Stochastic Analysis (Proceedings of the Taniguchi Symposimu 1982)*, pages 307–332. North-Holland, Amsterdam, 1984.
171. C. E. Leith, Stochastic backscatter in a subgrid-scale model: Plane shear mixing layer. *Phys. Fluids A* 2 (1990), 297–299.
172. Y.-J. Lee and H.-H. Shih, Analysis of generalized Lévy white noise functionals. *Functional Anal.* 211 (2004). 1–70.
173. X. M. Li, *Stochastic Flows on Noncompact Manifolds*. PhD thesis, University of Warwick, 1993.
174. W. Li and K. Lu, Poincaré theorems for random dynamical systems. *Ergodic Theory Dyn. Syst.* 25 (2005), 1221–1236.
175. W. Li and K. Lu, Sternberg theorems for random dynamical systems. *Comm. Pure Appl. Math.* 58 (2005), 941–988.
176. M. Liao, Lyapunov exponents of stochastic flows. *Ann. Prob.* 25 (1997), 1241–1256.
177. M. Liao, The Dirichlet problem of a discontinuous Markov process. *Acta Math. Sin. (New Ser.)* 5 (1989), 9–15.
178. G. M. Lieberman, *Second Order Parabolic Differential Equations*. World Scientific, River Edge, NJ, 1996.
179. T. M. Liggett, *Interacting Particle Systems*. Springer, New York, 1985.
180. C. C. Lin and L. A. Segel, *Mathematics Applied to Deterministic Problems in the Natural Sciences*. SIAM, Philadelphia, 1988.
181. J. W.-B. Lin and J. D. Neelin, Considerations for stochastic convective parameterization. *J. Atmos. Sci.* 59 (2002), 959–975.
182. Y. K. Lin and G. Cai, *Probabilistic Structural Dynamics*. McGraw-Hill, New York, 2004.
183. X. Liu, J. Duan, J. Liu, and P. E. Kloeden, Synchronization of systems of Marcus canonical equations driven by α-stable noises. *Nonlinear Anal.* 11 (2010), 3437–3445.
184. P. D. Liu and M. Qian, *Smooth Ergodic Theory of Random Dynamical Systems*. Springer-Verlag, Berlin, 1995.
185. M. Loève, *Probability Theory*. Vol. II, 4th ed. Springer-Verlag, New York, 1978.
186. H. Luschgy and G. Pages, Moment estimates for Lévy processes. *Elect. Comm. Prob.* 13 (2008), 422–434.
187. Z.-M. Ma, R. Zhu, and X. Zhu, On notions of harmonicity for non-symmetric Dirichlet form. *Sci. China (Math.)* 53 (2010), 1407–1420.
188. D. K. C. MacDonald, *Noise and Fluctuations: An Introduction*. Dover Mineola, New York, 2006.
189. R. S. MacKay and D. J. C. MacKay, Ergodic pumping: A mechanism to drive biomolecular conformation changes. *Phys. D* 216 (2006), 220–234.
190. R. S. MacKay, Langevin equation for slow degrees of freedom of Hamiltonian systems. In "*Nonlinear Dynamics and Chaos*", M. Theil, J. Kurths, M. C. Romano, G. Karolyi, and A. Moura, editors, pages 98–102. Springer, New York, 2010.
191. A. J. Majda, I. Timofeyev, and E. Vanden Eijnden, Models for stochastic climate prediction, *Proc. Natl. Acad. Sci. USA* 96 (1999), 14687–14691.

192. X. Mao, *Stochastic Differenntial Equations and Applications.* Horwood, England, 1997.
193. S. I. Marcus, Modelling and approximation of stochastic differenntial equations driven by semimaringales, *Stochastics* 4 (1981), 223–245.
194. M. Marsili, A. Maritan, F. Toigo, and J. R. Banavar, Stochastic growth equations and reparametrization invariance, *Rev. Modern Phys.* 68 (1996), 963–983.
195. J. C. Mattingly, N. Pillai, and A. M. Stuart, Diffusion limits of the random walk Metropolis algorithm in high dimensions. *Ann. Appl. Prob.* 22 (2012), 881–930.
196. B. J. Matkowsky and Z. Schuss, The exit problem for randomly perturbed dynamical systems. *SIAM J. Appl. Math.* 33 (1977), 365–382.
197. B. Matkowsky, Z. Schuss, and C. Tier, Asymptotic methods for Markov jump processes, *A. M. S. Lect. Appl. Math.* 27 (1991), 215–240.
198. R. C. McOwen, *Partial Differential Equantions.* Pearson Education, Mahwah, NJ, 2003.
199. R. Metzler and J. Klafter, The random walk's guide to anomalous diffusion: A fractional dynamics approach. *Phys. Rep.* 339 (2000), 1–77.
200. T. Mikosch, *Elementary Stochastic Calculus with Finance in View.* World Scientific, New Jersey, 1998.
201. S.-E. A. Mohammed and M. Scheutzow, The stable manifold theorem for stochastic differential equations. *Ann. Prob.*, 27 (1999), 615–652.
202. P. Mörters and Y. Peres, *Brownian Motion.* Cambridge University Press, Cambridge, 2010.
203. F. Moss and P. V. E. McClintock (Eds.), *Noise in Nonlinear Dynamical Systems.* 3 vol. Cambridge University Press, Cambridge 2007–2009.
204. J. R. Munkres, *Topology.* 2nd ed. Prentice-Hall, New Jersey, 2000.
205. T. Myint-U. *Partial Differential Equations for Scientists and Engineers.* 3rd ed. North-Holland, Amsterdam, 1987.
206. T. Naeh, M. M. Klosek, B. J. Matkowsky, and Z. Schuss, A direct approach to the exit problem, *SIAM J. Appl. Math.* 50 (1990), 595–627.
207. E. Nelson, *Dynamical Theories of Brownian Motion.* 2nd Ed., Princeton University, Press, Princeton, NJ, 2001.
208. J. P. Nolan, *Handbooks in Finance*, Vol. 1, *Modeling Financial Distributions with Stable Distributions.* Elsevier, Amsterdam, 2003.
209. I. Nourdin and T. Simon, On the absolute continuity of Lévy processes with drift. *Ann. Prob.* 34 (2006), 1035–1051.
210. D. Nualart and W. Schoutens, Chaotic and predictable representations for Lévy processes. *Stochastic Process. Appl.* 90 (2000), 109–122.
211. G. Di Nunno, B. Oksendal, and F. Proskea, White noise analysis for Lévy processes. *J. Funct. Anal.* 206 (2004), 109–148.
212. G. Ochs and V. I. Oseledets, Topological fixed point theorems do not hold for random dynamical systems. *J. Dyn. Differential Eqns.* 11 (1999), 583–593.
213. B. Oksendal, *Stochastic Differential Equations.* 6th ed. Springer, New York, 2003.
214. B. Oksendal, *Applied Stochastic Control of Jump Diffusions.* Springer, New York, 2005.
215. E. Olivieri and M. E. Vares, *Large Deviation and Metastability.* Cambridge University Press, Cambridge, 2004.
216. T. N. Palmer, G. J. Shutts, R. Hagedorn, F. J. Doblas-Reyes, T. Jung, and M. Leutbecher, Representing model uncertainty in weather and climate prediction. *Annu. Rev. Earth Planet. Sci.* 33 (2005), 163–193.

217. G. Papanicolaou, Asymptotic analysis of stochastic equations, MAA Studies No. 18: *Studies in Probability Theory*, Murray Rosenblatt, editor, pages 111–179. Mathematics Association of America, 1978.
218. G. Papanicolaou, Introduction to the asymptotic analysis of stochastic equations: Modern modeling of continuum phenomena, Lect. Appl. Math. 16 (1977) 109–147.
219. A. Papoulis, *Probability, Random Variables, and Stochastic Processes*. 2nd ed. McGraw-Hill, New York, 1984.
220. E. Pardoux and S. Peng, Adapted solution of backward stochastic equations. *Syst. Control Lett.* 14 (1990), 55–61.
221. C. Pasquero and E. Tziperman, Statistical parameterization of heterogeneous oceanic convection. *J. Phys. Oceanogr.*, 37 (2007), 214–229.
222. G. A. Pavliotis and A. M. Stuart, Multiscale methods: Averaging and homogenization. *Texts Appl. Math.*, 53 (2009).
223. H. L. Pecseli, *Fluctuations in Physical Systems*. Cambridge University Press, Cambridge, 2000.
224. S. Peng, Backward stochastic differential equations and applications to optimal control. *Appl. Math. Optim.* 27 (1993), 125–144.
225. S. Peng, Nonlinear expectation theory and stochastic calculus under Knightian uncertainty. In *Studies in Probability, Optimization and Statistics*, Vol. 5, *Real Options, Ambiguity, Risk and Insurance*, IOS Press, Amsterdam, pages 144–184.
226. L. Perko, *Differential Equations and Dynamical Systems*. Springer, New York, 2000.
227. S. Peszat and J. Zabczyk, *Stochastic Partial Differential Equations with Lévy Noises*. Cambridge University Press, Cambridge, 2007.
228. M. A. Pinsky, *Partial Differential Equations and Boundary Value Problems with Applications*, Waveland Press, Long Grove, IL, 2003.
229. L. I. Piterbarg and T. M. Ozgokmen, A simple prediction algorithm for the Lagrangian motion in 2D turbulent flows. *SIAM J. Appl. Math.* 63 (2002), 116–148.
230. V. Pipiras and M. S. Taqqu, Convergence of the Wererstrass-Mandelbrot process to fractinal Brownian motion. *Fractals* 8 (2000), 369–384.
231. J. Poirot and P. Tankov, Monte Carlo option pricing for tempered stable (CGMY) processes, *Asia-Pacific Financial Markets* 13 (2006), 327–344.
232. P. E. Protter, *Stochastic Integration and Differential Equations*. 2nd ed. Springer, New York, 2005.
233. H. Qiao and J. Duan, Topological equivalence for discontinuous random dynamical systems and applications. *Stochastic Dyn.* 14 (2014). doi: 10.1142/S021949371350007X
234. H. Qiao, X. Kan, and J. Duan, Escape probability for stochastic dynamical systems with jumps. *Springer Proc. Math. Stat.*, 34 (2013), 195–216.
235. H. Qiao and X. Zhang, Homeomorphism flows for non-Lipschitz stochastic differential equations with jumps. *Stochastic Process. Appl.* (2008), 2254–2268.
236. M. Renardy and R. Rogers, *Introduction to Partial Differential Equations*. Springer, New York, 1993.
237. D. Revuz and M. Yor. *Continuous Martingales and Brownian Motion*. 3rd ed. Springer, New York, 2005.
238. H. Risken, *The Fokker-Planck Equation: Methods of Solution and Applications*. Springer, New York, 1984.
239. A. J. Roberts, Normal form transforms separate slow and fast modes in stochastic dynamical systems. *Phys. A* 387 (2008), 12–38.
240. L. C. G. Rogers and D. Williams, *Diffusions, Markov Processes and Martingales*. Vol. 1, *Foundations*. 2nd ed. Cambridge University Press, Cambridge, 2000.

241. L. C. G. Rogers and D. Williams, *Diffusions, Markov Processes and Martingales*. Vol. 2, *Ito Calculus*. 2nd ed. Cambridge University Press, Cambridge, 2000.
242. S. M. Ross, *Simulation*. 4th ed. Elsevier, New York, 2006.
243. G. Samorodnitsky and M. S. Taqqu, *Stable Non-Gaussian Random Processes*. Chapman & Hall, New York, 1994.
244. K.-I. Sato, *Lévy Processes and Infinitely Divisible Distributions*. Cambridge University Press, Cambridge, 1999.
245. K. R. Schenk-Hoppé, Stochastic Hopf bifurcation: An example. *Int. J. Non-Linear Mech.* 31 (1996), 685–692.
246. D. Schertzer, M. Larcheveque, J. Duan, V. Yanovsky, and S. Lovejoy, Fractional Fokker–Planck equation for nonlinear stochastic differential equations driven by non-Gaussian Levy stable noises. *J. Math. Phys.* 42 (2000), 200–212.
247. M. Scheutzow, On the perfection of crude cocycles. *Random Comput. Dyn.* 4 (1996), 235–255.
248. B. Schmalfuss and K. R. Schneider, Invariant manifolds for random dynamical systems with slow and fast variables. *J. Dyn. Differential Eqns.* 20 (2008), 133–164.
249. B. Schmalfuss, Inertial manifolds for random differential equations. In *Probability and Partial Differential Equations in Modern Applied Mathematics*, pages 213–236. Springer, New York, 2005.
250. B. Schmalfuss, The random attractor of the stochastic Lorenz system. *Z. Math. Phys. (ZAMP)* 48 (1997), 951–975.
251. G. Schoner and H. Haken, The slaving principle for Stratonovich stochastic differential equations. *Z. Phys. B* 63 (1986), 493–504.
252. M. Schumaker, Center manifold reduction and normal form transformations in systems with additive noise. *Phys. Lett. A* 122 (1987), 317–322.
253. Z. Schuss, *Theory and Applications of Stochastic Differential Equations*, Wiley, New York, 1980.
254. Z. Schuss, *Theory and Applications of Stochastic Processes: An Analytical Approach*. Springer, New York, 2009.
255. L. Schwartz, *A Mathematician Grappling with His Century*. Birkhäuser, Berlin, 1997.
256. M. Shao and C. L. Nikias, Signal processing with fractional lower order moments: Stable processes and their applications. *Proc. IEEE*, 81 (1993), 986–1010.
257. H.-H. Shih, The Segal-Bargmann transform for Lévy white noise functionals associated with non-integrable Lévy processes. *J. Funct. Anal.*, 255 (2008), 657–680.
258. M. F. Shlesinger, G. M. Zaslavsky, and U. Frisch (Eds.), *Lévy Flights and Related Topics in Physics*, *Lecture Notes in Physics*, Vol. 450. Springer-Verlag, Berlin, 1995.
259. A. V. Skorokhod, *Asmptotic Methods in the Theory of Stochastic Differential Equations*. American Mathematical Society, Providence, RI, 1989.
260. W. Smith and G. S. Watson: Diffusion out of a triangle. *J. Appl. Prob.* 4 (1967), 479–488.
261. R. Song, Probabilistic approach to the Dirichlet problem of perturbed stable processes. *Prob. Theory Related Fields* 95 (1993), 371–389.
262. R. B. Sowers, Stochastic averaging near long heteroclinic orbits. *Stochastics Dyn.* 7 (2007), 187–228.
263. N. Sri Namachchivaya, Stochastic bifurcation. *Appl. Math. Comput.*, 38 (1990), 101–159.
264. E. Stone and P. Holmes, Random perturbations of heteroclinic attractors. *SIAM J. Appl. Math.* 50 (1990), 726–743.
265. W. A. Strauss, *Partial Differential Equations: An Introduction*. 2nd ed. John Wiley & Sons, Hoboken, NJ, 2008.

266. S. H. Strogatz, *Nonlinear Dynamics and Chaos – with Applications to Physics, Biology, Chemistry, and Engineering*. Addison-Wesley, New York, 1994.
267. D. W. Stroock and S. R. S. Varadhan, *Multidimensional Diffusion Processes*. Springer, New York, 2005.
268. X. Sun and J. Duan, Fokker-Planck equations for nonlinear dynamical systems driven by non-Gaussian Lévy processes. *J. Math. Phys.* 53 (2012), 072701.
269. X. Sun, J. Duan, and X. Li, An impact of noise on invariant manifolds in stochastic nonlinear dynamical systems. *J. Math. Phys.* 51 (2010), 042702.
270. X. Sun, X. Kan, and J. Duan, Approximation of invariant foliations for stochastic dynamical systems. *Stochastics Dyn.* 12 (2012), 1150011.
271. A.-S. Sznitman, *Brownian Motion, Obstacles, and Random Media*. Springer, Berlin, 1998.
272. K. Taira, *Diffusion Processes and Partial Differential Equations*. Elsevier, New York, 2004.
273. K. Taira, *Boundary Value Problems and Markov Processes*. 2nd ed. Springer, New York, 2009.
274. K. Taira, *Semigroups, Boundary Value Problems and Markov Processes*. Springer, New York, 2004.
275. D. Talay, Simulation and numerical analysis of stochastic differential systems: A review. In *Probabilistic Methods in Applied Physics*, Vol. 451, P. Kree and W. Wedig, editors, pp. 63–106. Springer, New York, 1995.
276. Y. Tang, H. Gao, W. Zou, and J. Kurths, Pinning noise-induced stochastic resonance. *Phys. Rev. E* 87 (2013), 062920.
277. R. Temam, *Infinite-Dimensional Dynamical Systems in Mechanics and Physics*. 2nd ed. Springer, New York, 1997.
278. M. Turcotte, J. Garcia-Ojalvo, and G. M. Suel, A genetic timer through noise-induced stabilization of an unstable state. *Proc. Natl. Acad. Sci.* 105 (2008), 15732–15737.
279. N. G. Van Kampen, How do stochastic processes enter into physics? *Lecture Notes Phys.* 1250 (1987), 128–137.
280. N. G. Van Kampen, *Stochastic Processes in Physics and Chemistry.* 3rd ed. Elsevier, New York, 2007.
281. M. Veraar, The stochastic Fubini theorem revisited. *Stochastics* 84 (2012), 543–551.
282. A. J. Veretennikov, Strong solutions and explicit formulas for solutions of stochastic integral equations. *Mat. Sb.* 111 (1980), 434–452.
283. G. M. Viswanathan, V. Afanasyev, S. V. Buldyrev, S. Havlin, M. Daluz, E. Raposo, and H. Stanley, Lévy flights in random searches, *Phys. A* 282 (2000), 1–12.
284. W. Wang and A. J. Roberts, Slow manifold and averaging for slow-fast stochastic differential system. *J. Math. Anal. Appl.* 398 (2013), 822–839.
285. T. Wanner, Linearization of random dynamical systems. In *Dynamics Report*, Vol. 4. Spring-Verlog, New York, 1995. pages 203–269.
286. P. Walters, *An Introduction to Ergodic Theory*. Springer, New York, 1982.
287. L. Wang, J. Xin, and Q. Nie, A critical quantity for noise attenuation in feedback systems. *PLoS Comput. Biol.* 6 (2010), e1000764. doi:10.1371/journal.pcbi.1000764
288. E. Waymire and J. Duan (Eds.), *Probability and Partial Differential Equations in Modern Applied Mathematics*. Springer, New York, 2005.
289. E. R. Weeks, T. H. Solomon, J. S. Urback, and H. L. Swinney, Observation of anomalous diffusion and Lévy flights. In *Lévy Flights and Related Topics in Physics*, M. F. Shlesinger, G. M. Zaslavsky, and U. Frisch, editors, pages 51–71. Springer, Berlin, 1995.
290. S. Wiggins, *Introduction to Applied Nonlinear Dynamical Systems and Chaos*. 2nd ed. Springer, New York, 2003.

291. D. S. Wilks, Effects of stochastic parameterizations in the Lorenz '96 system. *Q. J. R. Meteorol. Soc.* 131 (2005), 389–407.
292. P. D. Williams, Modelling climate change: The role of unresolved processes. *Philos. Trans. R. Soc. London, Ser. A* 363 (2005), 2931–2946.
293. E. Wong and B. Hajek, *Stochastic Processes in Engineering Systems*. Springer, New York, 1985.
294. W. A. Woyczynski, Lévy processes in the physical sciences. In *Lévy Processes: Theory and Applications*, O. E. Barndorff-Nielsen, T. Mikosch, and S. I. Resnick editors, pages 241–266. Birkhäuser, Boston, 2001.
295. J. Xiong, A stochastic log-Laplace equation. *Ann. Prob.* 32 (2004) 2362–2388.
296. C. Xu and A. J. Roberts, On the low-dimensional modelling of Stratonovich stochastic differential equations. *Phys. A*, 225 (1996), 62–80.
297. Y. Xu, R. Gu, H. Zhang, W. Xu, and J. Duan, Stochastic bifurcations in a bistable Duffing–Van der Pol oscillator with colored noise. *Phys. Rev. E.* 83 (2011), 056215.
298. Y. Xu, J. Duan, and W. Xu, An averaging principle for stochastic dynamical systems with Lévy noise. *Phys. D* 240 (2011), 1395–1401.
299. T. Yamada and S. Watanabe, On the uniqueness of solutions of stochastic differential equations. *J. Math. Kyoto Univ.* 11 (1971), 155–167.
300. J. Yan, *Lectures in Measure Theory* (Chinese). 2nd ed. Science Press, Beijing, 2004.
301. J. Yang and J. Duan, Quantifying model uncertainties in complex systems. *Progr. Prob.* 65 (2011), 221–252.
302. Z. Yang and J. Duan, An intermediate regime for exit phenomena driven by non-Gaussian Lévy noises. *Stochastics Dyn.* 8 (2008), 583–591.
303. K. Yosida, *Functional Analysis*. 6th ed. Springer, Berlin, 1980.
304. L.-S. Young, Some open sets of nonuniformly hyperbolic cocycles. *Ergod. Th. Dyn. Syst.*, 13 (1993), 409–415.
305. E. Zeidler, *Applied Functional Analysis: Applications to Mathematical Physics*. Springer, New York, 1995.
306. E. Zeidler, *Applied Functional Analysis: Main Principles and Their Applications*. Springer, New York, 1995.
307. X. Zhang, Strong solutions of SDES with singular drift and Sobolev diffusion coefficients. *Stochastic Process. Appl.* 115 (2005), 1805–1818.
308. X. Zhang, H. Qian, and M. Qian, Stochastic theory of nonequilibrium steady states and its applications. Part I: *Phys. Rep.* 510 (2012), 1–86. Part II: *Phys. Rep.*, 510 (2012), 87–118.
309. H. Zhao and Z. Zheng, Random periodic solutions of random dynamical systems. *J. Differential Eqns.* 246 (2009), 2020–2038.
310. M. Zhou, *Theory of Real Functions* (Chinese). Peking University Press, Beijing, 2001.
311. W. Zhu, *Nonlinear Stochastic Dynamics and Control* (Chinese). Science Press, Beijing, 2003.
312. R. Zwanzig, *Nonequilibrium Statistical Mechanics*. Oxford University Press, Oxford, 2001.

Index